Analysis of Evolutionary Processes

Princeton Series in Theoretical and Computational Biology
Edited by Simon A. Levin

Analysis of Evolutionary Processes, by Fabio Dercole and Sergio Rinaldi
Theories of Population Variation in Genes and Genomes, by Freddy Bugge Christiansen
Mathematics in Population Biology, by Horst R. Thieme
Individual-based Modeling and Ecology, by Volker Grimm and Steven F. Railsback

Analysis of Evolutionary Processes

The Adaptive Dynamics Approach and Its Applications

Fabio Dercole
Sergio Rinaldi

PRINCETON UNIVERSITY PRESS

PRINCETON AND OXFORD

Published by Princeton University Press, 41 William Street, Princeton, New Jersey 08540

In the United Kingdom: Princeton University Press, 3 Market Place, Woodstock, Oxfordshire OX20 1SY

Library of Congress Cataloging-in-Publication Number: 2007938907

ISBN: 978-0-691-12006-5 (acid-free paper)

British Library Cataloging-in-Publication Data is available

This book has been composed in LaTeX

The publisher would like to acknowledge the authors of this volume for providing the camera-ready copy from which this book was printed.

Printed on acid-free paper. ∞

press.princeton.edu

Printed in the United States of America

10 9 8 7 6 5 4 3 2 1

to Martina and her mamma

Contents

Preface

The aim of this book is to present Adaptive Dynamics (AD), a quantitative approach for the study of evolutionary processes that has recently received a great deal of attention.

Starting with the groundbreaking works of Charles Darwin and Gregor Mendel and passing through all the achievements of genetics, ecology, and modern molecular biology, we now accept the view that evolutionary change results from the mechanism of genetic mutation, introducing new variant forms of individuals in the community of coevolving populations, and from the demography of the community, shaping the dynamics of population abundances. The typical pattern of evolutionary change of an individual trait, such as body size or shape, can be imagined as a sequence of mutations, with small effects on the trait, that confer an advantage to affected individuals in terms of survival and reproductive success. After each advantageous mutation, the group of mutants has the potential to spread in the community and replace the resident group of similar individuals, thus leading to a small step in the evolution of the trait. Darwin first realized that those individuals best adapted to survive and reproduce should come to dominate the populations, all else being equal, namely for given and constant environmental conditions and in the absence of further mutations. He called selection the demographic process leading to the dominance of the best-adapted individuals. As a result, all individual traits affecting demography evolve through mutation-selection processes, by adapting to the local environmental circumstances. The evolutionary hypothesis is that the superposition of all such processes have led to the origin of all species starting from a common ancestral species more than 3 billion years ago, the process of "descent with modification," as Darwin called it.

Traditionally, the effects of demographic and evolutionary changes have been considered separately. This is due to the fact that mutations are typically rare events, so that demographic and evolutionary dynamics are often visible on contrasting timescales, the demographic timescale of everyday birth and death of individuals and the evolutionary timescale on which mutations occur. However, the real dynamics of populations integrate both demographic and evolutionary changes. In fact, individual adaptive traits influence the dynamics of population abundances, which, in return, determine the selective pressures operating on the community and ultimately the evolution of the traits. Thus, demographic and evolutionary dynamics are entangled in a feedback loop, a notion perhaps first introduced by Ford (1949) and subsequently conceptualised by Pimentel (1961, 1968), Stenseth and Maynard Smith (1984), and Metz et al. (1992), who later developed the key ideas behind AD (Metz et al., 1996; see also Dieckmann and Law, 1996; Dieckmann,

1997; Geritz et al., 1997, 1998).

AD explicitly takes into account the coupling between demography and evolution, describes the process of selection by means of deterministic demographic models, and stochastically characterizes the occurrence of mutations. By separating the demographic and evolutionary timescales, i.e., by looking at the limit case of extremely rare and small mutations, AD derives a deterministic approximation of the evolutionary dynamics of adaptive traits, in terms of an ordinary differential equation defined on the evolutionary timescale: the AD canonical equation (Dieckmann and Law, 1996). Such an equation is typically nonlinear, so that, in accordance with the theory of nonlinear dynamical systems, one can in principle expect evolutionary dynamics to be characterized by several long-term evolutionary regimes, reached, in the long run, from different ancestral conditions. Moreover, long-term evolutionary regimes can be stationary (evolutionary equilibria), as well as nonstationary (periodic or even more complex chaotic evolutionary regimes), and this formally supports the so-called Red Queen hypothesis proposed by Van Valen (1973), which suggests that evolutionary processes have the potential to sustain never-ending evolutionary change.

The timescale separation argument also allows AD to describe the evolution of diversity in the community. Diversity, abstractly measured by the number of coevolving groups of identical individuals, can change due to mechanisms that are exogenous to the community, like immigrations of new forms of individuals or the accidental extinction of resident groups. However, evolutionary change is endogenously responsible of the evolution of diversity. On one hand, selection pressures that favor trait values at the extremes of a range over those in the middle may split a resident group into two initially similar subgroups, which then diversify by following separate evolutionary paths. Although the formation of new species is still a highly debated topic in the biological community, this mechanism, called evolutionary branching, is considered among the possible causes of speciation (Maynard Smith, 1966; Dieckmann and Doebeli, 1999). On the other hand, evolutionary change may drive the community toward trait values at which some of the resident groups cannot persist, a counterintuitive drawback of evolution called evolutionary extinction. AD provides specific analytical conditions for the occurrence of evolutionary branching and extinction, and when these conditions are met new resident groups are added or removed from the AD canonical equation, whose dimension therefore grows through evolutionary branching and is pruned by evolutionary extinction.

The applicability of AD is virtually limited to asexual species, where individuals produce either perfect copies of themselves or mutants characterized by different trait values (clonal reproduction). While the use of asexual demographic models to describe the long-term evolutionary dynamics of sexual populations is somehow justified (Marrow and Johnstone, 1996; Williams, 1996), the evolutionary branching identified by AD does not amount to speciation unless reproductive isolation evolves between the branching populations. These and other minor technical limitations concerning the mutation and selection processes will be clearly identified and discussed. On the other hand, the power of the AD approach will be highlighted through a series of applications in which the core of AD, namely its canonical equation and the branching and extinction conditions, will be used in conjunction with numerical bifurcation analysis, the most effective tool for the study of nonlinear dynamical systems.

Until now AD has been applied in population biology. However, there are many other areas, in particular in social sciences, economics, and engineering, to which AD could virtually be applied. In fact, natural or artificial systems can often be modeled as composed of agents or units that compete on a short-term (demographic) timescale, in accordance with their characteristic features, and transmit such features to agents or units of the next generation. If agents or units with innovative features occasionally appear in the system, the long-term (evolutionary) dynamics of the characteristic features are driven by an innovation-competition process that can be studied through the AD canonical equation.

This book is composed of ten chapters and four appendixes. The first appendix contains an introduction to dynamical systems and bifurcation analysis that should be read by those who are not familiar with these topics. It also presents some nonstandard bifurcations relevant for the study of evolutionary extinction (see Section A.7), which may be of interest to expert readers. As for the chapters, we can divide them into three groups: introductory (the first two), basic (the third), and application oriented (the others).

In the first chapter we give an overview of evolutionary processes with emphasis on evolutionary biology. It is a simplified introduction, which runs the risk of generating the wrong impression that evolutionary processes are simple and well understood. Readers who are already familiar with basic evolutionary principles and know about genotypes, phenotypes, mutations, and selection may skip this chapter.

In the second chapter we survey the main quantitative approaches to evolutionary dynamics. We identify seven approaches (population genetics, individual-based evolutionary models, quantitative genetics, evolutionary game theory, replicator dynamics, fitness landscapes, and AD) and point out some of their advantages and limitations. All these approaches would deserve much more space for a satisfactory description and, in fact, entire books have been written on each of them except AD. Thus, Chapter 2 should be viewed as a rudimentary introduction to the model-based approaches to evolutionary dynamics.

By contrast, Chapter 3 is the key chapter of the book. There we sketch all the steps that lead to the formal derivation of the AD canonical equation and discuss the mechanisms of evolutionary branching and extinction (the mathematical details are hidden in Appendixes B, C, and D).

All remaining chapters are twofold. Each of them introduces a new feature of evolutionary dynamics viewed through the lenses of AD, but at the same time deals with a particular application that is of interest per se. We were therefore forced to make a decision in favor of one of the two aspects when we fixed the titles of the various chapters. Our choice has been to highlight the specific methodological issue introduced in each chapter. However, we have compensated this decision by mentioning the applications explicitly in the title of the book. Since some readers might be interested only in one or a few applications, we have often reported in these chapters useful summaries of the relevant AD principles.

The issues we discuss cover a wide spectrum: the origin of diversity (Chapter 4), viewed as a sequential series of evolutionary branchings; the existence of alternative long-term evolutionary regimes and the possibility of cyclic evolution (Chapter 5), emphasizing the role of ancestral conditions and the simplest form of Red Queen dynamics; catastrophes of long-term evolutionary regimes (Chapter 6), explaining the sudden loss of evolutionary persistence of one or more populations due to microscopic variations of environmental parameters; branching-extinction evolutionary cycles (Chapter 7), characterizing cyclical patterns of diversity in the community; evolutionary reversals due to population bistability (Chapter 8), giving rise to evolutionary cycles even in the extremely simple case of a single isolated population characterized by a single adaptive trait; the formation of evolutionary ridges (Chapter 9), namely the possibility that slow-fast demographic interactions canalize the evolution of the adaptive traits; and, last, the possibility of chaotic Red Queen dynamics (Chapter 10) in three-species coevolving assemblies. The general conclusion that emerges from these lessons is that evolutionary scenarios can be as complex and diverse as one can imagine. Evolution can build up complexity in two ways: by increasing indefinitely the number of resident groups in the community, or by keeping them in constant number but forcing the coevolving traits to vary in an apparently random fashion. However, evolution can also reduce complexity through avalanches of evolutionary extinctions, so that phases of diversity enrichment can alternate with phases of diversity impoverishment.

The nature of the applications is announced at the beginning of each chapter. The most simple applications deal with the dynamics of exploited renewable resources (Chapter 8) and with the emergence of technological variety due to innovations in markets with competing commercial products (Chapter 4). A number of applications refer to predation (exploitation) systems: the case of resource-consumer communities is studied in Chapters 5 and 9, while the more complex case of resource-consumer-predator communities is considered in Chapter 10. A variant of predation systems of quite significant interest in biology, namely that of cannibalistic populations, is studied in Chapter 7. Finally, the coevolution of mutualistic populations and the puzzle of the long-term persistence of cheaters are discussed in Chapter 6.

The book is addressed to students and professional researchers in the broad areas of applied mathematics and evolutionary biology, and can serve as textbook for courses at graduate level. Scholars in social sciences and economics may also find the book of interest for its systematic view of the Darwinian paradigm outside the biological realm. Undergraduate skill in mathematics is the only prerequisite. Advanced mathematical tools are introduced in a rigorous but not overwhelming style, and this makes the book essentially self-contained.

ACKNOWLEDGMENTS

This book is the result of five years of research on evolutionary dynamics and of a challenging proposal of Simon Levin who reacted with enthusiasm to the thesis of F. D. (2002). Now that we have accomplished our task, we express our warmest gratitude to Simon for suggesting this adventure to us, one that we would not have considered possible without his encouragement. We also thank Vickie Kearn, our editor at Princeton University Press, and her colleagues for their invaluable technical support and generous understanding of our delays.

A relevant part of the work has been carried out at the International Institute for Applied Systems Analysis (IIASA), Laxenburg, Austria, where S.R. is appointed and F.D. spent six months in the period 2000–2002. Our cooperation with IIASA developed within the Adaptive Dynamics Network (now Evolution and Ecology Program), an international project led first by Hans Metz and then by Ulf Dieckmann, who introduced us to the theory they had founded, transmitted to us their enthusiasm for the study of evolutionary processes, and guided us during the first stages of our research. In particular, Ulf Dieckmann helped us organize the first two chapters and Hans Metz gave us important advice on the proof of the invasion implies substitution theorem (see Appendix B).

During our visits to IIASA we had the chance to meet with members of the ADN project as well as with many guests and visitors attracted at Schloss Laxenburg by the scientific reputation of the core team. Among them we acknowledge Karl Sigmund, Régis Ferrière, Stefan Geritz, Eva Kisdi, and David Claessen for their stimulating ideas and suggestions.

All coauthors of our papers on AD and related topics are, of course, hidden contributors to this book. Régis Ferrière has, among them, the lead for his deep understanding of evolutionary processes and for his vigor and enthusiasm. Yuri Kuznetsov was strategically important for his expertise in bifurcation analysis, and Alessandra Gragnani, Jean-Olivier Irisson, and Michael Obersteiner were very effective on specific applications.

We also acknowledge Peter Abrams, Oscar De Feo, Odo Diekmann, Marino Gatto, and Cesare Marchetti for fruitful discussion and advice, and Carlo Matessi who helped us a lot in critically comparing the various quantitative approaches to evolutionary dynamics.

Finally, we are also pleased to thank the institutions that supported our work. First, our home institutions, IEIIT, CNR, and DEI, Politecnico di Milano and, then, IIASA, Laxenburg, the Mathematical Eco-Evolutionary Theory Group of the École Normale Supérieure, Paris, and the Department of Mathematics, Utrecht University, where we have always been welcomed with enthusiasm. Moreover, we also acknowledge the European Science Foundation, the National Project MIUR-FIRB 2001 on "Modeling and Control of Complex Systems," and the National Project MIUR-PRIN 2002 on "Nonlinear Dynamics and its Applications" for their financial support.

Analysis of Evolutionary Processes

Chapter One

Introduction to Evolutionary Processes

In this chapter we introduce the basic elements and the empirical evidence of evolutionary processes. Since the groundbreaking work *The Origin of Species* by Charles Darwin (1859), a great deal of effort has been dedicated to the subject (see, e.g., Fisher, 1930; Haldane, 1932; Dobzanski, 1937; Mayr, 1942, 1963, 1982; Wright, 1969; Dawkins, 1976, 1982, 1986; Cavalli-Sforza and Feldman, 1981; Maynard Smith, 1989, 1993; Maynard Smith and Szathmary, 1995, just to mention a few masterpieces). Our discussion on the origin of evolutionary theory is mainly taken from the introduction by Ernst Mayr (2001) to the seventeenth printing of Darwin's famous book, and from Dieckmann (1994, Chapter 1), Schrage (1995), Rizzoli-Larousse (2003), and the web pages of the University of California Museum of Paleontology. Throughout the exposition we emphasize that, even though the major scientists who developed evolutionary theory were stimulated by the study of nature, their ideas not only apply to the biological realm, but also capture many phenomena of self-organization encountered in social sciences, economics, and engineering.

1.1 ORIGINS OF EVOLUTIONARY THEORY

The idea that living organisms have been diversifying themselves through time, starting from a common origin, goes back to the Greek naturalistic philosophy. Among the precursors of evolutionary theory, as we define it today, we can mention Anaximander of Miletus (610–546 BC), Empedocles of Acragas (495–435 BC), and later some clergymen, such as Saint Augustine (354–430). The evolutionary conceptions of Greek philosophy were known during the Renaissance. However, no further contribution arose until the eighteenth century, when European scholars still believed that the universe was created in essentially its present and final state. During the eighteenth century, the work of intellectuals known as "encyclopedists" spread the Illuminism doctrine and, in particular, the results of pioneering research in systematic biology, aimed at hierarchically classifying organisms into groups that successively share more and more visible structural characteristics. Their work brought a better understanding of the concept of species and highlighted fundamental similarities between widely disparate organisms. Such similarities were in contrast with the hypothesis of creation in final state and prepared the ground for evolutionary theory. A considerable contribution came from Georges-Louis Leclerc Buffon (1707–1788), author of a compendium of biological history, and from Erasmus Darwin (1731–1802), Charles' grandfather, who first discussed the conjecture that life could have evolved from a common ancestor and posed the question of how a species could evolve into another. The first explicit evolutionary theory was formulated by Jean Baptiste Lamarck (1744–1829), disciple of Buffon, who introduced the notion of inheritance. The "Lamarckian" hypothesis, that simple life forms continually come into existence from dead matter and continually become more complex, was strongly criticized by most naturalists of the time. In particular, one of the most active antievolutionists, Georges Cuvier (1769–1832), paradoxically provided evidence to the evolutionary hypothesis with his research in systematic biology, comparative anatomy, and paleontology.

At this point Charles Darwin (1809–1882) and Alfred Russel Wallace (1823–1913) formulated the evolutionary theory that we still accept today. In their papers published in the same issue of the *Journal of the Proceedings of the Linnean Society* (Darwin, 1858; Wallace, 1858, often cited as a single paper with the title "On the Tendency of Species to form Varieties; and on the Perpetuation of Varieties and Species by Natural Means of Selection") they presented their theory of evolution by *natural selection*, arguing that

- there is individual variation in innumerable characteristics of populations, some of which may affect the individual ability to survive and reproduce;

- there is likely to be a hereditary component to much of this variation, but evolutionary change ultimately relies on the appearance of new variant forms of organisms, called *mutants*;

- generation by generation there is a natural selection of the characteristics associated with greater survival and reproductive success, whose frequencies in the populations increase over time;

- the cumulative effects of mutations and natural selection, over a long period of time, alter the characteristics of species from those of their ancestors;

- all living organisms have descended with modifications from a common ancestor, thus developing hierarchical patterns of similarities.

Darwin and Wallace combined empirical observations with theoretical insights gained from Malthus' (1798) work on competition and population growth. They had a precise idea of natural selection and realized the need of mutations. However, they were not aware of the laws of heredity, discovered seven years later by Gregor Mendel (1822–1884), who realized the discrete nature of heredity determinants, which we now call genes (Mendel, 1865). Darwin actually introduced the concept of natural selection and deserves, more than anyone else, the credit for having started, and firmly supported, the scientific and philosophical revolution from the dogma of creation and constancy of species to evolutionary theory.

The decisive event in Darwin's life was the five-year period spent as a naturalist on the vessel Her Majesty's Ship *Beagle* (from December 1831 to October 1836), in which he surveyed the coast of South America and the off-lying islands, collecting invaluable observations on the tropical forests of Brazil, on fossils in the Pampas of Argentina, on the geology of the Andes, and on the animal life of the

Galapagos Islands (Darwin, 1839). After the return of the *Beagle*, Darwin spent most of his time in the analysis and interpretation of his findings and became later more acknowledged than Wallace, thanks to his famous book *The Origin of Species* (1859).

A third important scientist in the development of evolutionary theory, though by far less acknowledged than Darwin and Wallace, is Patrick Matthew (1790–1874). In a letter to Charles Lyell (April 10, 1860) Darwin says: "In last Saturday Gardeners' Chronicle, a Mr. Patrick Matthew publishes long extract from his work on Naval Timber & Arboriculture published in 1831, in which he briefly but completely anticipates the theory of Natural Selection. — I have ordered the book, as some few passages are rather obscure but it is, certainly, I think, a complete but not developed anticipation." Matthew's evolutionary insights lie buried in an appendix of a book he wrote on raising trees of optimal quality for the Royal Navy (Matthew, 1831). In that appendix Matthew expressed his theory based on how tree species might vary in form and how artificial selection might improve cultivated trees.

Let us now listen directly to Darwin, Wallace, and Matthew. "The affinities of all the beings of the same class have sometimes been represented by a great tree. I believe this simile largely speaks the truth. The green and budding twigs may represent existing species; and those produced during each former year may represent the long succession of extinct species... The limbs divided into great branches, and these into lesser and lesser branches, were themselves once, when the tree was small, budding twigs; and this connexion of the former and present buds by ramifying branches may well represent the classification of all extinct and living species in groups subordinate to groups... From the first growth of the tree, many a limb and branch has decayed and dropped off, and these lost branches of various sizes may represent those whole orders, families, and genera which have now no living representatives, and which are known to us only from having been found in a fossil state... As buds give rise by growth to fresh buds, and these, if vigorous, branch out and overtop on all a feebler branch, so by generation I believe it has been with the Tree of Life, which fills with its dead and broken branches the crust of the earth, and covers the surface with its ever branching and beautiful ramifications" (Darwin, 1859).

"We have also here an acting cause to account for that balance so often observed in Nature — a deficiency in one set of organs always being compensated by an increased development of some others — powerful wings accompanying weak feet, or great velocity making up for the absence of defensive weapons; for it has been shown that all varieties in which an unbalanced deficiency occurred could not long continue their existence. The action of this principle is exactly like that of the centrifugal governor of the steam engine, which checks and corrects any irregularities almost before they become evident; and in like manner no unbalanced deficiency in the animal kingdom can ever reach any conspicuous magnitude, because it would make itself felt at the very first step, by rendering existence difficult and extinction almost sure soon to follow" (Wallace, 1858).

"As Nature, in all her modifications of life, has a power of increase far beyond what is needed to supply the place of what falls by Time's decay, those individuals who possess not the requisite strength, swiftness, hardihood, or cunning, fall prematurely without reproducing — either a prey to their natural devourers, or sinking under disease, generally induced by want of nourishment, their place being occupied by the more perfect of their own kind, who are pressing on the means of subsistence... a law universal in Nature, tending to render every reproductive being as the best possibly suited to its condition... There is more beauty and unity of design in this continual balancing of life to circumstance, and greater conformity to those dispositions of Nature which are manifest to us, than in total destruction and new creation. It is improbable that much of this diversification is owing to commixture of species nearly allied, all change by this appears very limited, and confined within the bounds of what is called species; the progeny of the same parents, under great differences of circumstance, might, in several generations, even become distinct species, incapable of co-reproduction" (Matthew, 1831).

After more than a century since the publication of Darwin's famous book, we can say that the impact it had on man's concept of himself and his activities has been dramatic and has gone far beyond biology. Reduced to the essential, evolutionary change can be described as a two-step process: the first step consists of *innovation*, namely the production of variations, while the second is ruled by *competition* and leads to the selection of the best-performing variants. This abstract paradigm can be applied to many processes of self-organization that drive the evolution of complex natural and artificial systems, composed of several interacting agents, or units, each characterized by individual traits that are transmitted, with possible modifications, to agents or units of next generation, and naturally or artificially selected by their effectiveness or by optimization criteria. In other words, after the Darwinian "revolution," networks of socio-cultural relationships, the global economy, and the design of several industrial processes can be interpreted and studied as evolutionary processes.

In particular, Richard Dawkins reformulated the paradigm of evolution independently of genetic inheritance to explain the evolution of culture. In his books *The Selfish Gene* (1976) and *The Extended Phenotype* (1982) he argued that cultures, namely the clouds of ideas, behavioral traits, and artifacts developed and produced by animal and human populations, compete, cooperate, mutate, and are transmitted as well as genetic traits. Thus, the constitutive elements of cultures do evolve, and actually coevolve with genetic traits in a whole biological-socio-cultural evolutionary process. Dawkins introduced the concept of a *replicator* as the minimum natural or artificial unit evolvable through a *mutation-selection* (or better *innovation-competition*) *process*. Replicators are characterized by four fundamental properties:

- *replication*: units generate or are replaced by new units;

- *transmission*: units are characterized by distinctive features that are passed to the units of next generation;

- *innovation*: transmission is not a perfect copy but allows for variations;

- *competition*: survival and effectiveness of units are regulated by the characteristic features of interacting units.

Dawkins' idea of a replicator was elaborated on just a few years after Francis Crick and James Watson's (1953) discovery of the double helix molecular structure of the *deoxyribonucleic acid* (DNA). Although this idea might simply seem an abstract summary of Darwin and Wallace's evolutionary theory, its conceptual impact is enormous because it offers a paradigm of evolution that can be scaled from DNA to macroscopic natural and artificial systems. In fact, even if Dawkins' work is focused on animal and, in particular, human biological and cultural coevolution, the concept of replicator is fully independent of biology and can in principle be applied to describe any innovation-competition process. However, for tradition, simplicity, and uniformity, we will mainly refer throughout this chapter to biological evolution, pointing out here and there, and, in particular, in Section 1.9, the analogy with the evolution of social and economic systems. Later, in Chapter 4, we will interpret and analyze the process of technological change as an evolutionary process, where replicators are commercial products competing in a market, which replicate in production and transmit, with possible innovations, technological characteristics to new generations of products. Through a stylized model we will show that the evolutionary interpretation of technological change supports the emergence of technological variety from a single ancestral technology.

In the next three sections we introduce the basic biological elements of evolutionary processes, namely the structure of the genetic material and the laws of heredity, underpinning the way of being and reproducing of all living organisms, the appearance of mutations, ultimate source of organisms variability, and the mechanisms of selection of successful mutants. Among the innumerable options, we closely follow Charlesworth and Charlesworth (2003), a lucid and concise introduction to evolutionary theory, presenting some of their contributions in a form appropriate for this book. Sections 1.5–1.8 describe the evolutionary patterns we might expect to emerge as long-term consequences of evolutionary processes, while the last section shows many examples of such patterns in biological as well as nonbiological contexts.

1.2 GENOTYPES AND PHENOTYPES

Starting with Mendel's (1865) work and passing through the achievements of modern molecular biology, we now know that the similarities between living organisms are not confined to visible structural characteristics, but are profound and extend to the smallest microscopic scale. All living organisms other than viruses (which are on the borderline of life) are composed of a single unit or an assembly of essentially similar units, the *cells*. In so-called *eukaryote* organisms, which include all multicellular species (animals, plants, and fungi) as well as some unicellular species, cells are delimited by a membrane and contain the *cytoplasm*, a gel with floating subcellular structures, and the *nucleus*, which carries the genetic material. The remaining unicellular organisms, called *prokaryotes* and including bacteria and similar organisms called *archaea*, are simpler cells in which the genetic material is floating in the gel with no subcellular structures and nucleus. Viruses are parasites that reproduce inside the cells of other organisms and consist of a chemical coat surrounding the genetic material.

Cells are miniaturized factories that produce the chemicals needed by the organism, generate energy from food, and provide body structures. Most chemical products of a cell, as well as the machinery employed for the production of chemicals and energy, are *proteins*. Some proteins are *enzymes* that take a chemical and perform a task on it, like breaking molecules in food into smaller pieces that can be assimilated by cells. Some other proteins have storage or transport functions, like binding to iron and storing it in the liver, or to oxygen and carrying it through the blood. Some others are communication proteins, like *hormones*, which circulate in the blood and control many functions, or signaling proteins, located on the cell surface and responsible of communications with other cells. Finally, there are structural proteins, e.g., those forming skin and bones. Proteins are large molecules composed of a chain of *amino acids* typically arranged in a suitable spatial configuration. There are only twenty different amino acids across the entire range of living organisms. This points out how similar all organisms are at a microscopic scale and indirectly supports the evolutionary hypothesis of a common ancestor.

The genetic material of all species is organized in so-called *chromosomes*, composed of a DNA molecule and a protein coat. As explained below, there are different types of chromosomes, which control different functions of the organism and are characterized by a different size and structure of their DNA molecule. The number of different types of chromosomes is strongly species-specific, e.g., 23 in humans.

A DNA molecule consists of two helices of alternating sugar and phosphate molecules, where each sugar binds to one of four possible molecules: *adenine* (A), *cytosine* (C), *guanine* (G), and *thymine* (T), called DNA *bases*. Each base in one helix binds to the base at the corresponding position of the other helix, but only the bindings A–T and C–G are possible, so that given the sequence of bases in one helix, the sequence in the other helix is simply obtained by exchanging A with T and C with G. Thus, a molecule of DNA is represented by a sequence of the "genetic letters" A, C, G, and T, whose length is specific to the considered type of chromosome.

Each type of chromosome is also characterized by a logical structure of its DNA molecule. In fact, within the sequence of genetic letters representing the chromosome there are particular subsequences, called *loci*, whose number, lengths, and positions are also specific of the type of chromosome. The portion of the DNA molecule corresponding to a locus is a *gene* and may take one of several forms, called *alleles*, corresponding to different subsequences of genetic letters that are possible for that gene. Chromosomes of the same type can therefore contain different alleles at the same loci.

The function of most genes is to code the structure of a protein. For this, each triplet of genetic letters identifies an amino acid, and the sequence of triplets identifies the amino acid chain of the protein. Since there are $4^3 = 64$ different triplets and only 20 amino acids, several triplets code the same amino acid. Other than genes, there are also DNA portions which control the activity of the cell. In particular, they determine the proteins to be produced, and this is the main mechanism of cell diversification.

The genetic material is transferred by parents to the progeny and this transmission is called heredity. There are two mechanisms of reproduction: *asexual* and *sexual*. In asexual species (all prokaryotes and a few eukaryotes) all cells are *haploid*, i.e., they contain only one chromosome for each type of chromosome characterizing the species. An individual produces an offspring by simply replicating its genetic material through the process of cell division (*mitosis*), where chromosomes are first duplicated and then the cell divides into two identical daughter cells.

In sexual species some cells are *polyploid* (e.g., *diploid*), i.e., they contain more than one (e.g., two) chromosome for each type (except for special chromosomes identifying sex), which form a group of so-called *homologous* chromosomes. There are fundamentally two mechanisms of sexual reproduction. In diploid species (animals and many plants) all cells are diploid except for eggs and sperms, called *gametes*, which are haploid. The production of gametes (*meiosis*) is a special type of cell division, where first each pair of homologous chromosomes exchanges the alleles at some loci (*recombination*), then all new chromosome pairs are duplicated, and finally the cell divides into four gametes, each with one chromosome for each type. The union of an egg and a sperm restores a diploid cell, the fertilized egg, which develops into a new individual by successive cell divisions (mitosis). The second mechanism of sexual reproduction characterizes sexual haploid species (most fungi, some plants, and some unicellular eukaryotes), where the fusion of haploid cells produces polyploid cells in which the recombination of homologous chromosomes takes place before cell division gives rise to new haploid cells. Asexual reproduction may also occur in some sexual haploid species, where it is either occasional or the predominant mechanism of reproduction. Notice that haploid cells are temporary in the first mechanism of sexual reproduction, while polyploid cells are temporary in the second. Moreover, in both cases, homologous chromosomes in polyploid cells are inherited one from each parent (one from the mother and one from the father in diploid species), so that the genetic material of the progeny is a mix of that of the parents.

The *genome* of a species is the set of all possible chromosomes characterizing an individual of the species and is therefore identified by all allelic forms of all genes of the species. By contrast, the *genotype* of an individual is a particular genome realization, given by the chromosomes carried by the individual. Notice that in asexual species and in sexual haploid species the genotype is identified by one chromosome for each type and is contained in all cells, while in diploid species the genotype is given by pairs of homologous chromosomes, i.e., by two chromosomes for each type, and is contained in all cells except gametes.

Any individual characteristic determined (to some extent) by the genotype is called a *phenotype* (or *phenotypic trait*) and is therefore a heritable characteristic from parents to the progeny. Almost every imaginable kind of characteristic is genetic-dependent, from physical traits, like body size or colors, to mental traits, like character, disposition, and intelligence. Phenotypic variability within populations, called *polymorphism*, may take the form of discrete differences, as for the number of limbs and blood type, or that of a continuous range of values measurable on a metric scale, as for body size and weight. Discrete phenotypes are typically controlled by differences in one or a few genes and are unaffected, or altered only

Table 1.1 Blood type genotype-to-phenotype map.

allelic pairs	blood type	allelic pairs	blood type	allelic pairs	blood type
aa++	A+	bb++	B+	ab++	AB+
aa+−		bb+−		ab+−	
ao++		bo++		ab−−	AB−
ao+−		bo+−		oo++	O+
aa−−	A−	bb−−	B−	oo+−	
ao−−		bo−−		oo−−	O−

slightly, by the environmental circumstances experienced by the individual. By contrast, continuous phenotypes are typically influenced by many genes and by the environmental conditions as well.

The map between genotypes and phenotypes can be astonishingly complex and in most cases it is not yet completely understood. Phenotypes can be controlled by many genes, and some genes control several phenotypes (so-called *pleiotropic genes*), so that different phenotypes can be controlled by different but overlapping sets of genes. The effect of a gene on a phenotype can be simply a direct contribution to the phenotypic value, but there are also genes whose effect is to alter or switch off the direct effect of other genes (so-called *epistatic genes*). A further complicacy in diploid species is the mechanism of dominance between alleles. An allele is *dominant* over another allele (called *recessive*) if the presence of the two alleles at the same locus of two homologous chromosomes gives only the phenotypic effect of the dominant allele. If the same allele is present at a certain locus of a homologous pair of chromosomes, the locus is said to be *homozygous* (as well as the genotype with respect to that locus); otherwise, it is *heterozygous*. Thus, the phenotypic effect of a dominant homozygous locus is the same as that of an heterozygous locus with a dominant and a recessive allele.

For example, there are two genes responsible of our blood type. One is coding for a protein present on the surface of red corpuscles, which can be of two different types or absent, corresponding to three allelic forms (say a, b, and o). The other is coding for the so-called Rh-factor, a protein that can be present in red corpuscles (allele +) or absent (allele −). Alleles a and b are dominant on o, and the positive Rh allele is dominant on the negative one, so that there are eight possible blood types, as summarized in Table 1.1. This is one of the rare examples in which the genotype-to-phenotype map is known. Notice, however, that such a map is not invertible, not only because of the mechanism of dominance in diploid species, but also because of the existence of genetic differences with no effect on phenotypes, as, for example, the existence of different triplets of genetic letters coding for the same amino acid.

1.3 MUTATIONS

As Darwin and Wallace first realized, evolutionary change relies on the appearance of new forms of organisms, called mutants, characterized by changes in their phenotypes with respect to their conspecifics. Such phenotypic changes reflect heritable changes in the organism genetic material, i.e., what biologists call *mutations*. Heritable phenotypic variations have been documented in many organisms and for all kind of characters, including aspects of intelligence and behavioral strategies, like the role of dominance in the social hierarchy and the altruistic propensity among nonrelated individuals. Variations in such characters may confer competitive advantages or disadvantages in terms of survival and/or reproductive success.

Genetic and phenotypic variability within a population does not necessarily require mutations. In fact, as we have seen in the previous section, different environmental conditions may shape the phenotypes of different individuals, and the mechanisms of sexual reproduction, in particular the recombination of homologous chromosomes, result in offspring genotypes and phenotypic values different from those of the parents. However, if we imagine that all individuals of a population experience the same environmental conditions and that the processes of chromosome recombination and duplication are error-free, then the genetic material of the progeny is a mix of that of the parents, so that no new phenotypic value can appear other than those already observed or potentially observable by suitably mixing individual genotypes. Over a sufficiently long period of time, the genetic and phenotypic variability would halt and all statistics of the genotypic and phenotypic distributions would not evolve anymore. Thus, mutations, i.e., miscopying errors in the duplication of DNA molecules, are the ultimate source of genetic variability and form the raw material upon which selection acts, as shown in the next section.

The simplest mutation consists of a base substitution in a gene. As a result, a wrong amino acid may be placed in the protein coded by the gene, and the protein may be unable to perform its task properly. Other types of observed mutations involve the modifications, insertions, or deletions of entire DNA portions and may thus alter the length and the logical structure of chromosomes. Mutations can cause serious diseases. For example, the development of a cancer is favored by mutations involving genes coding for proteins controlling cell division. But mutations can also be advantageous, for example, by increasing the resistance of animals and plants against diseases, as well as chemical resistance in pests and antibiotic resistance in bacteria.

The phenotypic effect of a mutation may vary greatly in magnitude. Some mutations have no phenotypic effect and are known only because it is nowadays possible to observe the genetic material directly. Most mutations affect a small portion of a single gene and therefore have small phenotypic effects, especially on phenotypes controlled by several genes. But even on phenotypes controlled by a few genes, like the blood type, a huge mutation replacing most DNA bases of a gene would be needed to change the phenotype. Large phenotypic effects, like a blood type not obtainable from the combination of parental genotypes or the heritable possession of an extra leg, are, in principle, possible but extremely unlikely to occur.

Large mutations are more plausible if we extend the evolutionary paradigm beyond the biological realm. As discussed in Section 1.1, cultures, behavioral strategies, companies, goods, technologies, and many other artificial replicators evolve through innovation-competition processes analogous to the mutation-selection process that drives the evolution of living organisms. Most of the time, innovations consist of slight modifications, but revolutionary ideas, like the invention of differential calculus by Isaac Newton, or relevant technological innovations, like the steam engine, are certainly instances of large mutations. However, not all important technological innovations can be associated with large mutations. For example, the first personal computer was huge and expensive like a big mainframe, and the first mobile phone was a heavy car phone, different from a traditional phone only for the presence of an antenna instead of a wire.

Finally, notice that a mutation is not the only way to inject a new variant into an evolving system, since similar or radically different traits can come from outside. Examples are immigration in biological systems, communication through fast and worldwide media such as the Internet in social systems, and import-export of goods in economic systems. New traits originated by mutations or external influences typically appear in a single replicator or in a tiny fraction of the population. Thus, a mutation can have a long-term impact on the system only if other processes can cause the new trait to increase in frequency within the population. This is the topic discussed in the following section.

1.4 SELECTION

From an evolutionary point of view, what matters is the effect of a mutation on the phenotypes of the mutants. In fact, the performance of an individual depends on its characteristic phenotypes and on the environmental conditions it experiences. The environment experienced by an individual is identified by all physical factors, such as climate, altitude, oceans level, and air or water pollution, which define the so-called *abiotic environment*, and by all individuals of the same or other species interacting with the considered individual, which form its *biotic environment* (see, e.g., Lewontin, 1983). The performance of an individual continuously involved in intra- and interspecific and environmental interactions therefore depends on its phenotypes, on the abundances and phenotypes of the interacting individuals, and on the abiotic environmental conditions, which typically fluctuate in time.

Individuals with phenotypic values different from those of the rest of the population may thus differ in their probability to survive to reproductive age (*viability*), in their mating success, or in their progeny abundance (*fertility*). As a result, if we imagine no mutations or injections of new phenotypes from outside in a constant abiotic environment, namely in the absence of mutations, immigrations, seasonalities, and other external perturbations, then the demography of populations (i.e., birth and death of individuals) must promote the best-performing phenotypic values, which come to dominate the populations. In other words, what Darwin called *selection* is demography in the absence of mutations and external influences.

In nonbiological contexts, the success of artificial replicators, like different behavioral strategies in social relationships or products competing in a market, depends on the characteristic traits of interacting replicators (the equivalent of phenotypes), on their abundances, and on the rules and conditions constraining the system (the equivalent of the abiotic environment in biological systems). Thus, the best-performing replicators become dominant if there are no further innovations and external influences.

Demography is the dynamical process that regulates population abundances. The state of a population is determined by the genotypic distribution, which gives the abundances of all genotypes present in the population at a given time, and the dynamics of the population are the changes in time of such a distribution that result from birth, death, and migration of individuals. Given the genotypic distribution of a population, the distribution of one of its phenotypes is, in principle, obtainable by applying the corresponding genotype-to-phenotype map. By contrast, the genotypic distribution cannot be inferred from the phenotypic distribution because, as we have seen in Section 1.2, the genotype-to-phenotype map is not invertible. Unfortunately, we have also seen that such a map is usually not known, so that the joint distribution of all relevant phenotypes must generally be measured or assumed to be known to completely characterize the population. Thus, the state of an evolving system is given by the genotypic and phenotypic distributions of all interacting populations, i.e., by the state of the entire community or, in other words, by the biotic environment. Given such a state at a given time, the future abundances are not predictable if there is no a priori information on mutations and possible external influences. In technical words, demography is a nonautonomous dynamical process, i.e., a process whose future is not solely determined by its current state. On the contrary, selection is an autonomous process (since it works in the absence of mutations and external influences) and, as such, drives the system toward a regime, which can be stationary as well as nonstationary (periodic or wilder, so-called *chaotic*, regime; see, e.g., Turchin, 2003a). Such regimes correspond to the so-called *attractors* of the dynamical process, since they attract nearby states.

When approaching an attractor, some of the phenotypic values may disappear from the community, because the groups of individuals bearing them are outcompeted by better-performing groups. Once the state of the community is close to the attractor, the genotypes and phenotypic values that coexist in the populations are called *resident*, as the groups of individuals bearing them. If a phenotype is characterized by a single resident value or, more weakly, if the resident phenotypic distribution is concentrated around its mean value, then the population is said to be *monomorphic* with respect to that phenotype. In reality, selection may be interrupted by a mutation, by the arrival of a new phenotype from outside, or by a perturbation of the abiotic environment, events that prevent the system from reaching an attractor. However, the time between successive mutations and external influences is often long enough to enable selection to define the resident groups of the community.

In this book, the characteristic timescale on which the process of selection drives the system toward one of its attractors is called the *demographic timescale*. Analogously, we respectively call *demographic dynamics* and *attractors* the population dynamics driven by selection and their possible attractors. In the biological literature the demographic timescale, dynamics, and attractors are often called "ecological" or "short-term," since ecology typically studies the relations between living organisms and their environment on a timescale that is so short that mutations and external influences can be neglected (see, e.g., Roughgarden, 1983b). As we will see in more detail in Chapter 8, there can be multiple demographic attractors, each one attracting a different set of states, which means that there can be different selection pressures acting on different genotypic/phenotypic distributions.

As pointed out by Dawkins (1976), selection is a selfish process, since each individual is moved by the ambition to maximize its impact on the next generation, by maximizing the transfer of its genes. Thus, it may seem difficult to explain altruistic behaviors and cooperation, which are commonly observed in many species, including the human one. Some biologists, starting with Wynne-Edwards (1962) (see also Wilson, 1980), have invoked a higher level of selection, called *group selection*, acting on groups rather than on individuals, hence promoting phenotypic values that are beneficial to the group at the expense of the single individual. However, the idea of group selection is founded on a weaker genetic basis and has been therefore severely criticized (see, e.g., Williams, 1966). Moreover, as we will show in Chapter 6, group selection is not necessary to explain the evolutionary emergence of altruistic behaviors, which can be the direct result of "individual selection." A related concept is that of *kin selection*, originally proposed by Darwin as an explanation of the existence of sterile individuals in various species of social insects (e.g., bees, wasps, and ants), where some of the females are workers who do not reproduce, thus apparently forgoing their ambition (Hamilton, 1963, 1964a,b; Eshel and Motro, 1981). However, members of a social group are typically close relatives, often sharing the same mother or father (e.g., all worker bees are daughters of the queen), so that the sacrifice of individual reproduction may considerably improve the reproductive success of relatives. Other altruistic behaviors toward relatives account for other details of animal societies. For example, young males guarding the nest, instead of attempting to mate, increase the viability of their relatives at the expense of their own reproductive success, possibly leading to an overall higher transfer of genes in the next generation. Thus, altruistic behaviors toward relatives can be interpreted as selfish behaviors, promoted by selection as well as phenotypic values improving individual survival and/or reproduction (Grafen, 1984).

For brevity, and to allow one to think in general terms, the word *fitness* is often used in the biological context to stand for overall ability to survive and reproduce. Quantitatively, the fitness of an individual is defined as the abundance of its progeny in the next generation or, equivalently, as the per-capita growth rate of the group of individuals characterized by the same phenotypic values (i.e., the abundance variation per unit of time relative to the total abundance of the group). These two definitions say that the abundance of individuals characterized by a given set of phenotypic values is increasing, at a given time, if the associated fitness is, respectively, larger than one or positive. As we will see in the next chapter, the quantitative approaches for modeling evolutionary dynamics make an extensive use of the concept of fitness, and adopt the first or the second of the above definitions depending on the fact that the description of the time is discrete, i.e., by generations, or continuous.

By definition, the fitness of an individual depends on both the abiotic and biotic components of its environment. In particular, the dependence on the biotic component, i.e., on the phenotypic distributions of all interacting populations, is often emphasized by saying that the fitness, or selection, is *frequency-* and/or *density-dependent* (see, e.g., Li, 1955; Lewontin, 1958; Haldane and Jayakar, 1963; Wright, 1969; and the discussion in Heino et al., 1998). For example, speed is crucial for antelopes living in savannas, especially at low abundances, since they cannot defend themselves by gathering in big groups, but it is not so important in other habitats where they find refuges from predators. Analogously, some technological innovations are useful in particular market sectors and conditions and not in others, and the benefit of behavioral strategies typically depends on the socio-cultural context.

As discussed in the previous section, mutations appear in a single individual, or in a tiny minority with respect to the resident groups. Disadvantageous mutations, e.g., those causing the malfunction of important proteins, reduce the survival and reproductive success of affected individuals, so that mutants will be underrepresented in the next generations and will eventually be eliminated by the competition with similar residents. One of the major roles of selection is indeed to keep undesired characteristics under control. By contrast, mutants that are at some advantage in terms of survival and reproduction with respect to resident individuals initially tend to grow in number. However, since they are few in number, they face the risk of accidental extinction and may fail to invade, i.e., to spread among the resident groups. The mutant fitness when mutants are just appeared is usually called *invasion fitness*, and its value gives a quantitative indication of whether the mutation gives some advantage or disadvantage to its bearers. As described in the next section, evolutionary change develops whenever there are phenotypic traits affecting the fitness of invading mutants that escape accidental extinction and become new resident individuals.

1.5 EVOLUTION

Intuitively, we would like to define evolutionary dynamics as the long-term dynamics of the phenotypic distributions of interacting populations. If there are several populations of different species, interacting directly like predator and prey and symbiotic partners, or indirectly like populations competing for common resources, most likely the evolution of one population affects the evolution of the others, so that the coupled evolution of all interacting populations, called *coevolution* (see, e.g., Futuyma and Slatkin, 1983; Thompson, 1994; Futuyma, 1998), must be studied.

As Darwin pertinently pointed out in the first chapter of *The Origin of Species* (1859), the evidence that the combination of the processes of mutation and selection leads to evolutionary change is given by artificial selection, the human activity aimed to the breeding of animals and plants with desirable phenotypic values. Artificial mutations have been achieved in the past by intentionally and suitably mixing the genotypes of wild species, and, recently, by directly modifying their genotypes, hence producing transgenic individuals. The obtained groups of mutants with desirable characteristics, like all kinds of dogs or many fruits and vegetables, are then intensively selected by favoring or allowing only intragroup reproduction. Artificial mutations and selection have produced striking evolutionary changes over a relatively short timescale, compared to the typical timescale of most natural evolutionary processes. In nature, mutations are often rare events on the demographic timescale (the typical probability of mutation per locus is of the order of one per million). Moreover, natural selection is the result of individual interactions such as predation, unidirectional or mutualistic symbiosis, competition for resources, and exploitation of environmental niches, which generally are not so intense as artificial selection can be.

So far, we have considered phenotypes affecting the demography of coevolving populations, i.e., affecting individual survival and/or reproductive success or, in a word, fitness, and we have seen that the combination of the processes of mutation and selection drives their evolution. Phenotypes with an effect on fitness are able to adapt to the environmental conditions experienced by individuals and are therefore said to be *adaptive*; they are often called *adaptive traits* and, most of the time, simply *traits* in the following chapters. In principle, there can also be phenotypes with no effect on fitness, as well as mutations with no effect on phenotypes. At each generation, the genes that control selectively neutral phenotypes, or that are altered by mutations with no phenotypic effect, are a sample of the genes present in the parental population, whose offspring are not filtered by any selection pressure. The accumulation of differences in such genes (and related phenotypes) produces an evolutionary change that biologists call *genetic drift*. Thus, we can distinguish between two sources of evolutionary change: mutation-selection processes involving phenotypes affecting fitness and the sole process of reproduction resulting in genetic or phenotypic differences with no effect on fitness. Notice, however, that individuals characterized by new phenotypic values (due to the recombination of parental genotypes and/or mutations) are initially few in number, so that if they are neutral to selection they have a much higher chance of being lost by accidental extinction than of spreading in the population. Thus, genetic drift is a very slow process, typically dominated by mutation-selection processes.

To focus on the basic dynamical features of mutation-selection processes, it is convenient to identify two contrasting timescales: the demographic timescale on which selection is acting and the timescale on which the cumulative effects of several mutations followed by the selection of successful mutants are noticeable. When these two timescales are strongly separated it is possible to qualitatively predict interesting features of evolution. In fact, as we have seen in the previous section, the time between successive mutations is in most cases long enough to allow selection to reach a demographic attractor and define the resident groups. More precisely, if a mutation confers some advantage to affected individuals and if mutants are able to escape accidental extinction, then selection brings the community in a new attractor. In the opposite case, mutants quickly disappear and the community is back to the previous attractor. Thus, in the idealized case of extremely rare mutations, we can define *evolutionary dynamics* as the sequences of attractors visited by the demographic dynamics. The timescale on which such a sequence develops is called in the following the *evolutionary timescale*. In this ideal frame, the demographic and evolutionary timescales are completely separated, i.e., any finite time on the evolutionary timescale corresponds to an infinite time on the demographic timescale. In the case of contrasting but not completely separated timescales, selection may not be able to reach an attractor before the occurrence of the next mutation, so that the difference between demographic and evolutionary dynamics is faded. As we will see in the next section, this problematic distinction has sometimes led to questionable interpretations of field and laboratory data.

The separation between demographic and evolutionary timescales not only allows a precise definition of demographic and evolutionary dynamics, but points out their coupling as well. In fact, demographic dynamics are determined by the biotic component of the environment, since they are defined under constant abiotic conditions in the absence of mutations and external influences. Immediately after a mutation, the biotic environment is given by the current attractor of the demographic dynamics, i.e., the current evolutionary state, since mutants are scarce and therefore have no effects. Thus, the current evolutionary state determines the demographic dynamics, which, in turn, define the new evolutionary state. Ford (1949) was perhaps the first to document that demographic and evolutionary changes are entangled in a "feedback loop," a concept that has been subsequently formalized by Pimentel (1961, 1968), Stenseth and Maynard Smith (1984), and Metz et al. (1992).

Other than being rare on the demographic timescale, mutations have often small effects on phenotypes, as we saw in Section 1.3, so that the evolutionary dynamics are slow and smooth. In fact, imagine a sequence of mutations with small phenotypic effects and assume that each time the mutant group is at advantage, it replaces the similar resident group, thus becoming itself the new resident. Then, we can picture a smooth evolutionary trajectory as follows. At a given time on the evolutionary timescale, define the trait space as the space with one axis for each adaptive trait characterizing each resident group. Then, the current evolutionary state corresponds to a point in trait space, which smoothly moves in the course of evolution. Notice that the dimension of the trait space changes each time a new resident group appears in the community or an old resident group disappears from it (see Sections 1.7 and 1.8).

As we will see in the next chapter, the assumption of rare mutations with small phenotypic effects is at the core of some of the mathematical descriptions of evolutionary processes, including the approach of adaptive dynamics. Although this assumption might seem a crude approximation of reality, we will see that it allows a mathematical description of the demographic and evolutionary dynamics that, at least qualitatively, provides useful insights on the real dynamics of the community.

As done in Section 1.4 for the demographic dynamics, we can now focus on the attractors of the evolutionary dynamics. However, the evolutionary process is intrinsically nonautonomous, since evolutionary dynamics depend on the particular sequence of mutations and on the fluctuations of the abiotic environment. Abiotic environmental conditions typically fluctuate in time as stationary processes or as (nonstationary) processes with relevant periodic components at particular frequencies, like seasonalities. Here and almost everywhere in the book we consider constant abiotic environments. This choice not only greatly simplifies the analytical treatment, but allows one to concentrate on the evolutionary dynamics endogenously generated by the mutation-selection process and not on those entrained by the variability of the abiotic environment. Of course, results obtained under this assumption remain qualitatively sound when the abiotic environmental fluctuations are sufficiently mild or when their characteristic frequencies are out from the frequency range of demographic and evolutionary dynamics. In conclusion, at least in a probabilistic sense, namely by averaging on all possible sequences of mutations, evolutionary change can be described as an autonomous dynamical process. Evolutionary dynamics are therefore characterized by stationary and nonstationary attractors, i.e., by points and periodic or chaotic evolutionary trajectories in trait space attracting nearby evolutionary states. Moreover, as we will better see in Chapter 5, evolutionary attractors can be multiple, so that different evolutionary trajectories may approach different attractors, making long-term evolutionary implications strongly dependent on the ancestral evolutionary state.

1.6 THE RED QUEEN HYPOTHESIS

So far, we have seen that evolutionary dynamics proceed by apparently random mutations followed by demographic selections of the individuals best adapted to the current environmental conditions. Thus, externally imposed perturbations of abiotic environmental factors, such as climate changes in biology, the availability of new communication media or transportation services in social systems, the opening of new markets, or the establishment of new trading rules or policies in economics, might be thought to be the ultimate driving forces of evolution. A very common idea is that in a constant environment mutation-selection processes should have the time to adjust the relevant traits to a state that confers the maximum fitness. As a consequence, little evolutionary change is expected, until the environment poses some new challenge. In other words, it is generally conjectured that in the absence of external forcing, evolutionary dynamics converge to a stationary attractor, namely to an evolutionary state at which mutants are always at disadvantage with respect to well-adapted residents. This belief, however, neglects the fact that the coevolving populations define the biotic component of the environment in which they live, so that the expression "evolution in a constant environment" actually becomes nonsense if both the abiotic and biotic components of the environment are taken into account. Adapting to an environment that is itself adaptive can thus prevent the evolutionary dynamics to reach a stationary regime.

The question as to whether the interaction between species may drive neverending evolutionary change in their phenotypes has been first posed by Van Valen (1973), who called this hypothetically endless evolutionary story *Red Queen dynamics* (see also Rosenzweig and Schaffer, 1978a,b; Stenseth and Maynard Smith, 1984; Rosenzweig et al., 1987). The name was inspired by the book *Through the Looking-Glass and What Alice Found There* by Lewis Carroll (1871), where Alice says: "Well, in our country, you'd generally get to somewhere else — if you ran very fast for a long time as we've been doing." And the Queen replies: "A slow sort of country! Now, here, you see, it takes all the running you can do, to keep in the same place."

The term the "Red Queen" is also associated in the biological literature with the evolution of sex, since chromosome recombination maintains genetic variability and phenotypic polymorphism and therefore tends to keep sexual populations "in the same place" (see, e.g., Bell, 1982; Lively, 1996). In particular, recombinations have the potential to combine favorable parental alleles in the loci of the progeny and to break up deleterious allelic pairs, thus conferring an evolutionary advantage to sexual versus asexual species. In this book, independently of sex, we refer to Red Queen dynamics as evolutionary dynamics that in the absence of external forcing, namely in a constant abiotic environment, lead to nonstationary evolutionary regimes.

Since Van Valen (1973), there have been many contributions in the literature on Red Queen dynamics (including our Chapters 5–10). However, there is essentially no neat empirical evidence of Red Queen evolutionary regimes, since data are often not available on a sufficiently long time interval or there is not enough information on the constancy of the abiotic environment to distinguish between externally imposed and autonomous evolutionary oscillations (Lythgoe and Read, 1998).

Some studies on living populations show interesting oscillations in population abundances and phenotypic traits, though the evidence is not conclusive. An example is provided by some African fish species, whose individuals eat scales from prey flanks and have developed an asymmetry toward left or right in the direction of mouth opening to improve the effectiveness of attacks from behind while approaching the right or left prey flank. Hori (1993; see also the commentary by Lively, 1993) reported oscillations in the abundance of left- and right-handed populations and concomitant oscillations of the antipredator traits (or behaviors) of the prey species, such as alertness to attacks from left or right. A second example comes from lizards, where oscillations of male reproductive strategies have been observed (Sinervo and Lively, 1996; Sinervo et al., 2000; see also Maynard Smith, 1996). Finally, Yoshida et al. (2003) recently showed oscillations in genetic characters in prey-predator (algal-rotifer) laboratory microcosms.

However, in all three cases, the observed oscillations develop on a demographic timescale and indicate the demographic nonstationary coexistence of different phenotypic values, rather than nonstationary evolutionary dynamics. Although some evolutionary biologists argue that in some species and under particular environmental conditions evolutionary change may be rapid and develop on a timescale comparable to the demographic timescale (see, e.g., Lively, 1993; Thompson, 1998; Zimmer, 2003), more investigations are needed. Recent and well-controlled experiments on host-parasite and bacterial evolution seem to be promising (see Dybdahl and Storfer, 2003, for a review), but at the moment the empirical evidence of Red Queen dynamics remains scant and it is plausible that Red Queen dynamics will remain, for a while, a conjecture.

1.7 THE EMERGENCE OF DIVERSITY

One of the major hypotheses of evolutionary theory is that all living organisms are the descendants of self-replicating molecules that were formed by chemicals more than 3 billion years ago. The successive forms of life have been produced by the

natural selection of successful mutations: the process of "descent with modification," as Darwin called it.

Throughout this chapter, we have often used the concept of *species* without giving a precise definition, but rather relying on an intuitive notion. However, in most
cases, a species can be defined as a group of morphologically and genetically similar individuals that, when reproduction is sexual, are capable of interbreeding and
reproductively isolated from other such groups (Mayr, 1942). Notice that while
two sexual species are kept separated by the lack of interbreeding, the distinction
between two asexual species critically depends on the considered measure of similarity. Biologists identify each species with two names, the *genus*, identifying
a group of extant or extinct species by a set of morphological and physiological
characters not shared by other genera, and the name of the species itself, both con-
ventionally written in italics, e.g., *Homo sapiens*. Genera are then grouped into
families, families into orders, orders into classes, classes into phyla, and phyla into
kingdoms, thus forming the so-called taxonomic classification of organisms, e.g.,
Hominidae, *Primates*, *Mammalia*, *Chordata*, *Animalia*, respectively, from family
to kingdom in our case.

The formation of new species, called *speciation*, is certainly one of the core
issues of evolutionary theory and a great deal of research has been devoted to it
(see, e.g., Hutchinson, 1959; Mayr, 1963; Maynard Smith, 1966; Felsenstein, 1981;
Kondrashov and Mina, 1986; Rice and Hostert, 1993; Kawecki, 1996; Dieckmann
and Doebeli, 1999; Higashi et al., 1999; Doebeli and Dieckmann, 2000, 2003;
Matessi et al., 2001; Dieckmann et al., 2004; Coyne and Orr, 2004; Gavrilets,
2004). Speciation occurs through the genetic and phenotypic divergence of conspecific populations that adapt to different environmental niches in the same or
different habitats. In the case of sexual reproduction, the populations must diverge
far enough to develop some barrier to interbreed, such as morphological or physiological incompatibilities, inviability or sterility of crossbred offspring (hybrids), or
simply different preferences in habitats or in mating periods, sites, and rituals.

The mechanism of speciation traditionally accepted by the scientific community
(and first proposed by Darwin) imagines that two populations of a given species
become geographically isolated and follow separate evolutionary paths. Since different selection pressures and different genetic drifts may act on different environments, the isolated populations may eventually become separate species. This form
of speciation is called *allopatric* when the two populations become geographically
separated by natural or artificial barriers, and *parapatric* when they evolve toward
geographic isolation by exploiting different environmental niches of contiguous
habitats. While allopatric and parapatric speciations are supported by the evidence
that variants of the same species, some of which may be candidate new species, often occupy different territories, their key ingredient, geographic isolation, remains a
somehow exogenous cause of speciation rather than an evolutionary consequence.

A different mechanism of speciation, put forward by Maynard Smith (1966) and
called *sympatric*, considers populations in the same geographic location. The key
ingredient here is a selection pressure favoring phenotypic values at the extremes of
a polymorphic range over those in the middle of the range. This so-called *disruptive selection* may result, for example, from the competition for alternative environ-

mental niches, where specialization for specific niches may be advantageous with respect to be generalist. Under the effect of disruptive selection a monomorphic population may turn dimorphic with respect to some relevant phenotypes, undergoing what is called an *evolutionary branching*. The monomorphic population splits into two initially similar resident groups, which then diverge by following separate evolutionary paths (branches), each one driven by its own mutations. Pictured in trait space, an evolutionary trajectory approaches an evolutionary state, called a *branching point*, at which the dimension of the trait space increases by gaining the adaptive traits characterizing the new resident group.

The potential of disruptive selection as a mechanism of speciation has been hotly debated. Some evolutionary biologists do not believe that sympatric speciation can generically occur in sexual species, because interbreeding would constantly produce phenotypes that are intermediate between the two incipient branches, thus contrasting any phenotypic difference that might appear under disruptive selection. Speciation would then require the evolution of mechanisms of *assortative mating* (i.e., the mating preference for similar phenotypes) disfavoring intermediate phenotypes. For example, morphological differences, reduced viability or fertility of crossbred offspring, or different reproductive timings within the season may accompany and therefore favor the phenotypic divergence fostered by disruptive selection. Based on these arguments, recent empirical and theoretical studies (see, e.g., Schliewen et al., 1994; Schluter, 1994; Feder, 1995; Doebeli, 1996a; Johnson et al., 1996; Kawecki, 1996; Dieckmann and Doebeli, 1999; Higashi et al., 1999; Matessi et al., 2001; Barluenga et al., 2006; see also the review by Gavrilets, 2003, and its commentary by Doebeli et al., 2005) suggest that sympatric speciation is, after all, not as rare as commonly believed.

Both geographic isolation and sympatric speciation consider the splitting of an existing population into two initially similar resident groups, which then genetically and phenotypically diverge and eventually become separate species. An alternative speciation route in sexual species, called *hybrid speciation* (Stebbins, 1959), occurs when offspring crossbred from two similar species are viable, fertile, and perform better than parental individuals in exploiting suitable environmental niches, where they can develop reproductive isolation and eventually form a new species. Although theoretically possible, hybrid speciation is regarded as rare and hence of little importance, especially in animals (Buerkle et al., 2000; Schwarz et al., 2005).

The last three centuries of biological studies have organized all known living organisms in a branching genealogy, called the *tree of life*. The root of the tree represents the hypothetical common ancestor, from which all species have been derived by subsequent speciations. All species ever derived are represented by the branches of the tree, whose leaves thus correspond to extant or extinct species. By dating all speciation events, the tree of life can be drawn on a vertical time axis with the present day at the top, as in Figure 1.1. Two species are more closely related at a given time, and therefore share more characteristics, the closer in time is their common ancestral species. Figure 1.1 is very schematic and shows only the major groups of organisms, namely prokaryotes (bacteria and archaea), unicellular eukaryotes (where the dotted branches stay for several not shown speciations), and

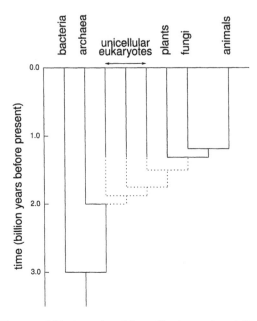

Figure 1.1 The tree of life (reproduced from Charlesworth and Charlesworth, 2003).

multicellular eukaryotes (animals, plants, and fungi) (see, e.g., Woese et al., 1990, and Wray, 2001, for an up to date and more detailed representation).

The classification of species and their genealogical relationships have been based for a long time on easily detectable morphological and structural characteristics, observed in living populations and in fossil records. Nowadays, it is possible to compare DNA and protein sequences among different extant species. Under the evolutionary hypothesis, the sequences of DNA bases of a given gene, or the chains of amino acids of a given protein, should be more similar for more closely related species than those of distantly related species. Thus, the times of speciations can be estimated under the assumption that the amount of differences increases proportionally with time since speciation. This is roughly true if mutational rates are constant and if the compared DNA and protein sequences have no effect on individual fitness, so that, as we have seen in Section 1.5, their evolutionary change consists of a random genetic drift solely driven by mutations. The accurate estimation of speciation times is therefore a complex problem and various techniques have been developed to provide better and better estimates (see, e.g., Wray, 2001). The result is that the evolutionary hypothesis is strongly supported, since the estimated times of speciations are in broad agreement with the times at which the major groups of animals and plants appear in fossil records. This confidence allows biologists to use molecular techniques to date the divergence between species for which there is no or little fossil evidence. By filling the gaps of fossil records, we are therefore getting closer and closer to the complete reconstruction of the tree of life.

As anticipated in Section 1.1, the evolutionary paradigm goes far beyond biology. The spontaneous emergence and maintenance of diversity are observable in many different areas of science and engineering and evolutionary theory may be used to identify the analogue of the biological mechanisms of speciation. For example, in Section 1.3 we have mentioned that the gradual innovation of technological products has led to different "product species," such as personal computers and mainframes, or fixed and mobile phones, which originated by a common ancestral product. As shown in Section 1.9 and formally in Chapter 4, the selection pressure exerted on the market by customers or other economic agents may be disruptive, so that evolutionary branching may occur and explain the emergence of product diversity. Goods that appear to us as radically different, like a horse carriage and a modern car, were very similar just after the branching. The socio-cultural history is also rich in gradual diversifications that we can describe through the mechanism of evolutionary branching. For example, different languages have common roots, as well as different political organizations and behavioral strategies.

In closing this discussion on the emergence of diversity in nature (*biodiversity*) as well as in other contexts, we like to emphasize the role of evolutionary branching, because it provides an autonomous evolutionary explanation for speciation, in the sense that it does not require exogenous triggers, like geographic isolation. However, notice again that an evolutionary branching only marks an increased polymorphism in a phenotypic distribution and thus only gives an indication for a possible speciation.

1.8 EVOLUTIONARY EXTINCTION

We have just seen how speciation mechanisms provide an evolutionary explanation to the diversity of life. It is estimated that roughly forty million species are living today on Earth. However, between five and fifty billion species are estimated to have existed at one time or another during the history of life (Wilson, 1988). These numbers reveal that the present diversity is the result of a small surplus of generation over loss of species, so that the causes of species extinction are as relevant as those of speciation.

Since the origin of life, diversity has generally increased over time, as shown for example in Figure 1.2, which reports the number of families of marine animals through the past 600 million years. Figure 1.2 also points out periods of diversity decline, where extinction was more intense than speciation, and abrupt extinction events, among which are the five so-called mass extinctions (vertical dotted lines, see Raup and Sepkoski, 1982). Minor extinction events, not visible at the scale of Figure 1.2, are noticeable in fossil records, so that biologists and paleontologists now believe that extinction and speciation have smoothly shaped the tree of life over the evolutionary timescale, with interruptions characterized by abrupt extinctions.

Clearly, abrupt extinction events, where many species living in different areas disappear in a relatively short time interval on the evolutionary timescale, are the fingerprints of some global catastrophe or rapid climate change that has little to do with evolution. In particular, the last mass extinction, where dinosaurs disappeared, thus making mammals and human evolution possible, is now well documented and associated with the impact of an asteroid in the Yucatan peninsula (Alvarez et al.,

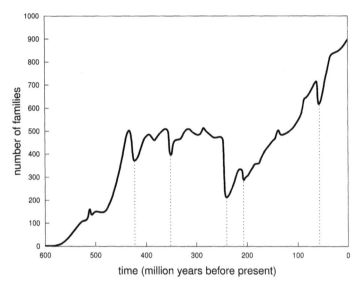

Figure 1.2 Evolution through the past 600 million years of the number of families of fossilized marine animals (reproduced from Sepkoski, 1984).

1980). The impact should have produced shock and tidal waves, darkness caused by dust with associated reduction of plant photosynthesis, acid rain, forest fires, heating due to the greenhouse effect, or cooling due to reduced sunlight. Other frequent causes of abrupt extinctions are glaciations and variations in oceans level, salinity, and oxygen concentration (Huey and Ward, 2005), while for minor extinction events the debate is much more open (see, e.g., Raup, 1981, 1991). External causes of minor extinctions may be several, e.g., earthquakes, volcanic eruptions, hurricanes, epidemic diseases, and even small-scale meteorite impacts (Raup, 1992).

Notice that the extinction of species is not a roulette-like random process, even when it is triggered by a catastrophe. In fact, a global change of the abiotic environment, occurring rapidly on the evolutionary timescale, begets an evolutionary transient in which species try to adapt to the new conditions. Some species may fail and go extinct, so that it is fully correct to refer to this evolutionary consequence as *evolutionary extinction*.

As we have repeatedly observed, the evolution of populations shapes the biotic environment in which they live, which in turn affects their evolution. Thus, as a change in the abiotic conditions may lead to the extinction of some of the coevolving populations, the same is, in principle, true for a change in the biotic conditions. In other words, evolution in a constant abiotic environment may drive the phenotypes of coevolving populations toward values at which some of the populations cannot persist. The fact that evolution may autonomously lead to the extinction of species was first claimed by Darwin, who observed that the mutation-selection process may favor phenotypic values that, in the long run, turn out to be inconvenient or even harmful to the survival of a population. Darwin concluded that if any part comes to be injurious, then either it will be modified by evolution, or the being will become extinct. Haldane (1932) noticed that fossil records are full of cases where enormous horns and spines have been the prelude to extinction and concluded that in some cases, one of which is presented in the next section, the species literally sank under the weight of their own armaments.

Much more recently, the possibility of evolution toward extinction in a constant abiotic environment has been theoretically investigated and three basic mechanisms have been identified (Matsuda and Abrams, 1994a; Ferrière, 2000; Gyllenberg and Parvinen, 2001; Dieckmann and Ferrière, 2004; Parvinen, 2005). The first, called *evolutionary runaway* by Matsuda and Abrams (1994a), is present when selection drives a population toward phenotypic values at which the population persists at low abundances, thus facing high risk of accidental extinction. For example, there are species in which a large, robust, or colorful body is accompanied by reduced fertility or increased predation risk and, hence, by low abundances of the population. However, selection between individuals may always favor larger and more robust or colorful bodies, because of their success in intraspecific competition for food or mates.

The other two mechanisms are based on the observation that the region of the trait space that allows the persistence of all coevolving populations is typically limited by an extinction boundary, since extreme phenotypic values are usually morphologically and/or physiologically incompatible with the abiotic and biotic environmental conditions. For example, a population may be able to persist up to a certain value of individual body size, which may depend on habitat morphology as well as on the phenotypes of coevolving predator or competing populations. In the course of evolution the abundance of the population may gradually vanish, when approaching the extinction boundary, or it may remain high even in the vicinity of the boundary but suddenly collapse (on the demographic timescale) when the boundary is reached, as is often the case for exploited renewable resources where a threshold abundance is needed for persistence. In the former case, the rate of phenotypic change of the population, being proportional to the number of mutants generated per unit of time, vanishes together with the population abundance. Thus, the extinction boundary is reached due to the evolution of other coevolving populations, which act as murderers, so that the extinction is called an *evolutionary murder*. By contrast, in the latter case, called *evolutionary suicide*, the population actively evolves toward self-destruction, i.e., mutants closer to the extinction boundary are at advantage with respect to resident individuals, even though they are unconsciously closer to extinction. In other words, as suggested by Darwin, what is advantageous for the individual may ultimately be disastrous for the population.

The subtle differences between evolutionary runaway, murder, and suicide are hard to observe based on empirical data, so that the debate on the possible causes of extinction remains wide open. In any case, we believe that theoretical investigations are important because in the lack of clear empirical evidence they conceptually show that evolution is able to autonomously destroy species. For example, in a recent review of fossil records Rohde and Muller (2005) have shown that diversity in marine animals, once detrended from environmental drivers, oscillated with a base period of about 62 million years, suggesting that speciation and extinction might have autonomously repeated. In Chapter 7 we will show that the mecha-

Figure 1.3 Icon of human evolution (reproduced from Wells, 2000).

nisms of evolutionary branching and extinction can indeed autonomously alternate if, after a branching, one of the two emerging lineages evolves to extinction while the remaining one evolves back to the branching point, thus forming a Red Queen evolutionary attractor called a *branching-extinction evolutionary cycle*.

1.9 EXAMPLES

We close this introductory chapter by reporting empirical evidence of evolutionary phenomena in a variety of fields, including biology, social sciences, economics, and engineering. In each example, we identify the basic characteristics and dynamical features discussed in the previous sections. Most examples present empirical data, some others discuss theoretical studies, and a few simply speculate on the applicability and descriptive power of the mutation-selection (innovation-competition) paradigm.

We start with our own evolutionary history, naively depicted in Figure 1.3. As we have seen in Section 1.7, we belong to the *Hominidae* family, also including great apes (bonobos, chimpanzees, gorillas, and orangutans) and extinct human-like species. No fossil data connecting great apes and humans were available when Darwin published *The Descent of Man* (1871), arguing on the basis of anatomical similarities that humans are closely related to gorillas and chimpanzees. Since then, many fossil remains have been found and accurately dated (see, e.g., the recent discovery of Moyà-Solà et al., 2004), showing that humans and great apes share a common ancestor 6 to 7 million years ago. Figure 1.4 sketches skulls of present-day gorillas and humans (A, F) and intermediate human-like species (B–E), while Figure 1.5 reports the evolution of brain size in the *Hominidae* family. All together, Figures 1.3–1.5 show the evolutionary dynamics of various human phenotypes, spinal curvature, hair abundance, skull shape, and brain volume, and confirm our speciation from great apes.

The causes of the evolutionary branching that led to humans and great apes have been largely discussed. An intriguing option comes from recent studies on the

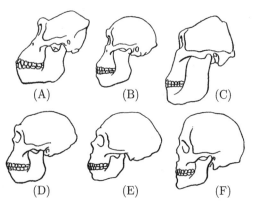

Figure 1.4 Skull evolution in the *Hominidae* family (reproduced from Radinsky, 1987): (A) present-day gorilla; (B, C) human ancestors from about 3 million years ago; (D, E) human ancestors from about 1.5 and 0.07 million years ago; (F) present-day human.

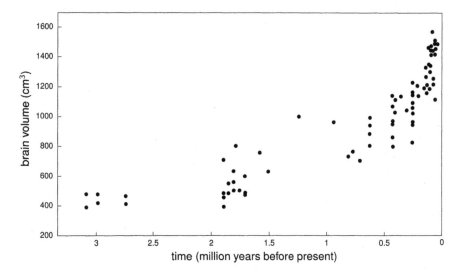

Figure 1.5 Brain size evolution in the *Hominidae* family (reproduced from McHenry, 1994).

evolution of language. The ability to develop articulate speech, and thus complex languages and cultures, relies on the fine control of the larynx and mouth, which is absent in great apes. Lai et al. (2001) showed that such an ability is regulated by a particular gene, while Enard et al. (2002) recently studied its evolution. Their results show that the gene is present in all great apes and even in other mammals, but that subtle differences are specific to the human lineage. From such human-specific mutations to the development of modern human languages the gap seems to be filled by cultural, rather than biological, evolution. In fact, the language is cul-turally transmitted and mutations are represented by the addition of new words or signals to the language. A more articulate language may confer several advantages to its speakers, especially when the language is still very poor, and such advantages may ultimately lead to a higher fitness, i.e., to a better reproductive success. This has been shown theoretically by Nowak et al. (2000, 2001), who studied the transition from nonsyntactic to syntactic communication, a step considered essential in the evolution of human languages (Hauser, 1996; Hauser et al., 2002). Animal communication is typically nonsyntactic, which means that there is a word or a signal for each event to be identified. By contrast, human languages are syntactic, i.e., events are identified by messages composed of words taken from a finite vocabulary. The fitness of an individual adopting a certain language is defined as the probability of successful communication with a randomly selected individual. The main result of these studies is that selection favors the emergence of syntax if the number of events to be identified exceeds a threshold value. This might explain why only humans evolved syntactic communication without requiring biological or genetic justification.

Many other examples of evolutionary dynamics may be found in fossil records, our only direct source of information along the evolutionary timescale, and actually the only source before the advent of the indirect methods of molecular biology that we have seen in Section 1.5. Fossilization is the replacement or molding of the body of a dead organism with a mineralized replica as minerals infiltrate or deposit around the body. Thus, fossilization is most likely to happen at the bottom of aquatic environments, where precipitation of minerals is regular, while for birds and other flying species we have almost no fossil record. New fossils continue to be found and systematically confirm the inferences previously made in accordance with the evolutionary hypothesis. In particular, there are cases of almost complete temporal sequences of fossils that provide examples of smooth evolutionary dynamics, as we have defined in Section 1.5. We now show five of such astonishing examples, three of them coming, not surprisingly, from columns of rock extracted from the bottom of oceans.

Figure 1.6 shows the evolution of body size of a zooplanktonic species over several million years. Planktonic species are unicellular eukaryotes floating in the ocean, and include algae, called phytoplankton, and organisms feeding on bacteria and algae, called zooplankton.

Figure 1.7, one of the two nonmarine examples, describes the coevolution of ungulate species and of their predators by showing two indexes positively correlated to running speed. These data provide an example of coevolution, where an increase of the prey speed is followed by an increase of the speed of some of the predators, though the evolution of predator indexes is less pronounced than that of prey indexes (we will focus on prey-predator (resource-consumer) coevolution in Chapters 5, 9, and 10).

The next three examples provide evidence for the three evolutionary patterns described in Sections 1.6–1.8, namely Red Queen dynamics, evolutionary branching, and evolutionary extinction. Figure 1.8 shows several millennia of reconstructed population abundances of two sardine species and points out repeated oscillations between commonness and rarity. Tens and even hundreds of consecutive generations at either high or very low abundance may have given time to contrasting

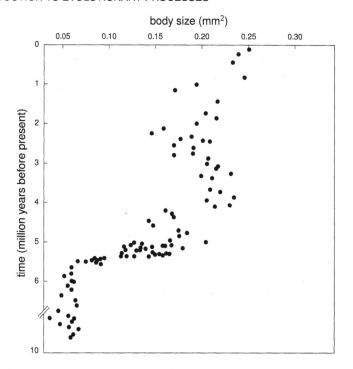

Figure 1.6 Body size evolution of the zooplanktonic species *Globorotalia tumida* (reproduced from Malmgren et al., 1983).

selective pressures to operate on phenotypes. For example, a physiological trade-off between competitive abilities and fertility often characterizes wild species (see, e.g., Roff, 1992; Stearns, 1992). An intuitive explanation is that at high population density individuals are engaged in many competitive contests, so that more competitive mutants are favored, even though characterized by a reduced fertility. By contrast, at low population density encounters are so rare that there is little to be gained from improved competitive abilities, so that one may expect advantageous mutants to be characterized by an increased fertility at the expense of competitive abilities. Thus, different selection pressures may have operated at high and low abundances on phenotypes related to competitive performance, and caused their evolutionary oscillations. Unfortunately, measures of such phenotypes are not available, so that the Red Queen dynamics remain a conjecture, that we formally support in Chapter 8.

Figure 1.9 illustrates evolutionary branching and speciation in the genus *Rhizosolenia* of (predominantly asexual) phytoplanktonic species. The evolutionary branching occurred about three million years ago and eventually led to two distinct species, *Rhizosolenia bergonii* (upper branch) and *Rhizosolenia praebergonii* (lower branch). The represented phenotype is the height of the region of the organism cell connecting the conical part with the top thin part (see sketch). Other phenotypes were measured and showed similar patterns of lineage splitting.

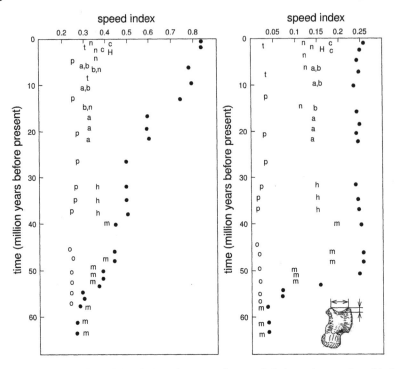

Figure 1.7 Coevolution of North American ungulates and their predators. Speed indexes: ratio metatarsal to femoral length (left); ratio astragular groove depth to trochlea width (right, see the sketched ankle joint). Dots refer to prey species (ungulates); letters refer to different predator species (reproduced from Bakker, 1983).

Our last example from fossil records documents the evolutionary extinction of the family *Brontotheriidae* of the order *Perissodactyla*, which includes modern horses, rhinoceroses, and tapirs. Brontotheres are known in fossil records of western United States from about 55 million years ago, when they were small and, in many respects, very similar to their contemporary relatives (primitive horses, rhinoceroses, and tapirs). Since then brontotheres underwent a diversification by evolving horns and attaining very large sizes, as illustrated in Figure 1.10. Both males and females showed this evolutionary pattern, though male horns were significantly more robust, since they probably conferred an evolutionary advantage in the competition for mating and food. The cause of brontotheres extinction is still debated among paleontologists, but most likely brontotheres were victims of their own armaments and their large and cumbersome bodies, thus suggesting a case of evolutionary suicide.

Evidence for evolutionary patterns may also be found in living populations. One of the most known and studied cases is that of Darwin's finches. During his visit to the Galapagos islands, Darwin noticed thirteen species of finches found nowhere else before. These species were different one from each other in the shape and size of the beak, in overall body size, and in resources exploited, e.g., seeds of differ-

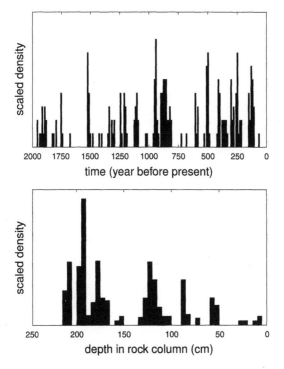

Figure 1.8 Population density time series of two species of Pacific sardines over the past millennia. (Top) *Sardinops caerulea* (reproduced from Soutar and Isaacs, 1974; Baumgartner et al., 1992). (Bottom) *Sardinops sagax*; the correlation between depth in rock column and time is roughly piecewise linear: 0–50 cm corresponds to 0–500 years before present; 60–140 to 1800–2500; 150–190 to 3000–3250; 195–220 to 11,400–11,700 (reproduced from De Vries and Pearcy, 1982).

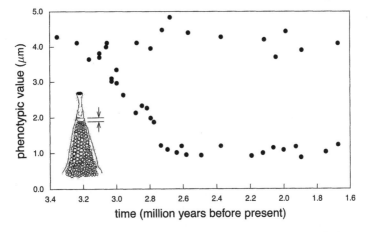

Figure 1.9 Speciation in the phytoplanktonic genus *Rhizosolenia* (source: Sorhannus et al., 1998, 1999; diagram reproduced from Benton and Pearson, 2001).

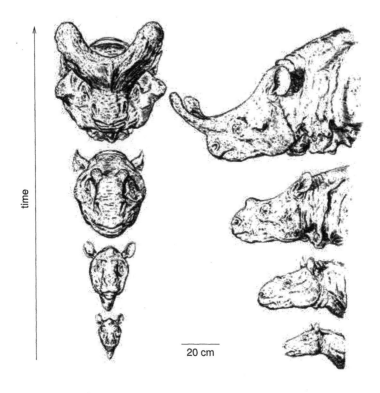

Figure 1.10 Sequence of dated skull sketches of the *Brontotheriidae* family: front view on
          left, side view on right. From top to bottom: late Eocene, about 35 million years
          ago; late middle Eocene, about 45 million years ago; early middle Eocene,
          about 50 million years ago; early Eocene, about 55 million years ago (repro-
          duced from Stanley, 1974).

ent size and hardness (see Figure 1.11). He then realized that those variations that
conferred survival advantages in a given environment come to dominate the population, all else being equal, and he argued that the drastic specialization of finches
was the result of speciation from a common ancestral species coming from Central or South America. Darwin suggested the geographic diversity of the islands,
characterized by different environments and food sources, as the cause of finches
speciation. Since then, Darwin's finches have provided a prototypical case study
for speciation (see, e.g., Grant et al., 1976, 1985; Schluter et al., 1985; Schluter,
1988; Grant and Grant, 2002; see also the book by Grant, 1986) and both morphological and genetic studies have confirmed that different species are derived from a
single ancestral finch (Petren et al., 1999; Sato et al., 2001). However, morphological and genetic differences are sometimes larger between species occurring on the
same island than between species on different islands, suggesting that competition
for different resources may have produced disruptive selective forces on the same
island. Thus, the finches example probably provides a case where both geographic

Figure 1.11 The finches drawings published by Darwin (1839) in the *Journal of Researches*.

Figure 1.12 (Left) Pelican, with four forward-pointing toes forming the paddle. (Right) Toucan, with two forward- and two backward-pointing toes (reproduced from Galis et al., 2001).

isolation and sympatric speciation have played a role in sexual populations (see, e.g., Benkman, 1999, 2003).

   Birds share a common ancestor with all other vertebrates about 500 million years ago, a fish-like creature from which the evolution of fins and limbs led to vertebrate fish and land reptiles. The further evolution of limbs into leathery and feathered wings led to flying reptiles and birds, while other evolutionary paths led to mammals about 300 million years ago. The evolution of limbs has also involved important speciations. Whale flippers, for example, though similar to fish fins, actually derive from mammals feet with increased number of fingers. The evolution of the number and morphology of fingers, recently reviewed by Galis et al. (2001), reveals another interesting case of speciation in birds. Modern techniques of molecular biology have shown that the ancestral position of toes in birds is that of one backward- and three forward-pointing toes. In aquatic birds, the three forward-pointing toes generally form the paddle, but in some species, like pelicans and cormorants, the backward-pointing toe has moved forward and is incorporated in the paddle, as shown in Figure 1.12 (left). The absence of a backward support for the foot explains the poor walking ability of pelicans, which have been subject, in the course

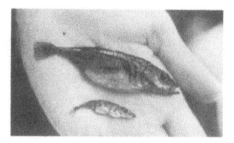

Figure 1.13 Adult individuals (females) of the large and small three-spined sticklebacks
          species living in southwestern British Columbia lakes (reproduced from Gib-
          bons, 1996).

of evolution, to a trade-off between swimming and walking. By contrast, in tree-
dwelling birds, like toucans, parrots, and woodpeckers, there has been a selective
advantage in having an extra backward-pointing toe, e.g., for climbing the trunk of
trees and grasping branches (see Figure 1.12, right).

Perhaps the best example of sympatric speciation is provided by species of fresh-
water fishes, commonly called three-spined sticklebacks (genus *Gasterosterus*),
deeply studied by Dolph Schluter and colleagues (Schluter and McPhail, 1992;
Schluter, 1994; Rundle et al., 2000; McKinnon and Rundle, 2002; see also Schluter,
2000). They have found two different species of three-spined sticklebacks in each
of five different lakes in southwestern British Columbia: a large species with a large
mouth primarily living on the lake bottom and feeding on large prey, and a smaller
species with a smaller mouth feeding on plankton in open water (see Figure 1.13).
DNA analysis (Taylor and McPhail, 1999) indicates that each lake was colonized
independently, presumably by a marine ancestor, about two million years ago, and
that the two species in each lake are more closely related to each other than to
any of the species in the other lakes. Moreover, the two species in each lake are
reproductively isolated, while individuals of the same species from different lakes
can interbreed and show mating preferences for similar size individuals (Rundle
and Schluter, 1998; Nagel and Schluter, 1998). Thus, the most likely explanation
is that in each lake an initially single population faced disruptive selection due to
competition for different resources and habitats. Such a competition favored fishes
at either extremes of body and mouth size over those closer to the mean. Disruptive
selection, coupled with assortative mating, split the population into two distinct
groups, exploiting different types of food in different parts of the lake, and eventually led to the two different species that we observe today.

Another remarkable example of evolutionary branching in fish populations is
provided by cichlid fish (family *Cichlidae*) in African lakes, in which the analysis of DNA sequences suggests that the hundreds of present species, differing in
size, colors, male courtship traits, and female preferences, have arisen within just
the last two million years (Meyer et al., 1990; Schliewen et al., 1994; Albertson
et al., 1999; Schliewen et al., 2001; Barluenga et al., 2006). The cause of this large
and rapid diversification is probably what Darwin first realized and called *sexual
selection*, namely selection for male traits associated with dominance in the social

hierarchy or with attractiveness to females. As selection for limited resources, sexual selection may become disruptive and lead to repeated speciations, as shown by Higashi et al. (1999).

In plant populations, speciation can be due to resource competition (see, e.g., Knox and Palmer, 1995) and to the coevolution with pollinator species in flower plants (see, e.g., Schemske and Bradshaw, 1999; Bradshaw and Schemske, 2003; Ramsey et al., 2003). In Chapter 6 we will show how the selection pressure governing the evolution of mutualistic interactions may turn disruptive.

While adaptation to resource consumption, to female mating preferences, and to the phenotypes of partner species certainly increases reproduction and therefore fitness, the alternative way to improve fitness is to reduce mortality, by adapting to environmental conditions and by evolving antipredator traits or behaviors. Thus, prey-predator, host-parasite, or, more in general, resource-consumer coevolution has often attracted the attention of evolutionary biologists and there is both empirical evidence and theoretical support that predation pressure may be a disruptive selection force leading to prey speciation. For example, Walker (1997) suggests that predation pressure may have contributed to the body size diversification of three-spined sticklebacks described above. Other examples of evolutionary branching induced by predation pressures are proposed by van Damme and Pickford (1995) and Stone (1998) in mollusks and by Chown and Smith (1993) in beetles, while other studies on resource-consumer coevolution have been recently reviewed by Abrams (2000). Further theoretical support will be given in Chapter 5.

Notice that, for all the above examples on living populations we have not shown figures representing evolutionary dynamics, as we have done for the examples from fossil records. The reason is simply that the length of significant evolutionary experiments is often impracticable and the experimental conditions are hardly controllable over long periods of time. However, these limitations are strongly attenuated for small and simple organisms, like bacteria, which typically have short generation times and large population sizes, so that their evolutionary dynamics develop on reasonably short timescales. Moreover, laboratory experiments can be easily controlled and repeated under the same or different conditions. Thus, laboratory experiments on simple organisms are best suited for testing theoretical evolutionary hypotheses and evolutionary models, as those described in this book (see the sequence of papers by Lenski and coauthors on long-term evolution in *Escherichia coli*, from Lenski et al. (1991) to Rozen et al. (2005), and references therein). However, though empirical evidence for evolutionary branching in bacteria is available (see, e.g., Rainey and Travisano, 1998), testing the Red Queen hypothesis is still a rather open problem.

Before closing this chapter, we now show that evolutionary phenomena not driven by genetic inheritance are relevant in biological as well as nonbiological contexts. As we first noted in Section 1.1, the mutation and selection processes do not necessarily concern genetically inherited phenotypes, but they can, as well, drive the evolution of culturally transmitted traits, like animal and human behavioral strategies (see, e.g., Cavalli-Sforza and Feldman, 1981; Boyd and Richerson, 1985; see also Le Galliard et al., 2005, for a recent theoretical study). The evolution of cooperation among nonrelated individuals is one of the most discussed problems in biology and social sciences and may have an evolutionary explanation (Axelrod and Hamilton, 1981). Sociologists investigate this problem by so-called public goods games, in which cooperative individuals contribute to a public resource from which both cooperative and cheating individuals benefit. Experimental results, typically obtained by means of interviews or by letting individuals play repeated games, show that, in the absence of a mechanism punishing cheaters, groups composed of cooperators perform better than groups composed of cheaters, though cheaters always outperform cooperators in mixed groups (Boyd and Richerson, 1992; Fehr and Gächter, 2000; Fischbacher et al., 2001; Fehr and Gächter, 2002). Thus, even if the cooperators' progeny tends to cooperate due to cultural transmission, the imitation of more successful cheating neighbors will eventually lead mixed groups to groups of cheaters and to the loss of the public goods. This is known as "The Tragedy of the Commons" (Hardin, 1968), which says that reciprocal altruism fails to provide an explanation for cooperation to get established, unless cheaters can be identified and punished.

In a recent theoretical study Hauert et al. (2002) draw attention to a simple mechanism to explain the evolution of cooperation. It consists in allowing individuals to quit the game (loner behavior) and get a low but safe income that does not depend on the strategy played by other individuals (see Doebeli et al., 2004; Hauert and Doebeli, 2004, for different mechanisms promoting or inhibiting the evolution of cooperation). Such risk-averse optional participation prevents cheating to become the dominant strategy and allows the evolutionary persistence of cooperation, even if individuals have no way of discriminating against cheaters. Mutations, namely the change of the strategy culturally learned from the family, and selection, favoring the switch to the better-performing strategies of the moment, drive the volunteering public goods game in a sort of "rock-scissors-paper" cycle, where individuals tend to quit the game in groups dominated by cheaters, to cooperate in groups dominated by loners, and start to cheat in groups dominated by cooperators. Hauert et al. (2002) concluded that volunteering is a "Red Queen mechanism" for the persistence of cooperation in public goods games. However, the game is characterized by three strategies that never change. What oscillates are the fractions of the population adopting the three strategies. Thus, variations occur on the demographic timescale and the game describes the cyclic coexistence of different strategies rather than their evolutionary change.

An example of evolutionary dynamics of a socio-cultural trait is shown in Figure 1.14, which reports the evolution of the skirt length of women's formal evening dresses reconstructed from the analysis of dress pictures appeared in fashion magazines over 200 years. The role of fashion is identity display. Two opposed selective forces are the tendency to imitate certain stereotypes with desirable characteristics and the tendency to diverge from them to proclaim an identity. The trade-off between imitation and personalization might induce complex evolutionary dynamics of fashion traits. For example, both diagrams in Figure 1.14 show an almost synchronous slow oscillating trend, with superposed year by year variations, suggesting the existence of a Red Queen evolutionary regime in which particular trait values periodically, or, better, erratically, become fashionable.

As already discussed in Section 1.1, Dawkins' replicators, the basic units able

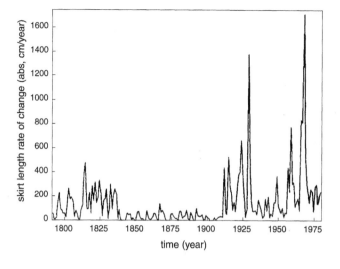

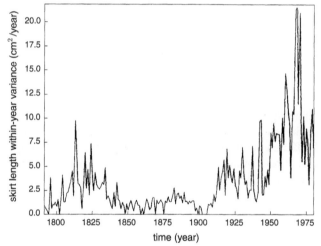

Figure 1.14 Skirt length of women's formal evening dresses (reproduced from Lowe and
Lowe, 1990). (Top) Rate of change (absolute value, cm/year). (Bottom) Within-
year variance (cm^2/year).

to evolve through innovation and competition processes, do not need to be living
organisms. All material products as well as ideas and social norms in use in our
societies are replicators in the sense of Dawkins. Not surprisingly, the evolution of
norms and, more in general, of the history of human societies have been recently addressed (see Ehrlich and Levin, 2005, and references therein, and Turchin, 2003b,
2006), while the technological evolution of commercial products will be described
in Chapter 4. Commercial products replicate each time a new product is purchased,
die out each time a product is dismissed, transmit their characteristic traits to newly
produced products, which from time to time happen to be new versions charac-

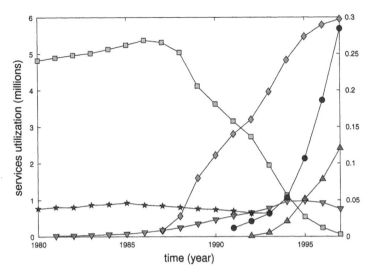

Figure 1.15 Swedish telecommunication services utilization. Squares [diamonds]: analog
[digital] fixed phones; triangles down [up]: analog [digital] mobile phone sub-
scribers; stars: public pay phones; circles: Internet hosts. Public pay phones
and Internet hosts are reported on the right vertical scale (source: Swedish National Post and Telecom Agency).

terized by slight innovations, and interact with each other in local or global markets. Market interactions are often competitive, as between same category cars or
watches of different producers and brands, but they can also be cooperative, as between cars and tires or watches and batteries. Technological change is therefore a
mutation-selection process, driven by technological innovations and by supply and
demand selective forces.

Figure 1.15 shows the case of the evolution of various telecommunication services in Sweden. The appearance of digital fixed phones (diamonds) has been a
successful innovation, which led in about ten years to the substitution of previous
analog fixed phones (squares). This is an example of trait substitution, the basic
step of evolution. In Section 1.5 we have seen that if substitutions are quick with
respect to the rate of occurrence of innovations, then the demographic timescale,
here better called market timescale, can be thought as separated from the evolutionary timescale, and this separation allows us to define the products resident in the
market, the evolution of their characteristic traits, and the corresponding evolution
of their market abundances, i.e., the evolution of the market share. In Figure 1.15,
however, as in all real situations, the demographic and evolutionary timescales are
not completely separated. Thus, we might interpret digital fixed phones as a mutant
group that replaced the resident group of analog fixed phones, or as a mutant group
that became resident by coexisting with other resident groups, thus marking an evolutionary branching in the group of fixed phones. In this second case, subsequent
mutation-substitution sequences in each resident group (not shown in Figure 1.15)
would have led to the evolutionary extinction (probably an evolutionary murder)

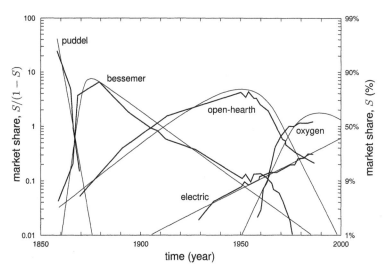

Figure 1.16 Evolution of the market share ($S \in [0, 1]$) among different production technologies in worldwide steel manufacturing. Thick lines connect historical data,
while thin lines are estimates based on a logistic substitution model (Fisher and
Pry, 1971). The adopted logarithmic scale renders the logistic grow (or decay)
linear; some corresponding share values are indicated on the right vertical axis
(source: Grübler, 1990b; diagram reproduced from Grübler, 1998).

of analog fixed phones. In fact, we might think that in ten years of development,
digital phones have gradually changed, e.g., improving the quality of speech, allowing faster and faster data communication, or simply changing design, begetting
trait substitution sequences in the telephone market, which eventually wiped out
analog phones. Analogously, we can see that digital mobile phones (triangles up)
are replacing analog mobile phones (triangles down), while they seem to coexist
with digital fixed phones and Internet hosts (dots). By contrast, public pay phones
(stars) are declining, possibly approaching an evolutionary extinction.

The somehow ambiguous interpretation of Figure 1.15 also applies to some of
the next figures, which show the evolution of the market share between several
technologies on a relatively long period of time, without reporting the concurrent
evolution of their characteristic traits. Figure 1.16 shows examples of technological
substitutions in steel manufacturing. Puddel steel was replaced by bessemer steel,
quite rapidly relative to the time required by the substitution of bessemer steel by
other technologies. Thus, we can reasonably interpret the bessemer technology
as a successful innovation that replaced the previously resident puddel technology.
By contrast, the advent of open-hearth and of more recent technologies are better
described as technological speciations, due to different technologies utilized in isolated markets or to the mechanism of evolutionary branching. Notice that while the
bessemer technology vanished gradually, the decline of the open-hearth technology has been much more sudden, two extinction patterns that we may interpret as
evolutionary murder and suicide, respectively.

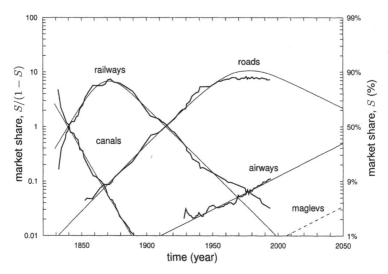

Figure 1.17 Evolution of the share ($S \in [0, 1]$) of the U.S. transport infrastructure by transportation system (logarithmic scale and right vertical axis as in Figure 1.16). Thick lines show historical data, thin lines are estimates based on a logistic substitution model (Nakicenovic, 1998), and the dashed line represents a model prediction (source: Grübler, 1990a, U.S. Bureau of the Census, and U.S. Department of Transportation; diagram reproduced from Ausubel et al., 1998).

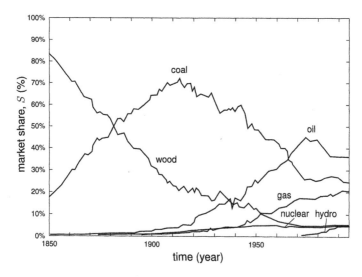

Figure 1.18 Evolution of the market share ($S \in [0, 1]$) among different energy sources in worldwide energy production (source: Nakicenovic et al., 1996; diagram reproduced from Grübler, 1998).

Analogous considerations apply to Figures 1.17 and 1.18, which show the evolution of the market share between different transport and energy production systems, while evolutionary dynamics of related traits are given in Figures 1.19–1.22.

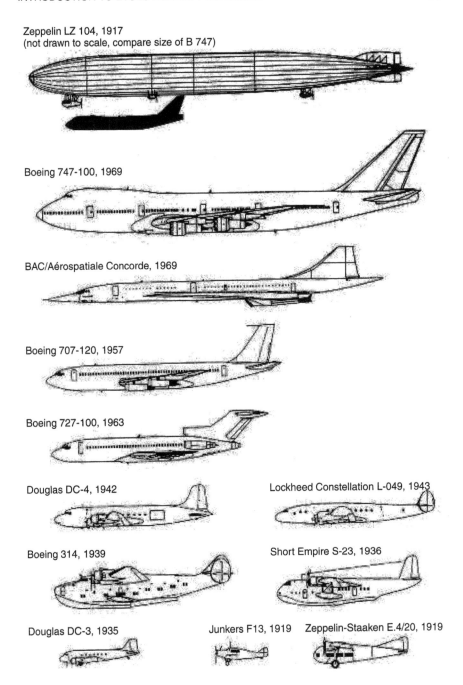

Zeppelin LZ 104, 1917
(not drawn to scale, compare size of B 747)

Boeing 747-100, 1969

BAC/Aérospatiale Concorde, 1969

Boeing 707-120, 1957

Boeing 727-100, 1963

Douglas DC-4, 1942

Lockheed Constellation L-049, 1943

Boeing 314, 1939

Short Empire S-23, 1936

Douglas DC-3, 1935

Junkers F13, 1919

Zeppelin-Staaken E.4/20, 1919

Figure 1.19 Evolution of the size of commercial passenger aircraft (reproduced from Hugill, 1993).

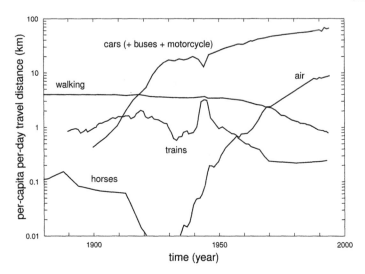

Figure 1.20 Evolution of U.S. per-capita per-day travel distance by transportation system (source: Grübler, 1990a, U.S. Bureau of the Census, and U.S. Department of Transportation; diagram reproduced from Ausubel et al., 1998).

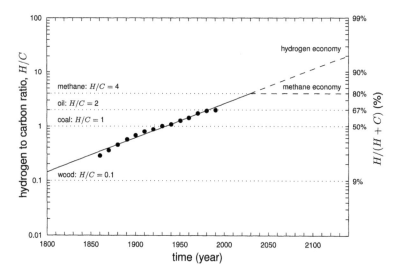

Figure 1.21 Evolution of the hydrogen (H) to carbon (C) ratio in worldwide energy production. Historical data (dots) are fitted by a logistic growth model (solid line) and two future scenarios (dashed lines) are predicted: a methane economy, in which the H/C ratio stabilizes at that of natural gas, and a hydrogen economy, in which energy is fully decarbonized (source: Ausubel, 1996; diagram reproduced from Ausubel et al., 1998).

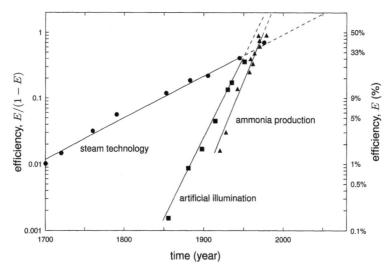

Figure 1.22 Evolution of thermodynamic efficiency ($E \in [0, 1]$) for three technologies. Efficiency is given by the ratio of the performance of the best available commercial machine to the maximum performance that is thermodynamically possible. Historical data (circles, squares, and triangles) are fitted by logistic growth models (lines) that extrapolate future projections (dashed lines) (reproduced from Marchetti, 1979).

The paradigm of mutation-selection evolution is also important in the design of complex machineries and computer programs. Engineers have often found that an efficient and effective way to find the optimal design is to successively make small and random changes, test the obtained design, and keep the version that perform better. Since the advent of modern and cheap powerful computers, this evolutionary approach has had an enormous impact on complex problem solving. Design or optimization algorithms inspired by the evolutionary analogy are generally called genetic algorithms and have been pioneered by Holland (1975; see also Goldberg, 1989). They are useful whenever we do not have a solution in mind for a complex problem but only the desired functional characteristics of the solution.

On the other hand, computer simulations are of interest to evolutionary biologists because they provide an artificial world in which experiments can be performed quickly and at low cost. As recognized by Maynard Smith (1992), "If we want to discover generalizations about evolving systems, we have to look at artificial ones." This is why the mathematical modeling of evolutionary systems is today more than ever an important and challenging problem.

As a last speculative example of the wide applicability and descriptive power of evolutionary paradigm, we report an intriguing problem that comes from the cosmologist Smolin (1997). Seeking an explanation of the physical constants of the universe, Smolin argued that universes themselves can be considered as replicators. His idea is that universes reproduce by means of black holes, which form when large stars collapse into a region of gravity so intense that nothing, not even light, can escape. New universes are then generated from black holes of parental universes, through expansions like the "Big Bang" that generated our universe. Each time a new universe is generated, its properties may vary slightly from those of the parent, thus providing a source of mutation. Thus, according to this hazardous evolutionary conjecture, universes most capable of reproducing, i.e., of forming black holes, would have been selected. In fact, the physical constants of our universe seem to be well adapted for the formation of black holes, and since the properties required for life are strongly correlated with those required for black holes, this conjecture would also explain why our universe is optimized for life.

The message emerging from this gallery of examples seems to be that the mechanisms of mutation (innovation) and selection (competition) are pervasive at each level of organization of natural and artificial systems. Dawkins (1989) even argued that evolvability is a trait that can be, and has been, selected for. In fact, the ability to be responsive to the surrounding environment through the mechanisms of mutation and selection is what allows living and artificial replicators to fit to their local conditions, and adaptation of structure to function is apparent in living organisms as well as in the human design of artificial systems. Not surprisingly, Ernst Mayr (2001) concludes his introduction to the seventeenth printing of Darwin's masterpiece by saying that the philosophical consequences of the Darwinian "revolution" have not yet been fully exploited.

Chapter Two

Modeling Approaches

In this chapter we survey the quantitative approaches proposed in the literature for modeling evolutionary dynamics. The evolution of biological as well as social and economic systems is determined by so many interacting factors that a detailed description is practically impossible. Any mathematical model takes into account some mechanisms (hopefully the relevant ones) and sacrifices the others. The trade-off between realism and mathematical tractability has produced, starting with Charles Darwin and Gregor Mendel, a rich variety of different studies, focused on different aspects of the evolutionary process. A sharp classification of these studies into well-confined modeling approaches is problematic because modeling hypotheses often overlap. Here we identify seven approaches, namely population genetics, individual-based evolutionary models, quantitative genetics, evolutionary game theory, replicator dynamics, fitness landscapes, and adaptive dynamics. However, different classifications have been proposed by other authors (see, e.g., Abrams et al., 1993a; Dieckmann, 1994; Marrow and Johnstone, 1996; Eshel et al., 1998; Abrams, 2001; Meszéna et al., 2001; Page and Nowak, 2002; Nowak and Sigmund, 2004; Day, 2005; Lessard, 2006). Throughout the chapter we do not pretend to be very precise and we describe each approach through its key ideas, virtues, and limitations. We follow, in particular, the scheme proposed by Dieckmann (1994, Chapter 2) and present the various approaches imagining that the reader is a newcomer to the theory of evolutionary dynamics with undergraduate skill in mathematics and basic notions of evolutionary biology.

2.1 OVERVIEW

Evolutionary change has been theoretically studied since Darwin (1859) proposed the theory of evolution of species by natural selection and Mendel (1865) used elementary mathematics to calculate the expected frequencies of genes in his experiments. After more than a century, we can say that Darwin's and Mendel's contributions concerned different aspects of evolution (Bishop, 1996), since the former ignores the detailed mechanisms of inheritance, while the latter disregards environmental selection forces.

The first mathematical investigations reconciling Darwinism with Mendelism started in the field of *population genetics*, founded by Fisher (1930), Wright (1931), and Haldane (1932) (see also Wright, 1969). Classical population genetics is focused on the change of the genotypic distribution of populations on a relatively short demographic timescale. However, as we have seen in Chapter 1, on a longer timescale individuals interact according to the value of various phenotypic traits, which are the macroscopic manifestation of their microscopic genetic structure (see Section 1.2). Interactions among individuals shape population abundances, which, in turn, affect reproduction rates and therefore genotype inheritance and mutations, the ultimate source of genetic variability. In other words, selection acts on phenotypes, so long-term predictions of genotypic distributions require knowledge of a map from genotypes into phenotypes. Unfortunately, such a map is, in most cases, extremely complicated and not yet fully understood.

But if the genotype-to-phenotype map is known (or conceptually hypothesized), one can make the step from demography to evolutionary dynamics. In fact, given the phenotypic distribution of populations, a stochastic demographic model can describe birth, death, and migration processes altering population abundances, while phenotypic variability is accounted by randomly generated genetic mutations and recombinations, which are quantitatively mapped into the corresponding phenotypic values. Such a description falls in the class of so-called individual-based models (see Bartlett, 1960; De Angelis and Gross, 1992; Grimm and Railsback, 2005; and, in particular, Dieckmann, 1994, for an evolutionary perspective), which are stochastic processes where each individual is explicitly represented, in this case, by its genotype. *Individual-based evolutionary models* represent the frontier of modern population genetics and provide the richest and more realistic framework for describing demography and evolutionary change. Other than the genetic information, they can virtually incorporate any further detail, like age, stage, or space distribution of populations and environmental fluctuations. Individual-based models are therefore particularly suited whenever the aim is to obtain long-term simulations of the stochastic process to be compared with field or laboratory data. However, the algorithms for the simulation of individual-based models typically require accurate tuning of several parameters and, in general, analytical analyses are too complex to be carried out (but see Bolker and Pacala, 1997; Dieckmann and Law, 2000; Grimm and Railsback, 2005).

The lack of genetic knowledge and the search for mathematical tractability have fostered the development of simpler approaches, called phenotypic approaches, aimed at describing the change of the phenotypic distribution of populations in the absence of genetic information. One route to bypass genetics is offered by *quantitative genetics* (Bulmer, 1980; Falconer, 1989), which considers phenotypes under the control of a large number of genes with small effect. Such phenotypes turn out to be almost continuously distributed across the population, so that they are called continuous or quantitative traits. Assuming that all possible genetic variability is always present in the population, mutations can be neglected, so that the evolution of the trait distribution is governed by the sole mechanisms of inheritance and natural selection and can be described, under suitable simplifying assumptions, independently of the genetic structure underneath (Lande, 1976).

A second route to bypass genetics is to simply ignore it, by assuming that individuals produce either perfect copies of themselves or mutants bearing slightly different trait values (asexual or clonal reproduction). In fact, there is general agreement that mutations are the ultimate source of phenotypic variability and that, in absence of mutation, the effects of inheritance on the distributions of quantitative traits average out on a relatively short timescale. By looking at the long-term evolutionary dynamics of quantitative traits, one can therefore imagine that trait distributions are concentrated around characteristic values. In other words, we can imagine that male and female resident populations are monomorphic, i.e., phenotypically homogeneous, and produce either identical males and females or mutants. Thus, given as granted that the sex ratio remains constant, as is common practice in ecological modeling (Roughgarden, 1979), one can measure population sizes in terms of males or females or, more abstractly, in terms of equivalent clonal organisms. The genetics of such organisms can be neglected, since they are completely characterized by their phenotypic values, which are inherited, with possible small mutations, by their progeny. In conclusion, if mutations are sufficiently rare on the demographic timescale and if one looks on a sufficiently long evolutionary timescale, the use of asexual demographic models in cases of sexual reproduction is qualitatively justified.

The first phenotypic models came from optimization theory and were supported by the assumption that, independently of the mechanisms of reproduction, under natural selection organisms adapt their phenotypes to maximize a suitable fitness function, i.e., a measure of their reproductive success (see, e.g., Cole, 1954; Cohen, 1966; Stearns, 1992). This is an intrinsically static approach that does not characterize evolutionary dynamics, but only the final evolutionary states as local fitness maxima. The lack of a dynamical description and the questionable freedom in the definition of an appropriate fitness measure motivated further theoretical developments. Moreover, fitness maximization is attained when the reproductive success of an individual depends only on its own phenotypic traits. Often, however, reproductive success also depends on the phenotypic distribution of the population, as well as on those of other interacting populations (so-called frequency-dependent selection, see Section 1.4). The correct maximization criterion in such cases is provided by *evolutionary game theory* (Maynard Smith and Price, 1973; Maynard Smith, 1982), which characterizes evolutionary equilibria as monomorphic phenotypic distributions that maximize the fitness of a small group of phenotypically different (e.g., mutants) individuals. Such distributions are therefore immune to invasion by any small group of dissidents and, for this reason, have been called evolutionarily stable (see the special issue of *Theoretical Population Biology*, vol. 69, pp. 231–350, edited by Lessard, 2006, for recent discussions). Evolutionary game theory is a revisitation of classical game theory, which was developed to study economic and social behavior (Nash, 1950, 1951; von Neumann and Morgenstern, 1953). This is not surprising, since asexual models fit with the case of evolving economic and social systems, where the spread of strategies (the analogue of phenotypes) occurs through learning, imitation, or other forms of cultural transmission (Fudenberg and Levine, 1998).

Although immunity to invasion became a fundamental concept, evolutionary game theory remains a static approach and cannot therefore explain if an evolutionary equilibrium will be asymptotically reached or not (i.e., the "real" stability). Indeed, only a dynamical theory can fully account for the evolutionary stability of evolutionary equilibria, as well as of nonstationary evolutionary regimes (Red Queen dynamics, see Section 1.6). A dynamical extension of evolutionary game theory is offered by *replicator dynamics* (Taylor and Jonker, 1978; Schuster and Sigmund, 1983), an approach dealing with the relative abundances of the strategies played in a game. However, the long-term evolution of strategies is not accounted for, since only the short-term dynamics of selection are described.

Long-term evolutionary dynamics have been pictured by many authors as a hill-climbing process on suitable *fitness landscapes* defined in the space of the relevant phenotypic traits (Levins, 1968; Leon, 1974). Phenotypic change is viewed as the result of a sequence of rare and random mutations followed by natural selection. The invasion success of a mutation is measured by the fitness function and invading mutations are assumed to replace similar resident traits. This simplified view is often referred to as *mutation-limited evolution* (Metz et al., 1992) and, in the limit of infinitesimal mutations, phenotypic change becomes smooth and described in terms of ODEs, thus accounting for evolutionary transients, steady and Red Queen evolutionary regimes and their stability.

The fitness maximization principles at the base of phenotypic approaches remained for a long time in contrast with population genetic models, which, taking the mechanisms of sexual reproduction into account, seemed to disprove any sort of maximization (Lewontin, 1974; Ewens, 1979). Although the discussion is far from conclusive (Gavrilets, 2004), a partial reconciliation is provided by the application of the paradigm of mutation-limited evolution to population genetics (Eshel and Feldman, 1984; Lessard, 1984; Matessi and Eshel, 1992; see also Eshel, 1996; Hammerstein, 1996a; Matessi and Di Pasquale, 1996; Weissing, 1996, published in the special issue of the *Journal of Mathematical Biology*, vol. 34, pp. 485–688, and the commentaries in Hammerstein, 1996b; Marrow and Johnstone, 1996; Williams, 1996). Using the short-term dynamics of genotype inheritance to test for mutant invasion and assuming rare mutations, it was shown (under some simplifying assumptions, e.g., frequency-independent selection) that genotypic distributions immune to invasion are characterized by a single dominant genotype with identical maternal and paternal genes (so-called homozygous genotype, see Section 1.2). At mating, such genotypes can only give rise to perfect copies of themselves or non-invading mutants, so that any phenotypic distribution remains monomorphic in the long run and reproduction can indeed be considered asexual.

The frontier of dynamical phenotypic approaches is represented by the more recent approach of *adaptive dynamics* (Dieckmann and Law, 1996; Metz et al., 1996; Geritz et al., 1997, 1998), which provides a particular fitness landscape by explicitly taking into account demographic resident-mutant interactions. In particular, the full investigation of demographic resident-mutant dynamics identifies the conditions under which a resident trait is replaced by an invading mutation. Moreover, the emergence of polymorphic trait distributions through the mechanism of evolutionary branching (coexistence of resident and mutant traits under opposite (disruptive) selection pressures, see Section 1.7) and the evolutionary extinction of resident traits (evolution toward the collapse of resident population abundances, see Section 1.8) are also formally explained.

In the following, each approach is described in a separate section and the chapter is closed by a comparative analysis, which also introduces the reader to the rest of the book.

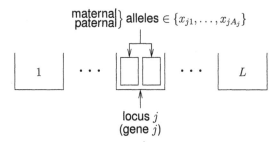

Figure 2.1 Diploid genotype model with L loci.

2.2 POPULATION GENETICS

Population genetics focuses on the change of genotype relative abundances, often called frequencies, but ignores, at least in its classical formulations, the change in genotype absolute numbers (see, e.g., Crow and Kimura, 1970; Ewens, 1979; Roughgarden, 1979). As we have seen in Section 1.2, the genotype of an individual is identified by its chromosomes. Here we consider diploid organisms (animals and most plants), in which the genotype is identified by pairs of so-called homologous chromosomes, one inherited from the mother and one from the father. As sketched in Figure 2.1, each pair of homologous chromosomes can be schematically represented as a sequence of positions (loci) occupied by two allelic forms of the corresponding gene, one for each of the two homologous chromosomes. Typically, the part of the genome that controls the characters of interest is small compared to the entire genome, so that only the relevant part must be considered. Let L be the number of loci of interest, C the number of involved chromosomes, L_i the number of loci in the ith chromosome ($i = 1, \ldots, C$ and $L_1 + \cdots + L_C = L$), and $\{x_{j1}, \ldots, x_{jA_j}\}$ the possible allelic forms of gene j ($j = 1, \ldots, L_1$ in chromosome 1, $j = L_1 + 1, \ldots, L_1 + L_2$ in chromosome 2, and so on).

Loci on the same chromosome are called *linked* loci, while loci on different chromosomes are said to be *unlinked*. Notice that the order between maternal and paternal alleles within loci is irrelevant. In fact, two individuals whose genotype models differ only in having all loci in a certain chromosome with exchanged first and second alleles cannot be distinguished, since they carry the same chromosomes. For example, with two loci, each characterized by two possible alleles (i.e., $L = 2$, $A_j = 2$, $j = 1, 2$), genotype models $((x_{11}, x_{12}), (x_{21}, x_{22}))$ and $((x_{11}, x_{12}), (x_{22}, x_{21}))$ are indistinguishable if the two loci are unlinked ($C = 2$, $L_i = 1$, $i = 1, 2$; individual chromosomes are identified by x_{11}, x_{12}, x_{21}, x_{22} in both models), while they represent different genotypes if the two loci are linked ($C = 1$, $L_1 = 2$; individual chromosomes are (x_{11}, x_{21}), (x_{12}, x_{22}) in the first model and (x_{11}, x_{22}), (x_{12}, x_{21}) in the second). Moreover, there are instances of the genotype model, such as $((x_{11}, x_{11}), (x_{21}, x_{22}))$ and $((x_{11}, x_{11}), (x_{22}, x_{21}))$, which are indistinguishable in both cases. As a consequence, the number G of different genotypes coded by the genotype model of Figure 2.1 is smaller than the number of different model instances obtained by considering as relevant the order

Table 2.1 Genotypes coded by two loci each with two alleles (different genotypes are progressively numbered in the # columns).

genotype model	linked loci		unlinked loci	
	chromosomes	#	chromosomes	#
$((x_{11},x_{11}),(x_{21},x_{21}))$	$(x_{11},x_{21}),(x_{11},x_{21})$	1	$x_{11},x_{11},x_{21},x_{21}$	1
$((x_{11},x_{11}),(x_{21},x_{22}))$	$(x_{11},x_{21}),(x_{11},x_{22})$	2	$x_{11},x_{11},x_{21},x_{22}$	2
$((x_{11},x_{11}),(x_{22},x_{21}))$	$(x_{11},x_{22}),(x_{11},x_{21})$	2	$x_{11},x_{11},x_{22},x_{21}$	2
$((x_{11},x_{11}),(x_{22},x_{22}))$	$(x_{11},x_{22}),(x_{11},x_{22})$	3	$x_{11},x_{11},x_{22},x_{22}$	3
$((x_{11},x_{12}),(x_{21},x_{21}))$	$(x_{11},x_{21}),(x_{12},x_{21})$	4	$x_{11},x_{12},x_{21},x_{21}$	4
$((x_{11},x_{12}),(x_{21},x_{22}))$	$(x_{11},x_{21}),(x_{12},x_{22})$	5	$x_{11},x_{12},x_{21},x_{22}$	5
$((x_{11},x_{12}),(x_{22},x_{21}))$	$(x_{11},x_{22}),(x_{12},x_{21})$	6	$x_{11},x_{12},x_{22},x_{21}$	5
$((x_{11},x_{12}),(x_{22},x_{22}))$	$(x_{11},x_{22}),(x_{12},x_{22})$	7	$x_{11},x_{12},x_{22},x_{22}$	6
$((x_{12},x_{11}),(x_{21},x_{21}))$	$(x_{12},x_{21}),(x_{11},x_{21})$	4	$x_{12},x_{11},x_{21},x_{21}$	4
$((x_{12},x_{11}),(x_{21},x_{22}))$	$(x_{12},x_{21}),(x_{11},x_{22})$	6	$x_{12},x_{11},x_{21},x_{22}$	5
$((x_{12},x_{11}),(x_{22},x_{21}))$	$(x_{12},x_{22}),(x_{11},x_{21})$	5	$x_{12},x_{11},x_{22},x_{21}$	5
$((x_{12},x_{11}),(x_{22},x_{22}))$	$(x_{12},x_{22}),(x_{11},x_{22})$	7	$x_{12},x_{11},x_{22},x_{22}$	6
$((x_{12},x_{12}),(x_{21},x_{21}))$	$(x_{12},x_{21}),(x_{12},x_{21})$	8	$x_{12},x_{12},x_{21},x_{21}$	7
$((x_{12},x_{12}),(x_{21},x_{22}))$	$(x_{12},x_{21}),(x_{12},x_{22})$	9	$x_{12},x_{12},x_{21},x_{22}$	8
$((x_{12},x_{12}),(x_{22},x_{21}))$	$(x_{12},x_{22}),(x_{12},x_{21})$	9	$x_{12},x_{12},x_{22},x_{21}$	8
$((x_{12},x_{12}),(x_{22},x_{22}))$	$(x_{12},x_{22}),(x_{12},x_{22})$	10	$x_{12},x_{12},x_{22},x_{22}$	9

between first and second alleles. This is explicitly shown in Table 2.1, where all the 16 model instances of the above example are listed in the first column, while the remaining columns list individual chromosomes and assign a progressive number to different genotypes (columns 2 and 3 for the linked case, where $G = 10$; columns 4 and 5 for the unlinked case, where $G = 9$).

The computation of G in the general case is now reported. However, it is not needed in the following, since we will deal only with simple cases. The noninterested reader can therefore skip this paragraph. Let G_i be the number of different genotypes coded by the loci in chromosome pair i, and I_i be the number of different instances of such loci in the genotype model. Then

$$I_i = \prod_{j=L_1+\cdots+L_{i-1}+1}^{L_1+\cdots+L_i} A_j^2,$$

because for each locus we can choose two alleles independently out of a set of A_j

elements, while G_i is given by I_i reduced by half of the number of instances with different first and second alleles at some loci (so-called heterozygous instances, in contrast with homozygous instances, where first and second alleles coincide at all loci, see Section 1.2). In fact, heterozygous instances are indistinguishable from those obtained from them by exchanging first and second alleles at all loci, and only one of the two indistinguishable instances must be counted. The number of heterozygous instances is given by I_i reduced by the number of homozygous instances, which is simply given by $\sqrt{I_i}$ (since we can choose only one allele per locus), so that

$$G_i = I_i - \frac{1}{2}\left(I_i - \sqrt{I_i}\right).$$

Finally, G is the product of all G_i, $i = 1, \ldots, C$, and after some algebra one gets

$$G = \prod_{i=1}^{C}\left(\frac{1}{2}\prod_{j=L_1+\cdots+L_{i-1}+1}^{L_1+\cdots+L_i} A_j\left(1 + \prod_{j=L_1+\cdots+L_{i-1}+1}^{L_1+\cdots+L_i} A_j\right)\right). \qquad (2.1)$$

Notice that if all loci are unlinked and characterized by two possible alleles (i.e., $C = L$, $L_i = 1$, $A_j = 2$, $i, j = 1, \ldots, L$), then (2.1) simplifies to $G = 3^L$, which is the formula reported in all elementary textbooks on genetics. However, linkage makes things a bit more complicated.

At mating, the process of meiosis (see Section 1.2) randomly splits the sequence of L loci into two allelic sequences, which represent the gametes. Each group of linked loci, identifying a pair of homologous chromosomes, is split into two allelic subsequences, which contribute to the corresponding positions of the two gametes. If no recombination takes place, all maternal alleles of the group go into the same subsequence (and all paternal into the other). However, since chromosomes split independently one from each other, maternal and paternal alleles of different groups of linked loci can go into the same gamete. Recombinations are modeled by characterizing each possible way of splitting a group of linked loci into two allelic subsequences with a suitable probability. Then, according to Mendel's laws, the genotype of the offspring is obtained by combining with equal probability one gamete from each parent.

In the following we consider the simplest case of a single locus with two alleles x_1 and x_2, so that there are three possible genotypes: (x_1, x_1), (x_1, x_2), and (x_2, x_2), and we denote by n_1, n_2, n_3 their respective population abundances (e.g., the number or density of individuals). Let p_1 be the frequency of the x_1-allele in the population, i.e., the number of x_1-allele over the total number of alleles, and $p_2 = 1 - p_1$ the frequency of the x_2-allele. In formulas:

$$p_1 = \frac{2n_1 + n_2}{2n}, \quad p_2 = \frac{n_2 + 2n_3}{2n},$$

where $n = n_1 + n_2 + n_3$ is the total population abundance, while, in words, p_i is the probability of randomly extracting the x_i-allele ($i = 1, 2$) from the gene pool. Other typical simplifying assumptions are that individuals randomly mate, that reproduction occurs in discrete and nonoverlapping generations, and that population abundances are sufficiently large, which yield the so-called Hardy-Weinberg law

(Hardy, 1908; Weinberg, 1908):

$$\frac{n_1(t)}{n(t)} = p_1^2(t),$$ (2.2a)

$$\frac{n_2(t)}{n(t)} = 2p_1(t)p_2(t),$$ (2.2b)

$$\frac{n_3(t)}{n(t)} = p_2^2(t),$$ (2.2c)

where t identifies the current generation. The equations in (2.2) simply say that the probability of sampling a genotype from the population, given by the corresponding genotype frequency n_i/n, is equal to the probability of sampling its alleles from the gene pool. As we will see in a moment, they allow the derivation of an autonomous recurrence

$$p_1(t+1) = F(p_1(t)),$$ (2.3)

which gives the change in allele frequencies from one generation to the next. The recurrence (2.3), together with the Hardy-Weinberg law (2.2), then gives the dynamics of the genotype frequencies n_1/n, n_2/n, n_3/n. Notice, however, that the dynamics of the genotype abundances n_1, n_2, n_3 are not given, unless one assumes that the total abundance n is constant.

We now show how the function F can be derived from the demographic dynamics of the three genotypes under the further assumption of initial identical genotype frequencies in both sexes. For example, is it easy to show that the average number of matings $(1, 2)$ per generation is proportional to $2n_1n_2$. In fact, such a number is given by the average number of matings (proportional to n^2) times the probability $2n_1n_2/n^2$ of a $(1, 2)$-mating (computed as the probability n_1/n of sampling a 1-male from the population times the probability n_2/n of sampling a 2-female, plus the probability of sampling a 1-female and a 2-male, i.e., again n_1n_2/n^2). As a result of $(1, 2)$-matings, half of the progeny has genotype 1 and half has genotype 2, because genotype 1 splits into two x_1-gametes, while genotype 2 splits into an x_1-gamete and an x_2-gamete, so that the offspring genotype results from combining an x_1-gamete (with probability 1) with an x_1- or x_2-gamete with probability $1/2$ each. Table 2.2 shows all possibilities. The genotype abundances in the next generation are therefore given by

$$n_1(t+1) = f_{11}n_1^2(t) + \frac{1}{2}f_{12}2n_1(t)n_2(t) + \frac{1}{4}f_{22}n_2^2(t),$$ (2.4a)

$$n_2(t+1) = \frac{1}{2}f_{12}2n_1(t)n_2(t) + f_{13}2n_1(t)n_3(t) + \frac{1}{2}f_{22}n_2^2(t)$$

$$+ \frac{1}{2}f_{23}2n_2(t)n_3(t),$$ (2.4b)

$$n_3(t+1) = \frac{1}{4}f_{22}n_2^2(t) + \frac{1}{2}f_{23}2n_2(t)n_3(t) + f_{33}n_3^2(t),$$ (2.4c)

where f_{ij}, $i \leq j$ is the fitness of the mating pair (i, j), defined as the product of viability (the probability that an offspring survives to reproductive age) and fertility (the average number of offspring produced per mating). Fitnesses of different mating pairs can be different, and these are the differences on which natural selection

Table 2.2 Mating table.

mating genotypes	average number of matings per generation	contribution to next generation		
		1	2	3
$(1,1)$	n_1^2	1	0	0
$(1,2)$	$2n_1n_2$	$1/2$	$1/2$	0
$(1,3)$	$2n_1n_3$	0	1	0
$(2,2)$	n_2^2	$1/4$	$1/2$	$1/4$
$(2,3)$	$2n_2n_3$	0	$1/2$	$1/2$
$(3,3)$	n_3^2	0	0	1

is acting on (Taylor, 1996). Fitnesses can also depend upon the relative or absolute abundance of genotypes, modeling so-called frequency or density dependencies (see Section 1.4). However, fitnesses depend on phenotypes rather than on genotypes, so that a correct definition of frequency- or density-dependent fitnesses must involve suitable genotype-to-phenotype maps. Moreover, the Hardy-Weinberg law (2.2) technically holds when the fitness f_{ij} of the mating pair (i,j) is the product of two terms that depend only on the mating genotypes i and j, respectively (this is the case, for example, if selection acts only on individual viability, see Bürger, 2000).

Substituting (2.4) into

$$p_1(t+1) = \frac{2n_1(t+1) + n_2(t+1)}{2n(t+1)},$$

one obtains the x_1-allele frequency in the next generation in terms of quadratic terms in $n_i(t)$, $i = 1,2,3$, at both numerator and denominator. Then, using (2.2) and canceling $n^2(t)$ between numerator and denominator, one finally gets a right-hand side composed of 4th-order terms in $p_1(t)$ and $p_2(t) = 1 - p_1(t)$, i.e., function F in (2.3).

Deterministic or stochastic recurrences on the relative frequencies of gametes (alleles in the single-locus case) or genotypes can be derived under more general assumptions. For example, the Hardy-Weinberg law (2.2) can be relaxed and more than two alleles per locus or multiple loci, with possible linkage between them and the effects of recombinations, can be considered, as well as mechanisms of assortative mating and complex fitness structures (see, e.g., Hartl and Clark, 1997; Christiansen, 2000; Bürger, 2000). Thus, in conclusion, population genetics provides a detailed model of the genome structure and of the mechanisms of inheritance, and describes the short-term dynamics of gamete and genotype frequencies. However, recurrences of the kind (2.3) do not account for the dynamics of population abundances, which are assumed to be sufficiently large.

Population genetics can also deal with long-term evolutionary predictions, where the processes of mutation and natural selection play a crucial role. Mutations are modeled as any change from one allelic form to another or as the introduction of a new allelic form. However, the description of natural selection requires the knowledge of the relevant phenotypes, according to which individuals interact with each other and with the environment, so that the genotype-to-phenotype map is strictly needed. As we have seen in Section 1.2, such a map can be extremely complicated due to the subtle mechanisms of dominance between alleles and to the presence of so-called epistatic genes, which act as inhibitors of other genes. In most cases, such a map is not known, so that it is hardly possible to quantitatively study long-term evolution at the DNA level. However, when the genotype-to-phenotype map can be identified, at least conceptually, the demographic interactions between individuals can be fully taken into account.

One way to address the long-term evolution of the relevant phenotypes is to consider mutations as rare events, so that the short-term demographic dynamics of genotype frequencies between two successive mutations can be described by recurrences of the kind (2.3). In particular, the asymptotic regimes of the genotypic distribution virtually reached in the absence of further mutations can be identified, as well as the invasion potentials of mutant genotypes injected at such regimes. As a result, long-term evolutionary dynamics are pictured as sequences of phenotypic distributions corresponding to sequences of short-term asymptotic regimes (see Section 1.5). As discussed in the previous section, the great merit of these mutation-limited genetic models is to (partially) support the view that long-term evolutionary predictions can be based entirely on phenotypic information (Marrow and Johnstone, 1996; Williams, 1996).

A different approach to long-term evolutionary dynamics, not restricted to the mutation-limited paradigm, describes both genetics and demography at the level of single individuals. Each individual of the community can be characterized by its genetic structure, which is then mapped into one or more phenotypes of interest. For example, in the case of diploid organisms, the multilocus model presented above can be used, with mating probabilities depending on individual phenotypes (modeling different mating mechanisms ranging from random to assortative mating) and on the phenotypic distributions of the community (modeling frequency- or density-dependent fitnesses). Once a couple of individuals are selected for mating, the genotype of the offspring can be stochastically composed by implementing random recombinations and mutations. Birth, death, and migration events can also be described as suitable stochastic processes depending on individual phenotypes (as well as on the phenotypic distributions of the community), taking into account any kind of ecological interactions such as competition for resources, mutualistic support, and predation. As a result, the evolutionary dynamics driven by the mutation-selection process or by the accumulation of genetic and phenotypic differences with no effect on fitness (genetic drift) are described by a stochastic individual-based process that brings together Darwin's theory of evolution by natural selection with Mendel's mechanisms of inheritance (see the pioneering work of Brues, 1954, and Cavalli-Sforza and Zei, 1967, and more recent studies in Doebeli, 1996a,b; Dieckmann and Doebeli, 1999; Kisdi and Geritz, 1999; van Dooren, 1999; Doebeli and Dieckmann, 2000; van Dooren, 2000; Geritz and Kisdi, 2000).

2.3 INDIVIDUAL-BASED EVOLUTIONARY MODELS

In this section we introduce individual-based evolutionary models, not necessarily genetically based, together with an algorithm for their simulation. Deterministic demographic dynamics, such as those described by differential or difference equations (see, e.g., (2.4)), are justified when population abundances are sufficiently large, so that even very small population densities correspond to populations far from being challenged by accidental extinction risks (Wissel and Stöcker, 1991). Despite infectious diseases and environmental threats, which can cause extinction at any population size (though most likely at low sizes), scarce populations face the accidental death of the last survivors as a further extinction risk, sometimes called *demographic stochasticity*.

To explore the consequences of finite population sizes, all the processes altering population abundances, such as birth, death, and migration, can be modeled as stochastic processes at the level of the single individual. The resulting class of models is said to be individual-based (De Angelis and Gross, 1992; Grimm and Railsback, 2005). Moreover, evolutionary processes have three other important sources of stochasticity (Kimura, 1983): first, genetic drift randomly alters genes and phenotypes with no effect on fitness; second, mutations introduce new genotypes and phenotypic values at random into the populations; third, mutants arise initially as single individuals and consequently are liable to accidental extinction.

Virtually any ecological, environmental, and genetic detail can be accounted for by individual-based models. For example, age or stage structures can be described by allowing individuals to grow from one stage to the next as time goes on or in accordance with their energy balance involving ingestion, metabolism, harvesting strategies, antipredator behavior, and reproduction. Heterogeneous environments can be modeled by dividing the space into homogeneous patches and allowing migration between patches, while external forcing due to seasons and climate change can also be easily considered (see, e.g., Dieckmann et al., 2000; Doebeli and Dieckmann, 2003). All interactions between individuals and with the environment depend upon both individual phenotypes and the whole phenotypic distribution of the community. Phenotypes can be the basic characteristics of individuals, or be determined by an underlying genetic structure, and the resulting evolutionary model is said *phenotypic* or *genotypic individual-based*, respectively. The mechanisms of reproduction (sexual versus clonal) described in Section 2.2 can be easily implemented and, finally, on the top of the whole process, mutations can be added as stochastic events occurring at birth, to make the step from demography to evolutionary dynamics (see Dieckmann, 1994, for a detailed treatment).

We now present a simple but representative example where reproduction is clonal and genetic details are omitted. Consider a community composed of a single species characterized by a continuous phenotypic trait x, living in a homogeneous environment. At time t there are $n(t)$ individuals characterized by a value x_i of the phenotype, $i = 1, \ldots, n(t)$. Thus, the state of the community is identified by the phenotypic distribution

$$p(x,t) = \sum_{i=1}^{n(t)} \delta(x - x_i),\qquad\qquad(2.5)$$

where δ denotes the Dirac impulsive function. The distribution (2.5) changes in time due to birth and death events. The general assumption behind individual-based models is that birth and death processes are independent Markov processes (i.e., the probabilities of birth and death events are determined by the present state (2.5); see, e.g., Bailey, 1964). This assumption allows one to simulate the dynamics of the distribution (2.5) by producing numerical realizations through the so-called minimal process method outlined below (Gillespie, 1976; see also Haccou et al., 2005; we follow, in particular, the Appendix of Dieckmann et al., 1995).

Let $b(x_i, p(x,t))dt$ and $d(x_i, p(x,t))dt$ be the probabilities that the ith individual, with phenotypic value x_i, will give a birth or die in the short time interval going from the current time t to $t + dt$. In other words, $b(x_i, p(x,t))$ and $d(x_i, p(x,t))$ are the per-capita birth and death rates and $b(x_i, p(x,t)) - d(x_i, p(x,t))$ is the per-capita growth rate of a population of x_i-individuals. The sum

$$e = \sum_{i=1}^{n(t)} (b(x_i, p(x,t)) + d(x_i, p(x,t)))$$

therefore gives the population events rate, namely the average number of birth and death events per unit of time and, consequently, the probability distribution of the waiting time Δt for the next event to occur is exponential, i.e.,

$$e \exp(-e\Delta t).\qquad\qquad(2.6)$$

Once Δt has been randomly selected in accordance with (2.6), the individual i involved in the event is picked up with probability $(b(x_i, p(x,t)) + d(x_i, p(x,t)))/e$ and, whether the event is a birth or a death, is selected according to probabilities $b(x_i, p(x,t))/(b(x_i, p(x,t)) + d(x_i, p(x,t)))$ and $d(x_i, p(x,t))/(b(x_i, p(x,t)) + d(x_i, p(x,t)))$, respectively. If a birth event involving individual i takes place, a mutation occurs with probability $\mu_i(x_i)$, so that the phenotype $x_{n(t)+1}$ of the newcomer is x_i with probability $(1 - \mu_i(x_i))$, and is otherwise randomly extracted from a suitable probability distribution centered in x_i. Finally, time t, total number of individuals $n(t + \Delta t)$, and the phenotypic distribution $p(x, t + \Delta t)$ are updated to $t + \Delta t$, $n(t) + 1$, $p(x,t) + \delta(x - x_{n(t)+1})$ in case of birth and to $t + \Delta t$, $n(t) - 1$, $p(x,t) - \delta(x - x_i)$ in case of death. The algorithm is then iterated until the desired final time is reached. Although many authors have developed their own code for the simulation of individual-based evolutionary models (see, e.g., the code available in Dieckmann, 1994), ZEN (Legendre, 2002) is a flexible and user-friendly available package.

The typical output of a simulation is a diagram like those shown in Figure 2.2. At time t, the distribution (2.5) is graphically represented by plotting $n(t)$ dots with coordinates (t, x_i), $i = 1, \ldots, n(t)$. Since individuals bearing the same trait generate the same dot, it is common to use a color intensity proportional to the frequency of the traits, e.g., a gray-scale where black represents the unit frequency and white

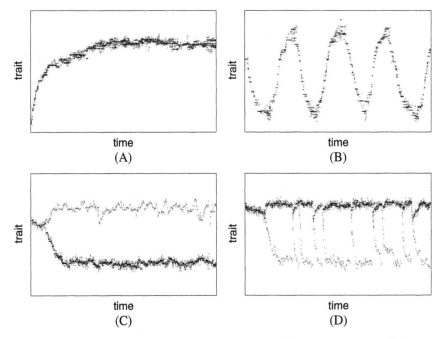

Figure 2.2 Stochastic simulations of phenotypic individual-based evolutionary models (reproduced from Doebeli and Dieckmann, 2000). See Chapter 6 for other examples.

the absence of individuals. Figure 2.2A shows a case in which the distribution (2.5)
converges to a sort of stable distribution (evolutionary equilibrium), while in Figure 2.2B the distribution never settles down, thus pointing out a Red Queen evolutionary regime (evolutionary cycle). After convergence in Figure 2.2A, individuals
are concentrated around a mean trait value and dissident individuals, bearing different trait values, are not able to invade. This marks the halt of evolutionary dynamics
and, as we will see in the next section using the terminology of evolutionary game
theory, the equilibrium mean trait value is said evolutionarily stable. By contrast,
in Figures 2.2C and D, selection turns disruptive and promotes the formation of a
bimodal trait distribution. In particular, Figure 2.2D shows recurrent branching and
extinction events (branching-extinction evolutionary cycle, see Section 1.8), a type
of Red Queen dynamics that we further discuss in Chapter 7: after each branching
the trait distribution becomes again unimodal through the evolutionary extinction
of the individuals composing the left peak of the distribution.

2.4 QUANTITATIVE GENETICS

Quantitative genetics models the effects of inheritance, natural selection, and environmental factors on the phenotypic distribution of populations by avoiding the
description of complex genetic mechanisms and intricate relations between geno-

Table 2.3 Genotype-to-phenotype map.

#	genotype	phenotypic value	H-W frequency
1	$((x_{11}, x_{11}), (x_{21}, x_{21}))$	2.2	1/16
2	$((x_{11}, x_{11}), (x_{21}, x_{22}))$	2.1	1/8
3	$((x_{11}, x_{11}), (x_{22}, x_{22}))$	2.0	1/16
4	$((x_{11}, x_{12}), (x_{21}, x_{21}))$	2.0	1/8
5	$((x_{11}, x_{12}), (x_{21}, x_{22}))$	1.9	1/4
6	$((x_{11}, x_{12}), (x_{22}, x_{22}))$	1.8	1/8
7	$((x_{12}, x_{12}), (x_{21}, x_{21}))$	1.8	1/16
8	$((x_{12}, x_{12}), (x_{21}, x_{22}))$	1.7	1/8
9	$((x_{12}, x_{12}), (x_{22}, x_{22}))$	1.6	1/16

types and phenotypes (Bulmer, 1980; Falconer, 1989). This is possible because quantitative genetics considers phenotypes controlled by a large number of loci whose alleles have small effects on phenotypes. In contrast with phenotypes controlled by one or a few loci, which assume discrete values, these phenotypes tend to assume values distributed on a continuum and can be measured on a metric scale. They are therefore called *continuous* or *quantitative traits* (sometimes simply *traits* in the following).

For example, imagine that the beak size of Darwin's finches is controlled by two unlinked loci with two alleles each. Beak size might have a background value, say 1 cm, plus the contribution of the two loci, with x_{11}, x_{12}, x_{21} and x_{22} adding 0.4, 0.2, 0.2, and 0.1 cm, respectively. Then, according to (2.1), there would be nine genotypes, mapped into seven different phenotypic values, as summarized in Table 2.3, which also shows the Hardy-Weinberg (H-W) genotype frequencies corresponding to equal allele frequencies. Table 2.3 can be graphically presented as in Figure 2.3, which shows that the distribution of the phenotype across the population is bell-shaped and indicative of a normal distribution.

This example has shown additive gene action. Other genetic mechanisms, such as dominance and epistasis (see Section 1.2), affect phenotypes and tend to reduce the number of different values that result from a given number of loci. For example, if dominance characterizes a locus, then the dominant homozygote and heterozygotes with dominant and recessive alleles will have the same phenotypic values, while if a locus is inhibited by an epistatic gene, then the alleles of that locus will have no effect on phenotypes. However, in the limit of a large number of loci with alleles adding small contributions on phenotypic values, additive gene action can be considered the most important phenotypic determinant, so that a quantitative trait x can be well described by a continuous probability distribution $p(x, t)$ across the population at generation t.

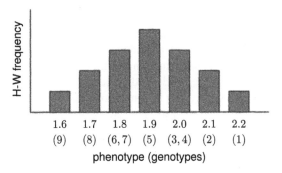

Figure 2.3 Phenotypic distribution of Table 2.3.

All the above factors are genetic in nature, but the environment also affects quantitative traits, in the sense that the additive allelic contributions may be influenced by environmental conditions. Thus, individuals of the same genotype growing in different environments may develop different trait values. Quantitative genetics takes into account both genetic and environmental factors and describes the change of the probability distribution $p(x, t)$ from one generation to the next. Mutations are not explicitly considered, but any value of x, namely all possible phenotypic variability, is always assumed to be present (though some values with negligible probability density). In other words, mutations are indirectly taken into account at each generation.

We now derive the simplest quantitative genetic model (Lande, 1976), which provides a recursive equation for the change in the mean trait

$$\bar{x}(t) = \int xp(x, t)dx,$$

under the assumption of discrete and nonoverlapping generations (see Roff, 1997, and references therein, for a more general and complete treatment). The trait x is assumed to be determined by two independent and additive contributions: the genetic one, g, and the environmental one, e. The relevance of genetic and environmental factors is measured by the *genetic variance* σ_g^2 and by the *environmental variance* σ_e^2, while the ratio

$$h^2 = \sigma_g^2/(\sigma_g^2 + \sigma_e^2), \tag{2.7}$$

called *heritability*, measures the predominance of genetic over environmental factors. In a constant and homogeneous environment all individuals are subject to the same environmental conditions, so that environmental factors do not alter the mean trait. Thus, the mean value $\bar{e}(t)$ of the environmental contribution e can be assumed to be equal to zero for all t (i.e., $\bar{x}(t) = \bar{g}(t)$ for all t). Genetic details are bypassed by assuming that the genetic value g of the progeny is the midpoint between parents genetic values, so that the mean genetic value in the next generation $\bar{g}(t + 1)$ is obtained by the mean genetic value $\bar{g}_s(t)$ of individuals in the current generation after selection. In formulas:

$$\bar{x}(t + 1) = \bar{g}(t + 1) = \bar{g}_s(t). \tag{2.8}$$

Selection acts on the trait, favoring the diffusion of the trait value x in proportion to its fitness $f(\bar{x}(t), x)$, defined, as in Section 2.2, as the product of viability and fertility. Notice that the fitness of an individual is not uniquely determined by its trait value x, since it also depends on the trait distribution through its mean value $\bar{x}(t)$. This models frequency-dependent effects. Once the fitness function is identified, the mean trait value $\bar{x}_s(t)$ at generation t after selection is given by

$$\bar{x}_s(t) = \frac{\int x f(\bar{x}(t), x) p(x, t) dx}{\int f(\bar{x}(t), x) p(x, t) dx},$$

(2.9)

and is related to $\bar{g}_s(t)$ through the heritability (2.7) as follows:

$$\bar{g}_s(t) - \bar{g}(t) = h^2 \left(\bar{x}_s(t) - \bar{x}(t) \right).$$

(2.10)

Then, substituting (2.10) and (2.9) into (2.8) (and recalling that $\bar{x}(t) = \bar{g}(t)$), one obtains

$$\bar{x}(t+1) = \bar{x}(t) + h^2 \frac{\int (x - \bar{x}(t)) f(\bar{x}(t), x) p(x, t) dx}{\int f(\bar{x}(t), x) p(x, t) dx}.$$

(2.11)

The recurrence (2.11) can be further simplified by assuming that the trait distribution is symmetric, unimodal (e.g., normal), and with a small variance $\sigma_x^2 = \sigma_g^2 + \sigma_e^2$. In fact, by expanding the fitness function in Taylor series

$$f(\bar{x}, x) = f(\bar{x}, \bar{x}) + \frac{\partial}{\partial x} f(\bar{x}, x) \bigg|_{x=\bar{x}} (x - \bar{x}) + \frac{1}{2} \frac{\partial^2}{\partial x^2} f(\bar{x}, x) \bigg|_{x=\bar{x}} (x - \bar{x})^2$$
$$+ O\left((x - \bar{x})^3 \right),$$

and substituting the above expansion in (2.11), one gets

$$\bar{x}(t+1) = \bar{x}(t) + h^2 \frac{\frac{\partial}{\partial x} f(\bar{x}(t), x) \bigg|_{x=\bar{x}(t)} \sigma_x^2 + O(\sigma_x^4)}{f(\bar{x}(t), \bar{x}(t)) + \frac{1}{2} \frac{\partial^2}{\partial x^2} f(\bar{x}(t), x) \bigg|_{x=\bar{x}(t)} \sigma_x^2 + O(\sigma_x^4)},$$

(2.12)

since for odd k the integral

$$\int (x - \bar{x}(t))^k p(x, t) dx$$

vanishes due to the symmetry of p around $\bar{x}$. Then, in the limit of small σ_x^2 and taking (2.7) into account, (2.12) turns out to be approximated by

$$\bar{x}(t+1) = \bar{x}(t) + \frac{\sigma_g^2 \frac{\partial}{\partial x} f(\bar{x}(t), x) \bigg|_{x=\bar{x}(t)}}{f(\bar{x}(t), \bar{x}(t))},$$

(2.13)

which is a deterministic recurrence in the mean value of the trait, independent of the trait distribution p (Lande, 1979; Abrams et al., 1993a; Vincent et al., 1993).

If there are S coevolving species, and each species is characterized by multiple traits, equation (2.13) easily extends to

$$\bar{x}^{(i)}(t+1) = \bar{x}^{(i)}(t) + \frac{\Sigma_g^{(i)} \frac{\partial}{\partial x^{(i)}} f^{(i)}(\bar{x}^{(1)}(t), \ldots, \bar{x}^{(S)}(t), x^{(i)}) \bigg|_{x^{(i)}=\bar{x}^{(i)}(t)}}{f^{(i)}(\bar{x}^{(1)}(t), \ldots, \bar{x}^{(S)}(t), \bar{x}^{(i)}(t))},$$

(2.14)

where $x^{(i)}$ is the vector of traits of species i, $f^{(i)}(\bar{x}^{(1)}, \ldots, \bar{x}^{(S)}, x^{(i)})$ is the fitness of the $x^{(i)}$-conspecific bearing trait values $x^{(i)}$ in an environment characterized by mean trait values $\bar{x}^{(1)}, \ldots, \bar{x}^{(S)}$, $\Sigma_g^{(i)}$ is the matrix of genetic covariances among traits $x^{(i)}$, and $\partial f^{(i)}/\partial x^{(i)}|_{x^{(i)}=\bar{x}^{(i)}(t)}$ is the fitness gradient which points in the best direction for species i (formally considered as a column vector), $i = 1, \ldots, S$ (Iwasa et al., 1991; Taper and Case, 1992).

Notice that the simplification leading to (2.13) and (2.14) requires that the trait distributions remain unimodal under the action of selection. Thus, (2.13) and (2.14) cannot explain the emergence of bimodal distributions, which would be indicative of a possible evolutionary branching (see Section 1.7).

In conclusion, quantitative genetics simplifies both maps from parents to off-spring genotypes and from genotype to phenotype, and allows one to consider the effect of natural selection on phenotypes independently of the details of genetics. In other words, by looking at quantitative traits and assuming that they spread in the environment according to bell-shaped continuous distributions, quantitative genetics describes the short-term demographic dynamics of such distributions (and their statistical indicators) under the effect of inheritance, natural selection, and environmental factors. Moreover, since mutations are indirectly taken into account at each generation (all possible phenotypes are always present), quantitative genetic models can also provide long-term predictions.

2.5 EVOLUTIONARY GAME THEORY

In evolutionary game theory individuals interact through repeated contests in which they take different actions with different probabilities. The number A of possible actions is fixed and each individual is characterized by a trait vector x, called strategy, whose A components $x_1, \ldots, x_A$ represent the probabilities according to which the individual takes the corresponding actions. Of course $x \in \Sigma_A$, where $\Sigma_A = \{x : x_1, \ldots, x_A \geq 0, \sum_i x_i = 1\}$ is the A-dimensional simplex.

Denote by S the number of different strategies $x^{(1)}, \ldots, x^{(S)}$ actually present in the community (S is unrelated to A) and by n_i the relative density of the population of $x^{(i)}$-strategists, $i = 1, \ldots, S$. Thus, $n = (n_1, \ldots, n_S) \in \Sigma_S$ and n_i is the probability of randomly picking an $x^{(i)}$-strategist out of the community. The strategy

$$m = \sum_{i=1}^{S} n_i x^{(i)} \tag{2.15}$$

is the mean strategy of the community, i.e., the probability vector according to which actions are taken on average.

The simplest formulation of game theory (von Neumann and Morgenstern, 1953), called *normal form game*, assumes that contests involve only two players and that p_{ij} is the payoff for a player taking the ith action against a player taking the jth action. The $A \times A$ matrix $P = [p_{ij}]$ is called the payoff matrix. Thus, the average payoff for an $x^{(i)}$-strategist against an $x^{(j)}$-strategist is given by $(x^{(i)})^T P x^{(j)}$, where the T superscript denotes transposition. Moreover, if encounters are random, as in large and well-mixed populations, the average payoff for an $x^{(i)}$-strategist is given by $(x^{(i)})^T Pm$ (see (2.15)).

For example, in the well-known prisoners' dilemma game (Axelrod and Hamilton, 1981), there are two possible actions ($A = 2$), namely cooperate and defect, thus the payoff matrix is given by

$$P = \begin{bmatrix} p_{11} & p_{12} \\ p_{21} & p_{22} \end{bmatrix}.$$

If both players cooperate, they receive a payoff p_{11} which is assumed to be larger than the payoff p_{22} obtained if they both defect. But if the antagonist player cooperates, it is better to defect and to exploit the other's effort ($p_{21} > p_{11}$), while if the antagonist player defects, it is also better to defect in order not to get exploited ($p_{22} > p_{12}$). In conclusion,

$$p_{21} > p_{11} > p_{22} > p_{12}.$$

The game got its name from the hypothetical situation in which two criminals, arrested under the suspicion of having committed a crime together, are kept isolated and offered the chance to be free if they provide evidence against the other, while the other does not. Each suspect can cooperate or defect by remaining silent or betraying the other suspect, respectively. Thus, p_{21} is freedom, p_{11} corresponds to remaining a suspect, p_{22} is a reduced punishment due to collaboration with the police, and p_{12} is the full punishment.

The most important notion in game theory is that of *Nash equilibrium* (Nash, 1950, 1951; see Nash, 1996, for a more recent survey): this is a strategy $\bar{x}$ that is a best reply to itself, i.e.,

$$x^T P\bar{x} \leq \bar{x}^T P\bar{x}, \quad \forall x \in \Sigma_A, x \neq \bar{x}. \tag{2.16}$$

A strategy $\bar{x}$ is a strict Nash equilibrium if it is the unique best reply to itself, i.e., if inequality (2.16) holds with the strict inequality sign. The idea behind the notion of Nash equilibrium is that if all individuals adopt such a strategy, no individual can improve its payoff by adopting a different strategy. However, imagine that the community is composed of two populations of relative densities ϵ and $(1 - \epsilon)$ and strategies x and $\bar{x}$, with $x^T P\bar{x} = \bar{x}^T P\bar{x}$ and $x^T Px > \bar{x}^T Px$. Thus, even though $\bar{x}$ is a Nash strategy, the average payoff for an x-strategist, i.e., $(1-\epsilon)x^T P\bar{x}+\epsilon x^T Px$, is greater than the average payoff for an $\bar{x}$-strategist, given by $(1 - \epsilon)\bar{x}^T P\bar{x} + \epsilon\bar{x}^T Px$. This means that a group of individuals might have an interest in adopting a strategy that differs from a Nash strategy. Of course, this is not possible in the case of a strict Nash equilibrium.

Evolutionary game theory defines a strategy as evolutionarily stable if, when all individuals of the community adopt it, no dissident group can have an advantage. Thus, a strategy $\bar{x}$ is an *evolutionarily stable strategy* (ESS) if

$$x^T P\bar{x} \leq \bar{x}^T P\bar{x}, \quad \forall x \in \Sigma_A, x \neq \bar{x}, \tag{2.17a}$$

and

$$x^T Px < \bar{x}^T Px \tag{2.17b}$$

whenever the equality sign holds in (2.17a). Clearly, a strict Nash equilibrium is an ESS, and an ESS is a Nash equilibrium. Notice that $\bar{x} = (0, 1)$, namely the strategy "always defect," is a strict Nash equilibrium for the prisoners' dilemma game. This formally explains the so-called tragedy of the commons that we introduced in Section 1.9, according to which the suboptimal use or even destruction of public resources (the "commons") is the result of private interests, whenever the best strategy for individuals conflicts with the common good (Hardin, 1968).

So far, we have assumed that the success of a strategy depends on the outcome of pairwise encounters with randomly chosen opponents. This needs not always be the case. Extensions of the normal form game, called *nonlinear games*, consider contests involving more than two players and/or non-purely random (e.g., assortative) encounter mechanisms. In these cases, the average payoff for a player taking the ith action is, in general, a nonlinear function $p_i(m)$ of the mean strategy m. Thus, the average payoff for an $x^{(i)}$-strategist is given by $(x^{(i)})^T p(m)$, where $p(m) = (p_1(m), \ldots, p_A(m))$.

The notion of ESS readily extends to nonlinear games as follows. If all individuals of the community adopt the strategy $\bar{x}$ except for a small dissident group of relative density ϵ adopting a strategy x, then $\bar{x}$ is an ESS if the average payoff for a dissident is lower than the average payoff for an individual adopting $\bar{x}$, i.e., if

$$x^T p((1 - \epsilon)\bar{x} + \epsilon x) < \bar{x}^T p((1 - \epsilon)\bar{x} + \epsilon x). \tag{2.18}$$

The strategy $\bar{x}$ is said to be a local ESS if condition (2.18) holds only for x close to $\bar{x}$, i.e., if no small dissident group with strategy x close to $\bar{x}$ can have an advantage. By expanding both sides of (2.18) in powers of $(x - \bar{x}, \epsilon)$ one gets, after some tedious but otherwise straightforward calculations,

$$(x - \bar{x})^T p(\bar{x}) + \epsilon(x - \bar{x})^T \left. \frac{\partial p}{\partial m} \right|_{m=\bar{x}} (x - \bar{x}) + O\left(\|(x - \bar{x}, \epsilon)\|^4\right) < 0, \tag{2.19}$$

where $\partial p/\partial m|_{m=\bar{x}}$ is the Jacobian matrix of $p(m)$ evaluated at $\bar{x}$. Thus, $\bar{x}$ is a local ESS if the first term in the left-hand side of (2.19) is negative, i.e., if

$$x^T p(\bar{x}) < \bar{x}^T p(\bar{x}),$$

or if such term is zero and the second term is negative. Taking into account that

$$\left. \frac{\partial p}{\partial m} \right|_{m=\bar{x}} (x - \bar{x}) \simeq p(x) - p(\bar{x}),$$

since x is close to $\bar{x}$, these conditions can be summarized by saying that $\bar{x}$ is a local ESS if

$$x^T p(\bar{x}) \leq \bar{x}^T p(\bar{x}), \quad \forall x \neq \bar{x} \text{ in a neighborhood of } \bar{x}, \tag{2.20a}$$

and

$$x^T p(x) < \bar{x}^T p(x) \tag{2.20b}$$

whenever the equality sign holds in (2.20a). Conditions (2.20) are therefore the generalization of conditions (2.17) for nonlinear games.

The notion of evolutionarily stable strategy was introduced by Maynard Smith and Price (1973) and Maynard Smith (1974). A similar notion was used by Hamilton (1967), while the notion of local ESS was first formulated by Pohley (1983).

A good survey of the development during the first decade of evolutionary game theory is offered by Maynard Smith (1982). For extensions to more general cases see Bomze and Pötscher (1989). Determining payoffs is usually not underpinned by a description of the demographic interaction between individuals, but see Reed and Stenseth (1984), Metz et al. (1992), and Rand et al. (1994) for interesting exceptions.

Although evolutionary game theory has the great merit of having introduced the notion of ESS, it is essentially a static approach which does not provide any description of evolutionary dynamics. Only a particular class of evolutionary equilibria, namely ESS, is pointed out by the theory, and evolutionary transients as well as Red Queen dynamics (see Section 1.6) are a priori excluded. Moreover, the word "stable" in the context of ESS is misleading since it does not refer to the standard notion of stability. Indeed, as we will see in Chapter 3 in the context of adaptive dynamics, ESS can be unstable, and therefore evolutionary unreachable, as first noticed by Eshel and Motro (1981) (see also Eshel, 1983; Nowak, 1990; Christiansen, 1991; Takada and Kigami, 1991; Eshel et al., 1997).

2.6 REPLICATOR DYNAMICS

Replicator dynamics describe the change of the relative densities (frequencies) $n_1, \ldots, n_S$ of S large and well-mixed interacting populations characterized by the strategies $x^{(1)}, \ldots, x^{(S)}$ of an underlying game. The relative rate of increase $\dot{n}_i / n_i$ of the ith population is a measure of its evolutionary success. Following the metaphor of the survival of the fittest, replicator dynamics express this success as the difference between the payoff $f_i(n) = (x^{(i)})^T p(m)$ for an $x^{(i)}$-strategist (see previous section) and the average payoff $f(n) = \sum_j n_j f_j(n)$. This yields the so-called *replicator equation*:

$$\dot{n}_i = n_i \left(f_i(n) - f(n) \right), \quad i = 1, \ldots, S, \tag{2.21}$$

which holds on the simplex Σ_S.

Figure 2.4 shows two examples of replicator dynamics in Σ_3, where three strategists, cooperators, defectors, and loners, play a public goods game (Hauert et al., 2002). Points $(1, 0, 0)$, $(0, 1, 0)$, and $(0, 0, 1)$ denote homogeneous populations of cooperators, defectors, and loners, and form a so-called "rock-scissors-paper" cycle, where defectors dominate cooperators (as in the prisoners' dilemma game described in the previous section), cooperators dominate loners, and loners dominate defectors. In Figure 2.4A the interior of the simplex consists of trajectories issued from and returning to $(0, 0, 1)$, i.e., brief intermittent bursts of cooperation interrupting long periods of prevalence of loners, while Figure 2.4B shows the periodic persistence of all strategies.

Let us stress that equation (2.21) does not consider explicitly the processes of mutation and selection underlying genuine evolutionary dynamics. In fact, equation (2.21) assumes that each individual is characterized by one of finitely many traits, namely the S strategies of a game, which do not vary in time. As clearly noticed by Nowak and Sigmund (2004), the replicator dynamics describe only se-

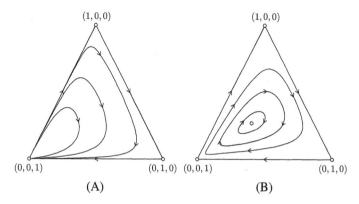

Figure 2.4 Examples of replicator dynamics (reproduced from Hauert et al., 2002).

lection, namely the short-term demographic dynamics of birth and death processes in absence of mutations.

An extension of the replicator equation, called the *replicator-mutator equation*, explicitly takes into account that the replicas of a strategist can be characterized by different strategies, belonging, however, to a fixed set of strategies. The replicator-mutator equation reads

$$\dot{n}_i = \sum_{j=1}^{S} n_j f_j(n) \mu_{ji} - n_i f(n), \quad i = 1, \ldots, S, \tag{2.22}$$

where μ_{ji} is the probability that replication of an $x^{(j)}$-strategist gives rise to an $x^{(i)}$-strategist. Mutations are represented by the fact that from time to time an individual generates miscopied replicas, while selection is taken into account by assuming that the replication rate of a strategist corresponds to its success in the underlying game. Thus, the replicator dynamics described by equation (2.22) can be interpreted at evolutionary timescale. Equation (2.22) has been also used in population genetics (Hadeler, 1981), game theory (Bomze and Bürger, 1995), and in a study of language evolution (Nowak et al., 2001), and has been recently extended to the case of a continuous distribution of strategies (Diekmann et al., 2005).

As already noticed in Section 1.9, in the social context the spreading of successful strategies is more likely to occur through cultural transmission, learning, or imitation than through inheritance. Cultural transmission can, for example, be described by the replicator-mutator equation (2.22) interpreted on a demographic timescale, where mutations represent the fact that new generations can adopt different cultures (e.g., religions) with respect to those learned from parents. The imitation of well-performing strategies is, for example, a possible source of mutations described by the replicator equation (2.21) (Hofbauer and Sigmund, 1998; Schlag, 1998). Suppose that occasionally an individual is picked out of the community and afforded the opportunity to change strategy. The selected individual samples another individual at random and adopts the sampled strategy with a certain probability. A description of such an imitation process is given by the input-output model

$$\dot{n}_i = \sum_{j=1}^{S} n_i n_j \left(f_{ji}(n) - f_{ij}(n) \right), \quad i = 1, \ldots, S, \qquad (2.23)$$

where $n_i n_j f_{ij}(n)$ is the flow of individuals switching from the ith to the jth population and $f_{ij}(n)$ is proportional to the probability of switching. Such a probability must depend on the current payoffs $f_i(n)$ and $f_j(n)$ for an $x^{(i)}$- and an $x^{(j)}$-strategist. Assuming that

$$f_{ij}(n) = \begin{cases} 0 & \text{if } f_j(n) \leq f_i(n) \\ f_j(n) - f_i(n) & \text{otherwise,} \end{cases}$$

which says that actions that perform better are more imitated, one can easily check that equation (2.23) reduces to equation (2.21).

The replicator equation (2.21) was introduced by Taylor and Jonker (1978) and generalized to nonlinear games by Palm (1984) and Ritzberger (1995). The term was coined by Schuster and Sigmund (1983) (see also Hofbauer and Sigmund, 1998) for the resemblance with the Dawkins' notion of replicator. The most interesting connection between the replicator equation and the underlying game is the following: a density vector $\bar{n}$ for which the corresponding mean strategy $\bar{m} = \sum_i \bar{n}_i x^{(i)}$ is an ESS is an asymptotically stable equilibrium of equation (2.21) (Hofbauer et al., 1979; Cressman, 1990, 1992).

In conclusion, the approach of replicator dynamics provides a dynamical framework for various evolutionary games (see Hofbauer and Sigmund, 2003, for a recent review). However, the evolutionary dynamics of strategies (traits) are not described. In fact, no mechanism is provided to generate new strategies, since mutations are considered as switches from a strategy to another already present in the community. Thus, replicator dynamics are more suited for describing selection processes rather than long-term evolutionary dynamics.

2.7 FITNESS LANDSCAPES

Evolution through natural selection has often been imagined to imply some sort of improvement and progress, so that the long-term evolutionary dynamics of a continuous trait x, like a vector of quantitative traits or the strategy of a game, can be pictured as an ascent on a so-called fitness landscape $F(x)$, which abstractly measures the advantage of bearing the trait value x (Wright, 1931, 1969). This suggests the simplistic view that the evolutionary rate of change $\dot{x}$ is given by the fitness gradient, i.e.,

$$\dot{x} = \frac{\partial F}{\partial x}. \qquad (2.24)$$

The evolutionary dynamics governed by (2.24) imply a continuous fitness increase, possibly up to a local peak of the fitness landscape, thus preventing the possibility of Red Queen evolutionary regimes. Finding the vector x that maximizes a function $F(x)$ is the standard problem of optimization theory. To solve this problem various numerical techniques can be used, among them the so-called genetic algorithms that we briefly discussed in Section 1.9.

The fitness function F can describe the reproductive success of a phenotype, as we have seen in Sections 2.2 and 2.4, or it can be derived from an underlying game, as in Sections 2.5 and 2.6, or from other plausibility arguments. The fitness function F summarizes the interactions of individuals with their environment and how such interactions select the most advantageous traits. However, (2.24) neglects the fact that the adaptation of a trait occurs through a sequence of mutations and substitutions of old trait values with new ones, and not by a synchronized decision of all individuals of a homogeneous population to change strategy or to reproduce progeny bearing the same mutation. What really matters is the advantage $f(x, x')$ that a scarce population of mutants with trait x' has in competing against a resident population of individuals with trait x. In other words, the fitness landscape is often determined by a family of functions of the mutant trait x', where the particular function of the family experienced by mutants is identified by the resident trait x (frequency-dependent selection). Thus, the change of x reshapes the fitness landscape experienced by the next mutants and the fitness landscape can be seen as an "adaptive" landscape. As x evolves, new peaks and valleys appear in the fitness landscape and this determines the further evolution of x.

Three biological assumptions are needed to formally support the idea that evolutionary dynamics can be obtained through an *adaptive fitness landscape*. Synthetically:

1. mutations are rare,

2. mutations are small,

3. invasion implies substitution.

As already discussed in Section 2.1, if mutations are rare events on the demographic timescale (so-called mutation-limited evolution), population genetics supports the idea that sexual reproduction can be neglected, so that one can look at long-term evolutionary dynamics of quantitative traits by using, at least for qualitative purposes, asexual demographic models, where individuals are characterized by their phenotypes. Since the demographic and evolutionary timescale are kept separated, each time a mutation occurs in one of the relevant traits the resident and mutant populations have plenty of time to interact and define the new structure of the community, before the next mutation occurs. In particular, if the mutant population replaces the resident population, the trait undergoes an evolutionary step (on the evolutionary timescale), while if mutants go extinct the trait does not change. Moreover, if mutations are small (as is the case most of the time, in particular for quantitative traits), evolution proceeds by small steps in trait space and, in the limit of infinitesimal mutations, one can pretend to describe evolutionary trajectories by means of ODEs.

However, each time the community experiences a mutation, the acceptance of the evolutionary step in the involved trait virtually requires to study the competition between the resident and mutant populations, for example, by considering the replicator dynamics in the augmented space of resident and mutant populations (see Section 2.6). As already said, in the case of substitution the step is accepted, while it is rejected if mutants go extinct. But if residents and mutants coexist, then a new resident population is formed and an evolutionary branching may occur (though the mechanisms of sexual reproduction here play a crucial role, see Section 1.7). The identification and analysis of resident-mutant competition models has often been avoided by assuming that if a mutant has an advantage in terms of fitness with respect to a similar resident, then the mutant population will invade and oust the resident population (assumption 3), thus becoming the new resident (Rand et al., 1994; Dieckmann and Law, 1996; Metz et al., 1996). This assumption has tradi- tionally been called the "invasion implies fixation" principle, but here we prefer the more accurate expression "invasion implies substitution."

Under assumptions 1–3, mutants invade and therefore substitute if

$$\frac{\partial}{\partial x'} f(x, x') \bigg|_{x'=x} (x' - x) > 0,$$

so that, in the end, the evolutionary dynamics are given by

$$\dot{x} = \frac{\partial}{\partial x'} f(x, x') \bigg|_{x'=x}. \tag{2.25}$$

The right-hand side of (2.25) is often called the *selection gradient* or *selection derivative*. Notice that the selection gradient is not the gradient of the fitness func- tion $F(x) = f(x, x)$. In other words, the evolutionary rate of change $\dot{x}$ does not point in the direction that is best for the whole population, but in the direction that is best from the mutant point of view. Thus, the evolutionary dynamics governed by (2.25) do not describe a hill-climbing process on a constant fitness landscape (like (2.24)) and can give rise to cyclic or chaotic Red Queen evolutionary regimes (Khibnik and Kondrashov, 1997). In particular, stable evolutionary equilibria $\bar{x}$ of (2.25) can be local maxima as well as minima of the mutant fitness $f(x', \bar{x})$, re- spectively corresponding to terminal points of the evolutionary dynamics at which no mutant invasion is possible (see the notion of ESS introduced in Section 2.5) and points at which disruptive selection may lead to evolutionary branching (Chris- tiansen, 1991; Brown and Pavlovic, 1992; Metz et al., 1992; Abrams et al., 1993b).

The extension to the case of S coevolving species with traits $x^{(1)}, \ldots, x^{(S)}$ is straightforward. If $f^{(i)}(x^{(1)}, \ldots, x^{(S)}, x')$ is the fitness of a mutant in the ith species with trait x', then the evolutionary dynamics read

$$\dot{x}^{(i)} = K^{(i)}(x^{(1)}, \ldots, x^{(S)}) \frac{\partial}{\partial x'} f^{(i)}(x^{(1)}, \ldots, x^{(S)}, x') \bigg|_{x'=x^{(i)}}, \tag{2.26}$$

where $K^{(i)}$, $i = 1, \ldots, S$, are suitable matrices which take into account the corre- lations between the trait components of a species and scale the rate of evolutionary change between different species.

Notice that the selection gradients of equations (2.25) and (2.26) are formally equivalent to those of equations (2.13) and (2.14) of quantitative genetics (Sec- tion 2.4), though they model evolutionary processes on different timescales: long- term mutation-limited evolution and short-term phenotypic variability, respectively.

Such a formal similarity has induced Dieckmann and Law (1996) to consider the selection gradient as a sort of "canonical" ingredient of evolutionary models.

The notion of fitness landscape goes back to Wright (1931) and phenotypic models of trait-related interactions can be traced back to Levins (1962a,b) and Leon (1974). The limitations of the optimization view to evolutionary dynamics have been discussed in Lewontin (1979, 1987) and Emlen (1987), while the idea of adaptive fitness landscapes has been independently proposed by various authors (see Brown and Vincent, 1987b; Hofbauer and Sigmund, 1990; Christiansen, 1991; Vincent et al., 1993; Metz et al., 1996). Evolutionary dynamics of the kind (2.25) or (2.26) have frequently been used in applications (see, e.g., Brown and Vincent, 1987a, 1992; Rosenzweig et al., 1987; Iwasa et al., 1991; Takada and Kigami, 1991; Abrams, 1992a; Taper and Case, 1992; Abrams et al., 1993b; Marrow et al., 1992; Marrow and Cannings, 1993; Matsuda and Abrams, 1994a,b; Marrow et al., 1996; Abrams and Matsuda, 1997) and have recently been formalized as "Darwinian dynamics" by Vincent and Brown (2005). The major limitations of the fitness landscapes approach are assumption 3 and the rather vague identification of the scaling factor $K^{(i)}$ in (2.26). By suitably shaping the functions $K^{(i)}$, $i = 1, \ldots, S$, one can in fact get from (2.26) arbitrary evolutionary dynamics. Moreover, the mechanisms of evolutionary branching and extinction are not explained by fitness landscapes (see, however, Gavrilets, 2004). As we will see in the next section, the approach of adaptive dynamics overcomes these limitations.

2.8 ADAPTIVE DYNAMICS

With the term "adaptive dynamics" authors generically refer to the long-term evolutionary dynamics of quantitative characters driven by the processes of mutation and selection. However, here we refer to the particular approach developed by Dieckmann and Law (1996), Metz et al. (1996), and Geritz et al. (1997, 1998).

As in the approach of fitness landscapes, genetic details are ignored by exploiting the timescale separation argument that justifies the use of asexual demographic models (see assumption 1 in the previous section). The competition between resident and mutant populations is explicitly described by a so-called *resident-mutant model*, an ODE system that describes the dynamics of resident and mutant population densities for any given value of the traits, thus including frequency- and density-dependent effects. In the absence of mutants, such a model degenerates into the *resident model*, which defines the region of the trait space where all resident populations coexist on the demographic timescale. If the evolutionary dynamics drive the traits toward the boundary of this region, at least one resident population goes extinct.

The use of deterministic demographic models is justified when the populations are abundant, so that they are not subject to the effect of demographic stochasticity (see Section 2.3). Of course this is not the case for an initially scarce mutant population and, consequently, the resident-mutant model can be used only to ascertain the fate of a mutant population that has escaped accidental extinction. Mutations are described as random events (as in individual-based evolutionary models, see

Section 2.3) and once a mutation has occurred the probability of avoiding accidental extinction is measured by the probability that the mutant population reaches a threshold abundance. Thus, evolutionary dynamics are nothing but a sequence of trait substitutions and can be depicted as random walks in trait space. The emergence of new resident traits occurs when a mutant population escapes accidental extinction and coexists with the resident populations, thus becoming itself a resident population. By contrast, an evolutionary extinction occurs when the evolutionary dynamics drive the extinction of a resident population, thus reducing the number of coexisting resident traits. Moreover, the realizations of a phenotypic individual-based evolutionary model, defined by the same stochastic description of mutations and by the birth and death rates of the resident-mutant model, have been proved to converge to the trait random walks of adaptive dynamics for vanishing mutation probabilities (Champagnat et al., 2001). This confirms that the random walks of adaptive dynamics represent a reasonable description of long-term mutation-limited evolution.

In the case of continuous traits (see Section 2.4), the limit to infinitesimal mutations can be considered, and the result is that the trait random walks converge to the trajectories of an ODE system of the form (2.26), where the terms $K^{(i)}$ and $f^{(i)}$ take explicit forms derived from the stochastic description of the mutation process and from the resident-mutant model (Dieckmann and Law, 1996). Thus, the "canonical" selection gradient introduced in Section 2.7 again plays an important role, so that Dieckmann and Law (1996) called their ODE system the *canonical equation of adaptive dynamics*. The canonical equation therefore describes long-term mutation-limited evolution in the case of infinitesimally small mutations. Its derivation and properties are the core of next chapter.

As we will see, the derivation of the canonical equation requires the invasion implies substitution principle. However, the conditions under which such a principle holds can now be derived by studying the resident-mutant model, thus transforming the principle into a theorem. So far, such a theorem has been proved under the following technical conditions:

a. spatial heterogeneity and physiological structures (e.g., age, stage, and energy reserves) of resident and mutant populations are not described, so that the resident-mutant model is composed of one equation for each population,

b. stationary coexistence of the resident populations,

c. the community is composed of (c') a single resident population characterized by mutations with effects on many traits, or of (c'') many resident populations characterized by mutations which only influence a single (scalar) trait.

Case c' has recently been proved by Geritz (2005), while case c'' is proved in Appendix B (see Dieckmann, 1994, Chapter 7, and Eshel et al., 1997, for proofs under particular resident-mutant dynamics). Case c' is relevant in situations in which several traits may be affected by a single mutation in accordance with a multivariate probability distribution. This happens when the traits are controlled by different but overlapping sets of genes that control more characters (pleiotropic genes, see Section 1.2). In fact, if the traits are controlled by nonoverlapping sets of genes, they mutate independently, so that each mutation affects only a single trait. By contrast, if the traits are controlled by the same genes, their concomitant mutations are so strongly correlated that one can define a one-dimensional equivalent trait and the results for case c'' can still be applied.

In both cases c' and c'' the conditions under which invasion implies substitution are not valid at an equilibrium of the canonical equation. Thus, once the evolutionary dynamics have found a halt at a stable equilibrium of the canonical equation, a further investigation of the resident-mutant model is required. Geritz et al. (1998) found specific analytical conditions to establish if resident-mutant coexistence is possible and, if so, whether similar resident and mutant trait values are subject to opposite selection pressures (disruptive selection). If resident-mutant coexistence is not possible or selection is not disruptive, we are back to the notion of ESS (see Section 2.5), while in the opposite case the equilibrium is a branching point: the mutant population becomes a new resident and the two initially similar resident populations diverge in trait space under the disruptive selection described by a higher-dimensional canonical equation.

At this point it is important to remark that the canonical equation describes the long-term evolution of reproductively isolated monomorphic populations, a description that fits with the case of asexual populations (where reproduction is clonal) and with that of noninterbreeding sexual populations, i.e., those belonging to different species. Thus, after an evolutionary branching, the higher-dimensional canonical equation is justified for sexual species only if a speciation event actually takes place at the branching point (see Section 1.7), an issue that can only be verified experimentally or theoretically by means of population genetic models. This is the main limitation of adaptive dynamics, as well as of all phenotypic approaches to long-term evolution. As described by adaptive dynamics, evolutionary branching is associated with disruptive selection, under which a sexual monomorphic population turns dimorphic, while speciation occurs provided reproductive isolation evolves between the two morphs. Describing the further evolution of two interbreeding morphs with a higher-dimensional canonical equation can be nonsense (Matessi and Gimelfarb, 2006). This is best explained by the following example. Imagine that a sexual, diploid-one-locus, population is dimorphic with respect to a given trait, which takes a high value x_1 if the locus is homozygous and a low value x_2 if the locus is heterozygous. Then, a two-dimensional canonical equation could pretend to describe the evolution of x_1 and x_2, possibly arriving to the absurd conclusion that one of the two morphs goes extinct, while sexual reproduction guarantees that each of the two morphs is able to sustain the other.

So far, the canonical equation of adaptive dynamics has been used even outside its ascertained domain of validity (conditions a–c), just taking the invasion implies substitution principle as granted. For example, Parvinen (2002) and Ernande and Dieckmann (2004) consider spatially structured populations, physiological structures are described by Diekmann (2004) and Ernande et al. (2004) (see also Durinx et al., 2007, for recent advancements), and van Doorn and Weissing (2006) investigate the evolutionary consequences of sexual conflicts. Moreover, Dieckmann and Law (1996) address both mutations with effects on many traits and nonstationary coexistence of the resident populations, while Parvinen et al. (2006) and Dieckmann et al. (2006) study the evolution of functional (i.e., infinite-dimensional) traits (see also Ernande and Dieckmann, 2004; Ernande et al., 2004). Except for Chapter 9, where we consider a case in which resident populations coexist either at equilibrium or on a cycle, depending upon trait values, we restrict ourselves to the conditions a, b, and c″ above. In our opinion, the full understanding of the competition between residents and mutants, under general conditions on the structure of the community and on the resident coexistence regime, constitutes the major challenge for the future development of adaptive dynamics (see Geritz et al., 2002, for a promising direction).

Summarizing, the approach of adaptive dynamics describes the long-term evolutionary dynamics of continuously varying traits, by explicitly taking into account the resident-mutant demographic interactions underlying the process of selection. In the limit of infinitesimal mutations, adaptive dynamics defines a particular fitness landscape and provides the evolutionary dynamics in terms of an ODE system, growing in size through evolutionary branching and pruning through evolutionary extinction.

2.9 A COMPARATIVE ANALYSIS

All the classifications of quantitative approaches to evolutionary dynamics that we are aware of, including the one proposed in this chapter, are different. However, the differences do not concern the substance of the problem, but only the meaning associated with a few strategic keywords, like "adaptive dynamics." While there is general agreement that all models explicitly describing genotypes fall in population genetics, the classification of the phenotypic approaches is much more faded. The long-term phenotypic changes resulting from the processes of mutation and selection are often generically mentioned as adaptive dynamics (see Waxman and Gavrilets, 2005, for a review, and the 15 commentaries appeared in the same issue of the *Journal of Evolutionary Biology*, vol. 18, pp. 1139–1373; see also McGill and Brown, 2007 for a more recent review). Thus, phenotypic individual-based evolutionary models (Section 2.3), fitness landscapes (Section 2.7), and what we have called adaptive dynamics (Section 2.8) can be seen as different versions of a very general approach called adaptive dynamics. Hofbauer and Sigmund (1990) were the first to use the term adaptive dynamics for describing the evolutionary dynamics of the strategies of a game (see also Meszéna et al., 2001; Page and Nowak, 2002; Nowak and Sigmund, 2004), in a way that we have here classified as fitness landscapes. Dieckmann (1994), Doebeli and Dieckmann (2000), and Kisdi and Geritz (2001) used the term adaptive dynamics for describing phenotypic individual-based evolutionary models and extended the approach of adaptive dynamics to the case of sexual reproduction by adding genetic details to their models (Dieckmann and Doebeli, 1999; Kisdi and Geritz, 1999; Doebeli and Dieckmann, 2000; Geritz and Kisdi, 2000), thus obtaining models that we would have here classified within population genetics. Abrams (2001) does not mention fitness landscapes, but considers them as a by-product of quantitative genetics that, as we have seen in Section 2.4, arrives at the same formal equation of the fitness

Table 2.4 Comparison of modeling approaches to evolutionary dynamics.

approach	geno- type	pheno- type	short- term	long- term	evol. tree	det. mod.	stoch. mod.
population genetics*	✓		✓			✓	✓
individual-based evolutionary models	✓	✓	✓	✓	✓		✓
quantitative genetics		✓	✓			✓	
evolutionary game theory		✓				✓	
replicator dynamics		✓	✓			✓	
fitness landscapes		✓		✓		✓	
adaptive dynamics		✓		✓	✓	✓	✓

*classical formulation only: short-term evolution of genotypic distributions.

landscapes approach. Finally, Page and Nowak (2002) and Nowak and Sigmund (2004) refer to evolutionary game theory as a general paradigm including all phenotypic approaches, unified by the common feature of considering some sort of local optimization through the eyes of a small group of mutant dissidents.

Despite the somehow irrelevant problem of terminology, lucidly pointed out by Lewontin (1982), the choice of the approach to be used depends on the desired level of description, on the particular questions one likes to answer, and on the technical tools of analysis available for such purposes. From a biological point of view, the main distinctive aspects of a modeling approach are its genotypic or phenotypic foundation, the relevant timescale (the short-term demographic timescale, on which the mechanisms of inheritance and natural selection are predominant on mutations, or the long-term evolutionary timescale on which repeated selections of successful mutations are the dominant evolutionary driver), and the capability of predicting the evolutionary tree of the community, whose branches generate through the emergence of new resident traits and are pruned by evolutionary extinctions. By contrast, from a technical point of view, the main discriminant is between stochastic and deterministic models. A comparison of the seven approaches presented in this chapter with respect to these descriptive and technical aspects is shown in Table 2.4, where the difficulties of a sharp classification are of course reflected. In particular, we have seen that population genetics includes, on one hand, short-term demographic descriptions of the detailed mechanisms of reproduction and, on the other hand, long-term stochastic descriptions which find maximum flexibility and realism in genotypic individual-based evolutionary models. These two different approaches have been kept separated in Table 2.4, to avoid the misleading message that population genetics fulfills all requirements. Also the evolutionary tree checkmark attributed to adaptive dynamics deserves the usual comment that in sexual species disruptive selection amounts to speciation provided reproductive isolation evolves between the two branching phenotypes.

The aim of this book is to show how one can produce a (possibly complete) catalog of qualitatively different long-term evolutionary regimes in a prescribed region of the parameter space. We do not focus on the quantitative differences between two similar evolutionary regimes reached for slightly different parameter settings, but rather on the qualitative characteristics of the regime, be it stationary, cyclic, or chaotic and involving the branching and/or the pruning of the evolutionary tree. In other words, we want to partition the parameter space into niches where the characteristics of the long-term evolutionary regime do not change. The boundaries of such niches are of crucial importance for the formulation of environmental and socio-economic control policies and for the evaluation of the effects of human activities on the long-term behavior of ecosystems.

In mathematical jargon, a qualitative change of the long-term regime of a dynamical system observed in response to a small parameter perturbation is called a *bifurcation*. In principle, one might imagine being able to detect bifurcations by simulating the system for various parameter values and initial conditions, and by checking for each simulation the characteristics of the obtained regime. However, this "brute force" approach rarely works in practice, because, as explained in the tutorial on bifurcation theory reported in Appendix A, a bifurcation is typically related to a loss of stability of the long-term regime, so that the length of the simulations would need to be dramatically increased while approaching the bifurcation. In other words, there is no hope of accurately detecting a bifurcation through repeated simulations, since close to the bifurcation there is no way of knowing if the long-term regime is still qualitatively the same or if we have just not waited enough.

Numerical bifurcation analysis (see Appendix A) is thus the ideal tool for our purpose. However, it is well developed, together with effective software packages, only for deterministic systems (i.e., for differential and difference equations; see, e.g., Kuznetsov, 2004). This rules out individual-based evolutionary models and restrains our choice to fitness landscapes and adaptive dynamics. Moreover, we are also interested in predicting the structure of the evolutionary tree, so that our final choice is the canonical equation of adaptive dynamics (Section 2.8). As we will see in more detail in the next chapter, it explicitly takes into account the underlying mutation and selection processes and provides a description of the dynamics of adaptive traits in terms of an ODE system, whose size can vary through evolutionary branching and extinction, thus tracking the structure of the evolutionary tree of the community.

The canonical equation of adaptive dynamics is derived under conditions 1 and 2 of Section 2.7 and a–c of Section 2.8. They might seem to be so many and so crude that one can wonder about the significance of deriving and analyzing the canonical equation in real situations, where many (if not all) of these conditions are likely to be violated. To which extent the canonical equation may reveal the essence of the evolutionary process is a rather questionable issue that no one is in a position to address. As discussed in Sections 2.1 and 2.7, conditions 1 and 2 taken together, i.e., rare and small mutations, can be exploited to sweep genetic details under the carpet and go from short-term to long-term evolutionary predictions. However, this step is not fully justified in cases of sexual reproduction, in particular as far as evolutionary branching is concerned. Moreover, though the separation between demographic and evolutionary timescales is a commonly accepted assumption (Roughgarden, 1983b), recent studies have documented cases of rapid evolution, in which demographic and evolutionary changes occur on comparable timescales (Lively, 1993; Thompson, 1998; Albertson et al., 1999; Yoshida et al., 2003). In such cases, as well as when conditions a–c are relaxed, or in general to validate the predictions obtained through the canonical equation, one may use a suitable genotypic individual-based evolutionary model (which, however, requires the description of the genotype-to-phenotype map) to run selected simulations guided by the bifurcation analysis of the canonical equation.

On the other hand, the canonical equation is the ideal tool for a sound mathematical analysis of the dependence of the long-term evolutionary regime upon physiological, environmental, and control parameters. This is why we strongly believe that the conclusions one can obtain by applying the theory and tools of nonlinear dynamics to the canonical equation of adaptive dynamics may serve at least as a qualitative guide for further modeling investigations and for designing field surveys and laboratory experiments.

Since its conception, the approach of adaptive dynamics has been used in many applications, ranging from infectious diseases (see Dieckmann et al., 2002, and references therein) to the evolution of dispersal rates in heterogeneous environments (Doebeli and Ruxton, 1997; Parvinen, 1999; Gyllenberg et al., 2002; Kisdi, 2002; Parvinen, 2002, 2006; Dercole et al., 2007b), and from the evolution of predation, mutualistic interactions, and competition (Dieckmann et al., 1995; Marrow et al., 1996; Kisdi, 1999; Doebeli and Dieckmann, 2000; Kisdi and Geritz, 2001; Ferrière et al., 2002; Le Galliard et al., 2003, 2005; Ferdy and Godelle, 2005) to seed size, seedling, and germination strategies (Geritz et al., 1999; Levin and Muller-Landau, 2000; Mathias and Kisdi, 2002), just to cite a few examples. Chapters 4–10 extend this list and gradually emphasize several methodological aspects of the analysis. In particular, Chapter 4 presents the first application of adaptive dynamics outside the biological realm, by studying an economic system in which traits measure technological characteristics of products, mutations represent product innovations, selection is exerted by customers, and the environment is the market in which products compete.

Further reading on adaptive dynamics is available on the web pages of the Evolution and Ecology Program (formerly known as Adaptive Dynamics Network) at the International Institute for Applied Systems Analysis (IIASA), where a list of related papers is constantly kept updated, and in the book series Cambridge Studies in Adaptive Dynamics. The five available titles (Dieckmann et al., 2000, 2002; Ferrière et al., 2004a; Dieckmann et al., 2004; Haccou et al., 2005) minimally overlap with the present manuscript. They are focused on particular evolutionary issues and do not push the analysis of the canonical equation so far. Finally, we like to mention that interesting forthcoming books are in preparation (see http://www.iiasa.ac.at).

Chapter Three

The Canonical Equation of Adaptive Dynamics

In this chapter we derive the canonical equation of Adaptive Dynamics (AD). As anticipated in Chapter 2, the AD approach focuses on the long-term evolutionary dynamics of continuous (quantitative) adaptive traits and bypasses genetic details by using asexual demographic models. This is justified under contrasting demographic and evolutionary timescales, i.e., if mutations are rare events at birth (see discussion of condition 1 in Section 2.7). We consider physiologically unstructured populations coexisting at equilibrium in spatially homogeneous environments and characterized by traits that mutate independently (conditions a, b, and c″ of Section 2.8). If mutations are rare, the resident populations are challenged by one mutation at a time and are at equilibrium when a mutant appears. We show that if mutations are also small (condition 2 of Section 2.7), invading mutants generically replace the corresponding residents, so that the final consequence of a successful mutation is a small variation of the trait. This result allows one to consider the limit case of infinitesimal mutations and to approximate a stochastic sequence of successful mutations by a smooth evolution of the traits governed by a system of ODEs, one ODE for each trait, i.e., the AD canonical equation. The evolutionary trajectories produced by the AD canonical equation may lead to stationary evolutionary regimes, which may be terminal points of evolutionary dynamics or sources of diversity (evolutionary branching, see Section 1.7), to nonstationary evolutionary regimes (Red Queen dynamics, see Section 1.6), or to evolutionary extinction (see Section 1.8). Most of the material presented in this chapter and in related Appendixes B–D is adapted from the founding papers of Dieckmann and Law (1996), Metz et al. (1996), and Geritz et al. (1997, 1998).

3.1 THE EVOLVING COMMUNITY

We consider a community composed of several resident populations with individuals characterized by one or more continuous adaptive traits. We imagine that each trait can be described by a real variable mapped into the actual phenotypic value through a suitable scaling. This allows us to consider dimensionless traits.

Each population is a homogeneous group of individuals belonging to the same species and bearing the same trait values. Some of the resident populations may be conspecific, but characterized by different trait values or *morphs*. Species present in a single morph are said *monomorphic*, while species present in several morphs are said *polymorphic* (*dimorphic* in the case of two morphs). Thus, monomorphic species are represented by a single resident population, while polymorphic species are composed of several resident populations, one for each morph.

Evolutionary dynamics result from the processes of innovation (mutation), introducing in the community new populations characterized by new trait values, and competition (selection), shaping the dynamics of population abundances on the demographic timescale. Darwin (1858) first realized that those populations best adapted to survive and reproduce should come to dominate the community, all else being equal, and he used the term *natural selection* for the demographic process leading to the dominance of the best-adapted populations (see Section 1.4).

As discussed in Section 2.8, we describe natural selection by means of deterministic demographic models where sex, space, and physiological details, such as age, stage, and energy reserves, do not appear. Thus, all individuals of a population are identical and uniformly distributed in a homogeneous habitat, so that each population is identified by its abundance and by the values of all its characteristic traits. Deterministic demographic models typically define population abundances as positive real numbers that represent the actual number of individuals (or the corresponding population biomass or density per unit of space). Deterministic demographic models are justified only when the actual number of individuals in each population is sufficiently large to avoid accidental extinction risks (the so-called demographic stochasticity discussed in Section 2.3). In the following we assume that even small abundances correspond to relatively large numbers of individuals.

We consider mutations that have a small effect only on one trait, i.e., the case in which the traits characterizing the populations are controlled by nonoverlapping sets of genes, each with a small effect on the controlled trait (see discussion of case c'' in Section 2.8). The mutant population is therefore identified by the same trait values carried by the resident population in which the mutation has occurred (simply called resident population in the following), except for the trait affected by the mutation, which is slightly different from the corresponding resident value. After each mutation, the mutant population is initially very scarce (the mutation occurs in one or a few individuals), but it has the potential to spread in the community and replace the resident population. Since mutations are rare events on the demographic timescale, demography has plenty of time between successive mutations to define the resident populations of the community. This allows us to focus on the consequences of a single mutation.

Let us now fix the notation used throughout the chapter and the rest of the book. Discriminating between species, characteristic traits, and morphs within each species is possible but would be rather cumbersome and force the reader to pay attention to notational details rather than to the logical steps of the derivation of the AD canonical equation. Moreover, as summarized in Table 3.1, all applications described in Chapters 4–10 consider communities characterized by at most three adaptive traits, so that the use of a general formalism able to describe all possible community structures is not really needed.

We therefore use the simplest possible notation by sacrificing the detailed description of the community structure. For this we focus on the trait affected by the mutation and denote by x and x' its resident and mutant values, by n and n' the abundances of the corresponding resident and mutant populations, and we pack all other traits characterizing the community and all other resident population abun-

Table 3.1 Community structures considered in Chapters 4–10.

chapter	number of species	number of traits per species	number of morphs per species	total number of traits
4	1	1	1 and 2	2
5	2	1,1	1,1	2
6	2	1,1	1,1	2
7	1	1	1 and 2	2
8	1	1	1	1
9	2	1,1	1,1	2
10	3	1,1,1	1,1,1	3

dances in two vectors X and N. Notice that the dimension of X can be different from the dimension of N since some populations can be characterized by more than one trait, while others may not be endowed with adaptive traits. However, in all applications considered in Chapters 4–10 each resident population is characterized by a single trait (see Table 3.1 third column), so that X and N have the same dimension.

The chapter is organized as follows. In the next section we formally describe the dynamics of population abundances, n, n', and N, on the demographic timescale, namely for frozen values of the traits x, x', and X. Then, before going into the details of the derivation of the AD canonical equation, which will take place in Sections 3.4–3.7, we present in Section 3.3 an example of evolving community composed of a resource (prey) and its consumer (predator). The evolution of resource-consumer communities based on this example will be fully investigated in Chapters 5 and 9. As we will see, the derivation and the analysis of the AD canonical equation requires basic notions of nonlinear dynamical systems, reviewed in Appendix A. Section 3.8 highlights in particular the role of bifurcation analysis, while Section 3.9 relaxes some of the conditions under which the AD canonical equation is derived.

3.2 THE RESIDENT-MUTANT MODEL

The demography of populations is regulated by the interactions of single individuals with their surrounding environment, meant as all biotic and abiotic factors affecting them (see Section 1.4). The biotic environment is the whole community and is therefore identified by the abundances n, n', and N of resident and mutant populations and by a set of demographic parameters characterizing the interactions between individuals of the same or different populations. Such parameters can be affected by the traits of interacting individuals and by abiotic environmental conditions which may fluctuate in time. However, for simplicity, but also for focusing on the evolutionary dynamics endogenously generated by mutation and competition processes and not on those entrained by exogenous influences, we assume a constant abiotic environment (e.g., no seasons), hence described by means of constant parameters.

The rates of growth $\dot{n}$, $\dot{n}'$, and $\dot{N}$ of the resident and mutant populations are therefore functions of all abundances n, n', and N and of all traits x, x', and X, as well as of constant demographic and environmental parameters which are not explicitly pointed out in the following. Since a population is composed of identical individuals, the common practice is to define the growth rate of a population when its abundance is equal to one, called per-capita (or per-unit) growth rate, and then scale it into the population growth rate.

We denote by $f(n, n', N, x, x', X)$ the per-capita growth rate $\dot{n}/n$ of the resident population. Denoting by $b(n, n', N, x, x', X)$ and $d(n, n', N, x, x', X)$ the per-capita birth and death rates, i.e., the abundance gain due to births and the abundance loss due to deaths per unit of time and per unit of abundance, then the per-capita growth rate results from their balance, i.e.,

$$f(n, n', N, x, x', X) = b(n, n', N, x, x', X) - d(n, n', N, x, x', X). \quad (3.1)$$

Given f, the per-capita birth and death rates b and d are not uniquely defined, since a common function of the community variables (n, n', N, x, x', X) can be added to both b and d without affecting f. Since mutations occur at birth, one might conjecture that different evolutionary dynamics of the trait x arise for the same function f but for different functions b and d. Surprisingly, in Section 3.5, such a conjecture will be disproved.

Once the functions b and d are available, the per-capita birth, death, and growth rates of the mutant population can be immediately obtained. In fact, residents and mutants are conspecific individuals that differ in only one of their traits, so that they are involved in the same intra-, interspecific, and environmental interactions described by the functions b, d, and f. In other words, either one of the two populations can be considered as mutant, provided the other is considered as resident, so that the mutant per-capita birth, death, and growth rates are derived from those of the resident population by exchanging resident and mutant abundances and trait values, i.e., they are given by $b(n', n, N, x', x, X)$, $d(n', n, N, x', x, X)$, and $f(n', n, N, x', x, X)$.

As for the remaining resident populations, we could introduce their per-capita growth rates but this would require us to be explicit on the various components of the abundance vector N, by spanning them through an index i and by defining each per-capita growth rate $\dot{N}_i/N_i$. To avoid this notational inconvenience, we pack the growth rates $\dot{N}_i$ of all remaining populations in the vector $F(n, n', N, x, x', X)$. Thus, the demographic dynamics of the community are described by

$$\dot{n} = n f(n, n', N, x, x', X), \quad (3.2a)$$
$$\dot{n}' = n' f(n', n, N, x', x, X), \quad (3.2b)$$
$$\dot{N} = F(n, n', N, x, x', X), \quad (3.2c)$$

where the abundances n, n', and N are the demographic state variables and the traits x, x', and X play the role of constant parameters. Model (3.2) is called the *resident-mutant model*.

We now point out a few properties of functions f and F that will be useful in the following to study the resident-mutant model (3.2). Function f, by definition, gives the per-capita growth rate of the population whose abundance and trait appear as first and fourth argument. Thus, if the second argument vanishes, then f does not depend anymore on its fifth argument, since the growth of a population cannot be affected by the trait of a nonexisting population, i.e.,

$$f(n, 0, N, x, x', X) = f(n, 0, N, x, \cdot, X), \tag{3.3a}$$

where the dot ($\cdot$) stands for any trait value. By contrast, if the first argument vanishes, the value of f is still affected by the fourth argument, since the potential growth of an absent population depends on its trait values.

A second property of f concerns the case of identical resident and mutant populations, i.e., $x' = x$. Then, there is actually only one population, characterized by the trait x, with abundance $n+n'$ and per-capita growth rate $f(n+n', 0, N, x, \cdot, X)$. If one virtually splits the population into two subpopulations of abundances n and n', it must therefore result

$$(\dot{n} + \dot{n}')|_{x'=x} = nf(n, n', N, x, x, X) + n'f(n', n, N, x, x, X)$$
$$= (n + n')f(n + n', 0, N, x, \cdot, X),$$

which implies that the two subpopulations are characterized by the same per-capita growth rate, i.e.,

$$f(n, n', N, x, x, X) = f(n', n, N, x, x, X),$$

for any value of n and n', or equivalently

$$f(n, n', N, x, x, X) = f((1 - \phi)(n + n'), \phi(n + n'), N, x, x, X), \tag{3.3b}$$

for any $0 \le \phi \le 1$. Of course, similar properties hold for the per-capita birth and death rates b and d.

Analogously, for function F we have

$$F(n, 0, N, x, x', X) = F(n, 0, N, x, \cdot, X), \tag{3.4a}$$

$$F(n, n', N, x, x, X) = F((1 - \phi)(n + n'), \phi(n + n'), N, x, x, X). \tag{3.4b}$$

Moreover,

$$F(n, n', N, x, x', X) = F(n', n, N, x', x, X), \tag{3.4c}$$

since the growth of a population interacting with two conspecific populations cannot depend upon the order in which these populations are considered.

In the absence of the mutant population, the resident-mutant model (3.2) degenerates into the so-called *resident model*

$$\dot{n} = nf(n, 0, N, x, \cdot, X), \tag{3.5a}$$
$$\dot{N} = F(n, 0, N, x, \cdot, X), \tag{3.5b}$$

which identifies the demographic attractor of the resident populations before the appearance of the mutant population. We assume that in suitable ranges of the

traits x and X model (3.5) has a stable and strictly positive equilibrium at which
the resident population abundances are constant and given by

$$n = \bar{n}(x, X), \quad N = \bar{N}(x, X), \tag{3.6}$$

i.e.,

$$f(\bar{n}(x, X), 0, \bar{N}(x, X), x, \cdot, X) = 0, \tag{3.7a}$$
$$F(\bar{n}(x, X), 0, \bar{N}(x, X), x, \cdot, X) = 0. \tag{3.7b}$$

We also assume that the equilibrium (3.6) is the only strictly positive attractor of
model (3.5). This condition is not necessary but simplifies the discussion and will
be relaxed in Section 3.9. In other words, we assume that the equilibrium (3.6) is
the only demographic attractor at which the resident populations can coexist. Thus,
the evolutionary dynamics of the community can be defined only in the open set $\mathcal{X}$
of the trait space (x, X) in which the equilibrium (3.6) exists. Such a set is called
the *evolution set* of the community.

The equilibrium (3.6) is a point in the demographic state space (n, N) that is
stable and strictly positive for each $(x, X) \in \mathcal{X}$, while crossing the boundary of $\mathcal{X}$ it
loses at least one of its properties, namely existence, stability, and positivity. More
precisely, the equilibrium (3.6) may disappear by colliding with a strictly positive
saddle, lose stability while giving rise to a small stable limit cycle, or lose positivity
(and stability) by crossing one of the faces of the state space (n, N). Since such
faces are invariant for the resident model (3.5) (i.e., trajectories of model (3.5)
starting on one face remain there forever), the only way in which the equilibrium
(3.6) may cross one face is therefore through the collision with a saddle lying on
the face.

Technically, each one of the above degeneracies can be easily revealed by a corresponding degeneracy of the eigenvalues of the Jacobian matrix of the resident
model (3.5) evaluated at the equilibrium (3.6), i.e.,

$$J_r(x, X) = \begin{bmatrix} \bar{n}(x,X) \frac{\partial}{\partial n} f(n,0,\bar{N}(x,X),x,\cdot,X) & \bar{n}(x,X) \frac{\partial}{\partial N} f(\bar{n}(x,X),0,N,x,\cdot,X) \\ \frac{\partial}{\partial n} F(n,0,\bar{N}(x,X),x,\cdot,X) & \frac{\partial}{\partial N} F(\bar{n}(x,X),0,N,x,\cdot,X) \end{bmatrix}_{\substack{n=\bar{n}(x,X) \\ N=\bar{N}(x,X)}} . \tag{3.8}$$

In fact, $J_r(x, X)$ has a vanishing eigenvalue if the equilibrium (3.6) collides with
a saddle and a pair of purely imaginary eigenvalues if the equilibrium loses stability and gives rise to a small stable limit cycle, while all eigenvalues of $J_r(x, X)$
generically have negative real part for $(x, X) \in \mathcal{X}$.

3.3 THE EXAMPLE OF RESOURCE-CONSUMER COMMUNITIES

As an example of evolving community, we consider a two-species community
composed of two resident populations, a resource (prey) and a consumer (predator) harvesting on it, with abundances n_1 and n_2, respectively. The most standard model describing resource-consumer demographic interactions is the so-called
Rosenzweig-MacArthur model (Rosenzweig and MacArthur, 1963). The biological
assumptions needed to justify the model are many. First of all, as assumed from

the beginning of this chapter, the abiotic environment is supposed to be constant and spatially homogeneous, and the two populations are assumed to be randomly distributed in their habitat and characterized by the absence of relevant sex, age, or stage structures. Moreover, if consumers are not present, the resource population, when scarce, is assumed to grow exponentially (so-called *Malthusian model*), i.e.,

$$\dot{n}_1 = rn_1,$$

where r is the *net* (or *intrinsic*) *growth rate* (per capita), i.e., the difference between basic (intrinsic) natality and mortality per capita. However, when the resource is abundant, i.e., when n_1 is high, various phenomena, like scarcity of nutrients, scarcity of suitable niches for reproduction, and epidemics, become relevant and give rise to lower birth rates and/or higher death rates. All these phenomena, usually classified as *intraspecific competition*, can be taken into account by modifying the Malthusian model into the so-called *logistic model*

$$\dot{n}_1 = rn_1 - cn_1^2,$$

where c is a constant coefficient of intraspecific competition. The logistic model has two equilibria, namely the trivial equilibrium $n_1 = 0$, which is unstable, and a positive equilibrium $n_1 = r/c$, which is stable and represents the abundance of the resource in the absence of consumers. This equilibrium abundance, called *carrying capacity*, is universally denoted by K and the most popular form of the logistic model is actually

$$\dot{n}_1 = rn_1 \left(1 - \frac{n_1}{K}\right).$$

To complete the resource equation, one must add to the logistic model the rate of resource consumption due to the presence of consumers. This can be done in a simple way by assuming that the amount H of resource harvested by each consumer in a unit of time (called the *functional response* of the consumer) is an increasing and saturating function of resource abundance, i.e.,

$$H(n_1) = \frac{an_1}{1 + a\tau n_1},$$

where a is the *attack rate*, namely a measure of the aggressiveness of the consumer, and τ is the *handling time*, measuring the time needed by each consumer to capture, transport, eat, and digest a unit of resource. The function H specified above is called the *Holling type II functional response* (Holling, 1965), to distinguish it from other functional responses proposed in the literature. It is worth noticing that the *maximum consumption rate* of each consumer (obtained when resources are very abundant) is $H(\infty) = 1/\tau$ (i.e., the inverse of the handling time), while the so-called *half-saturation constant* h, defined as the resource abundance at which consumption is half maximum, is $h = 1/(a\tau)$. Actually, the Holling type II functional response is often written in terms of maximum consumption rate and half-saturation constant as

$$H(n_1) = \frac{\tau^{-1} n_1}{h + n_1}.$$

However, the expression in terms of attack rate and handling time is more easily extended to the case of two or more resources (e.g., resident and mutant resource populations) (Murdoch, 1969; Chesson, 1983).

Finally, the Rosenzweig-MacArthur consumer equation is obtained by assuming that the consumer birth rate (per capita) is given by a constant basic natality plus a contribution proportional, through a constant conversion factor e called *consumer efficiency*, to the functional response, while the *net death rate* (per capita) d (basic mortality minus basic natality) is constant.

In conclusion, the Rosenzweig-MacArthur model is

$$\dot{n}_1 = rn_1 - cn_1^2 - \frac{an_1}{1 + a\tau n_1} n_2, \tag{3.9a}$$

$$\dot{n}_2 = e\frac{an_1}{1 + a\tau n_1} n_2 - dn_2, \tag{3.9b}$$

and is often written in terms of carrying capacity as follows:

$$\dot{n}_1 = rn_1 \left(1 - \frac{n_1}{K}\right) - \frac{an_1}{1 + a\tau n_1} n_2, \tag{3.10a}$$

$$\dot{n}_2 = e\frac{an_1}{1 + a\tau n_1} n_2 - dn_2. \tag{3.10b}$$

Model (3.9) has six positive demographic parameters and degenerates for $\tau \to 0$ into the classical Lotka-Volterra prey-predator model (Lotka, 1920; Volterra, 1926). It is important to notice that two parameters (resource growth rate r and competition coefficient c) depend upon the resource, one (consumer death rate d) depends upon the consumer, while the three others (a, τ, and e), despite their names, depend upon both populations. The number of parameters of model (3.9) could be reduced to three through rescaling. However, we do not follow this option because it would complicate the biological interpretation of the dependence of the parameters upon resource and consumer adaptive traits. Moreover, to have a meaningful problem, one must assume that $e > d\tau$, because, otherwise, the consumer population cannot grow even in the presence of an infinitely abundant resource population. The reader interested in more details on the biological interpretation of the parameters can refer to Muratori and Rinaldi (1992).

Under suitable parametric conditions, fully specified in Section A.4, model (3.9) has a stable and strictly positive equilibrium

$$(\bar{n}_1, \bar{n}_2) = \left(\frac{d}{a(e - d\tau)}, \frac{e\left(ar(e - d\tau) - cd\right)}{a^2(e - d\tau)^2}\right), \tag{3.11}$$

or

$$(\bar{n}_1, \bar{n}_2) = \left(\frac{d}{a(e - d\tau)}, \frac{er\left(aK(e - d\tau) - d\right)}{a^2 K(e - d\tau)^2}\right), \tag{3.12}$$

in terms of carrying capacity.

To study the evolution of a resource trait x_1 and of a consumer trait x_2, one must first transform model (3.9) into a resident model of the form (3.5), by specifying the dependence of the demographic parameters upon the traits, i.e., $r(x_1)$, $c(x_1)$, $d(x_2)$, $a(x_1, x_2)$, $\tau(x_1, x_2)$, $e(x_1, x_2)$. Moreover, one must determine the evolution set $\mathcal{X}$, namely the pairs (x_1, x_2) for which the equilibrium (3.11) exists and is stable and strictly positive.

Then, extra information must be added to the resident model to obtain two resident-mutant models in the form (3.2), one in which a mutant resource population, characterized by abundance n_1' and trait x_1', is also present in the community, and one in which the mutant population is a consumer population with abundance n_2' and trait x_2'. This step typically requires the introduction of new trait-dependent demographic parameters describing the intraspecific competition between the resident and mutant populations. For example, denoting by $\gamma(x_1, x_1')$ the competition coefficient characterizing reduced birth rate and/or increased death rate in the resource resident population due to competition with the resource mutant population, where necessarily $\gamma(x_1, x_1) = c(x_1)$, we can write the resource resident-mutant model as

$$\dot{n}_1 = r(x_1)n_1 - c(x_1)n_1^2 - \gamma(x_1, x_1')n_1 n_1'$$
$$- \frac{a(x_1, x_2)n_1}{1 + a(x_1, x_2)\tau(x_1, x_2)n_1 + a(x_1', x_2)\tau(x_1', x_2)n_1'}n_2, \qquad (3.13a)$$

$$\dot{n}_1' = r(x_1')n_1' - \gamma(x_1', x_1)n_1'n_1 - c(x_1')n_1'^2$$
$$- \frac{a(x_1', x_2)n_1'}{1 + a(x_1, x_2)\tau(x_1, x_2)n_1 + a(x_1', x_2)\tau(x_1', x_2)n_1'}n_2, \qquad (3.13b)$$

$$\dot{n}_2 = e(x_1, x_2)\frac{a(x_1, x_2)n_1}{1 + a(x_1, x_2)\tau(x_1, x_2)n_1 + a(x_1', x_2)\tau(x_1', x_2)n_1'}n_2$$
$$+ e(x_1', x_2)\frac{a(x_1', x_2)n_1'}{1 + a(x_1, x_2)\tau(x_1, x_2)n_1 + a(x_1', x_2)\tau(x_1', x_2)n_1'}n_2$$
$$- d(x_2)n_2. \qquad (3.13c)$$

Introducing the trait-dependent carrying capacity $K(x_1) = r(x_1)/c(x_1)$, equations (3.13a) and (3.13b) read

$$\dot{n}_1 = r(x_1)n_1\left(1 - \frac{1}{K(x_1)}\left(n_1 + \alpha(x_1, x_1')n_1'\right)\right) + \cdots,$$
$$\dot{n}_1' = r(x_1')n_1'\left(1 - \frac{1}{K(x_1')}\left(\alpha(x_1', x_1)n_1 + n_1'\right)\right) + \cdots,$$

where the (nondimensional) function

$$\alpha(x_1, x_1') = \frac{\gamma(x_1, x_1')}{c(x_1)}$$

$(\alpha(x_1, x_1) = 1)$ is called the *competition function* (MacArthur, 1969, 1970; Gatto, 1990).

Competition between two populations is said to be *symmetric* if $\alpha(x_1, x_1') = \alpha(x_1', x_1)$, i.e., if the competition function α is symmetric with respect to the diagonal $x_1' = x_1$ of the resource trait space (x_1, x_1'). Competition is said to be *asymmetric* otherwise. Notice that symmetric competition neither requires nor implies the symmetry $\gamma(x_1, x_1') = \gamma(x_1', x_1)$ of the competition coefficient. Cases of asymmetric resident-mutant competition will be studied in Chapters 4, 6, and 8.

The consumer resident-mutant model is easier to derive, because the Rosenzweig-MacArthur model has no intraspecific competition in the consumer population (the per-capita growth rate $\dot{n}_2/n_2$ is independent of n_2). Thus, no new ingredients are needed to specify the consumer resident-mutant model, which using the carrying capacity $K(x_1)$ reads

$$\dot{n}_1 = r(x_1)n_1 \left(1 - \frac{n_1}{K(x_1)}\right)$$

$$-\frac{a(x_1, x_2)n_1}{1 + a(x_1, x_2)\tau(x_1, x_2)n_1}n_2 - \frac{a(x_1, x_2')n_1}{1 + a(x_1, x_2')\tau(x_1, x_2')n_1}n_2',$$

$$\tag{3.14a}$$

$$\dot{n}_2 = e(x_1, x_2)\frac{a(x_1, x_2)n_1}{1 + a(x_1, x_2)\tau(x_1, x_2)n_1}n_2 - d(x_2)n_2, \tag{3.14b}$$

$$\dot{n}_2' = e(x_1, x_2')\frac{a(x_1, x_2')n_1}{1 + a(x_1, x_2')\tau(x_1, x_2')n_1}n_2' - d(x_2')n_2'. \tag{3.14c}$$

If we like to explicitly rewrite model (3.13) with the notation introduced in the previous sections, then $x = x_1$, $x' = x_1'$, $X = x_2$, $n = n_1$, $n' = n_1'$, $N = n_2$, and

$$f(n, n', N, x, x', X) = r(x)\left(1 - \frac{1}{K(x)}(n + \alpha(x, x')n')\right)$$

$$-\frac{a(x, X)}{1 + a(x, X)\tau(x, X)n + a(x', X)\tau(x', X)n'}N,$$

$$F(n, n', N, x, x', X) = e(x, X)\frac{a(x, X)n}{1 + a(x, X)\tau(x, X)n + a(x', X)\tau(x', X)n'}N$$

$$+ e(x', X)\frac{a(x', X)n'}{1 + a(x, X)\tau(x, X)n + a(x', X)\tau(x', X)n'}N$$

$$- d(X)N.$$

Similarly, model (3.14) can be rewritten with $x = x_2$, $x' = x_2'$, $X = x_1$, $n = n_2$, $n' = n_2'$, $N = n_1$, and

$$f(n, n', N, x, x', X) = e(X, x)\frac{a(X, x)N}{1 + a(X, x)\tau(X, x)N} - d(x),$$

$$F(n, n', N, x, x', X) = r(X)N\left(1 - \frac{N}{K(X)}\right)$$

$$-\frac{a(X, x)N}{1 + a(X, x)\tau(X, x)N}n - \frac{a(X, x')N}{1 + a(X, x')\tau(X, x')N}n'.$$

Notice that in both cases functions f and F satisfy properties (3.3) and (3.4).

3.4 DOES INVASION IMPLY SUBSTITUTION?

In this section we study the competition between resident and mutant populations. In particular, we specify the conditions under which the mutant population spreads in the community (*invasion*) and those under which it replaces the resident population (*substitution*), thus leading to a small step, from x to x', in the evolution of

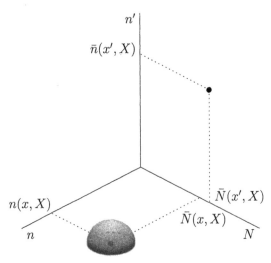

Figure 3.1 The resident-mutant demographic state space (n, n', N) sketched in three dimensions. Immediately after a mutation, the state of the resident-mutant model (3.2) is close to the equilibrium (3.15) (shaded volume). The equilibrium (3.17) is also represented.

the trait affected by the mutation. However, the conclusions drawn from the analysis of a deterministic demographic model, such as model (3.2), apply only if the mutant population, initially composed of one or a few individuals, escapes accidental extinction. Thus, even if the conditions of invasion are satisfied, the mutant population may disappear, leaving the trait unchanged. The effects of accidental extinction are quantified in the next section and in Appendix C.

Just before the occurrence of the mutation, the resident populations are at the equilibrium (3.6) of the resident model (3.5), or close to it. Thus, just after the mutation, the resident and mutant populations are close to the equilibrium

$$(n, n', N) = (\bar{n}(x, X), 0, \bar{N}(x, X)) \tag{3.15}$$

of model (3.2) (shaded volume in Figure 3.1).

The faces $n' = 0$ and $n = 0$ of the resident-mutant demographic state space (n, n', N) are obviously invariant for the resident-mutant model (3.2). In particular, on the face $n' = 0$ model (3.2) degenerates into the resident model (3.5), while on the face $n = 0$ it reduces to

$$\dot{n}' = n' f(n', 0, N, x', \cdot, X), \tag{3.16a}$$
$$\dot{N} = F(0, n', N, \cdot, x', X), \tag{3.16b}$$

which, by virtue of (3.4c), is simply model (3.5) with n and x replaced by n' and x'. Thus, provided $(x', X) \in \mathcal{X}$, the equilibrium $(n', N) = (\bar{n}(x', X), \bar{N}(x', X))$ is the only strictly positive attractor of model (3.16) and its associated eigenvalues are those of the Jacobian matrix $J_r(x', X)$ obtained from (3.8) by replacing x with x'. Moreover, the point

$$(n, n', N) = (0, \bar{n}(x', X), \bar{N}(x', X)), \tag{3.17}$$

shown in Figure 3.1, is another equilibrium of the resident-mutant model (3.2).

The stability of equilibrium (3.15) can be discussed through linearization (see Section A.1). By suitably ordering the demographic state variables n, n', and N, one can easily verify that the Jacobian matrix of the resident-mutant model (3.2) evaluated at equilibrium (3.15) is given by

$$J(x, x', X) = \begin{bmatrix} J_r(x, X) & \cdots \\ 0 & f(0, \bar{n}(x, X), \bar{N}(x, X), x', x, X) \end{bmatrix}. \quad (3.18)$$

Thus, due to the block-triangular structure of (3.18), the eigenvalues associated with equilibrium (3.15) are those of the matrix $J_r(x, X)$, which have negative real part, and the eigenvalue

$$\lambda(x, x', X) = f(0, \bar{n}(x, X), \bar{N}(x, X), x', x, X), \quad (3.19)$$

which is nothing but the growth rate of a very scarce mutant population. If this eigenvalue is positive the abundance of a scarce mutant population initially increases, i.e., the mutant population invades. For this reason $\lambda(x, x', X)$ is called the *invasion eigenvalue*.

The invasion eigenvalue gives the initial exponential rate of growth of the mutant population appeared in an environment set by the resident populations. Such a quantity is traditionally called *invasion fitness* (or simply *fitness*) of the mutant population (Metz et al., 1992) and function (3.19) will be often called the *fitness function* in the following. Positive fitness values characterize mutations that confer to mutant individuals some advantage, in terms of survival or reproductive success, in the competition against similar resident individuals, while negative fitness values characterize disadvantageous mutations.

Similarly, the eigenvalues associated with the equilibrium (3.17) are those of matrix $J_r(x', X)$ and the invasion eigenvalue $\lambda(x', x, X)$. In conclusion, since matrices $J_r(x, X)$ and $J_r(x', X)$ have eigenvalues with negative real part, the stability of equilibria (3.15) and (3.17) is related only to the sign of the invasion eigenvalues.

Notice, first, that $\lambda(x, x, X) = 0$. In fact, if the resident and mutant populations are identical ($x' = x$), we can use (3.3a), (3.3b) with $\phi = 0$, and (3.7a) to obtain

$$\lambda(x, x, X) = f(0, \bar{n}(x, X), \bar{N}(x, X), x, x, X)$$
$$= f(\bar{n}(x, X), 0, \bar{N}(x, X), x, \cdot, X) = 0. \quad (3.20)$$

Moreover, (3.3), (3.4), and (3.7) imply that

$$f((1 - \phi)\bar{n}(x, X), \phi\bar{n}(x, X), \bar{N}(x, X), x, x, X) = 0,$$
$$F((1 - \phi)\bar{n}(x, X), \phi\bar{n}(x, X), \bar{N}(x, X), x, x, X) = 0,$$

for $0 \leq \phi \leq 1$, i.e., all points

$$(n, n', N) = ((1 - \phi)\bar{n}(x, X), \phi\bar{n}(x', X), (1 - \phi)\bar{N}(x, X) + \phi\bar{N}(x', X)) \quad (3.21)$$

of the segment connecting equilibria (3.15) and (3.17) are equilibria of the resident-mutant model (3.2) when $x' = x$. All such equilibria are neutrally stable (i.e., they are not unstable, since $(n + n', N)$ converges to the equilibrium (3.6) of the resident model (3.5) when $x' = x$, but they do not attract all nearby trajectories) and hence have one vanishing associated eigenvalue, namely $\lambda(x, x, X)$, all other eigenvalues being those of $J_r(x, X)$.

Second, the invasion eigenvalues $\lambda(x, x', X)$ and $\lambda(x', x, X)$ are, in general, of opposite sign, so that if equilibrium (3.15) is stable, then equilibrium (3.17) is unstable, and vice versa. In fact, for any given X, the function $\lambda(x, x', X)$ vanishes for $x' = x$. This means that if we exclude nongeneric cases we have

$$\left.\frac{\partial}{\partial x'}\lambda(x, x', X)\right|_{x'=x} \neq 0, \tag{3.22}$$

so that the function $\lambda(x, x', X)$ has opposite sign for $x' > x$ and $x' < x$ (x' close to x).

If we look at the eigenvectors of the Jacobian matrix $J(x, x', X)$ we can again take advantage of its block-triangular structure (3.18) and conclude that all eigenvectors of $J_r(x, X)$, lying in the (n, N) space, become eigenvectors of $J(x, x', X)$ by simply adding the component $n' = 0$. By contrast, the eigenvector associated with the invasion eigenvalue $\lambda(x, x', X)$ (called the invasion eigenvector in the following) is almost aligned with segment (3.21) and tends to it as x' tends to x (see Figure 3.2). In fact, for $x' = x$, all points of segment (3.21) are equilibria of the resident-mutant model (3.2) characterized by a null eigenvalue with associated eigenvector lying on segment (3.21). Analogously, the eigenvectors of $J_r(x', X)$, lying in the (n', N) space, become eigenvectors of $J(x', x, X)$ by adding the component $n = 0$, and the invasion eigenvector (associated with the invasion eigenvalue $\lambda(x', x, X)$) is almost aligned with segment (3.21).

Now, expanding $\lambda(x, x', X)$ in Taylor series around $x' = x$, one gets

$$\lambda(x, x', X) = \lambda(x, x, X) + \left.\frac{\partial}{\partial x'}\lambda(x, x', X)\right|_{x'=x} (x' - x) + O(|x' - x|^2),$$

with $\lambda(x, x, X) = 0$. Thus, for x' sufficiently close to x, if

$$\left.\frac{\partial}{\partial x'}\lambda(x, x', X)\right|_{x'=x} (x' - x) \tag{3.23}$$

is positive, i.e., if

$$\left.\frac{\partial}{\partial x'}\lambda(x, x', X)\right|_{x'=x} > 0 \quad \text{and} \quad x' > x$$

or

$$\left.\frac{\partial}{\partial x'}\lambda(x, x', X)\right|_{x'=x} < 0 \quad \text{and} \quad x' < x,$$

the invasion eigenvalue $\lambda(x, x', X)$ is positive and equilibrium (3.15) is unstable, while equilibrium (3.17) is stable. Vice versa, if (3.23) is negative, equilibrium (3.15) is stable, while equilibrium (3.17) is unstable (see Figure 3.2). Moreover, the absolute values of the two invasion eigenvalues are of order $O(|x' - x|)$, and are therefore smaller than the absolute values of the real parts of all other eigenvalues of equilibria (3.15) and (3.17). This means that in the vicinity of equilibria (3.15) and (3.17) the dynamics of the resident-mutant model (3.2) in the direction of the invasion eigenvector are slower than those along all other directions. In other words, nearby trajectories quickly converge toward the trajectory associated with

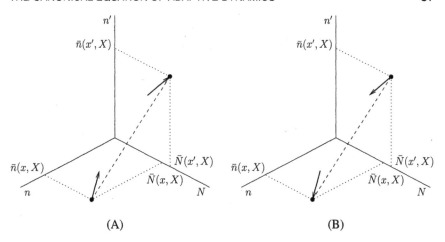

(A)   (B)

Figure 3.2 Stability of equilibria (3.15) and (3.17) of the resident-mutant model (3.2). (A) (3.23) is positive: equilibrium (3.15) is unstable and equilibrium (3.17) is stable. (B) (3.23) is negative: equilibrium (3.15) is stable and equilibrium (3.17) is unstable. Arrows point in the direction of the resident-mutant dynamics along the invasion eigenvectors (dashed line: segment (3.21)).

the invasion eigenvector and then move along it, by approaching the corresponding equilibrium or going away from it depending on its stability.

So far we have established through linearization that if the first-order term (3.23) is positive the mutant population invades, while if (3.23) is negative the mutant population goes extinct. However, the question whether invasion with positive (3.23) implies the substitution of the resident population cannot be answered through a local analysis of the resident-mutant model (3.2), but requires, in principle, the study of its global behavior. The fact that invasion with positive (3.23) implies the substitution of the resident population has been assumed true since the first formulation of the AD canonical equation (Dieckmann and Law, 1996) and even before, when the notion of adaptive fitness landscapes was introduced (see Section 2.7). Such an assumption, called the "invasion implies substitution" principle, was later turned into a theorem, which is proved in Appendix B. What is interesting is that the invasion implies substitution theorem guarantees the substitution of the resident population under the same conditions assumed so far, namely

– $(x, X) \in \mathcal{X}$,

– x' close to x,

– (3.23) positive,

– initial condition of the invasion transient close to equilibrium (3.15).

In particular, Appendix B shows that under the first two conditions there exists an invariant tube that connects equilibria (3.15) and (3.17) (see Figure 3.3). If (3.23) is positive (case A in the figure), all trajectories starting in the tube tend

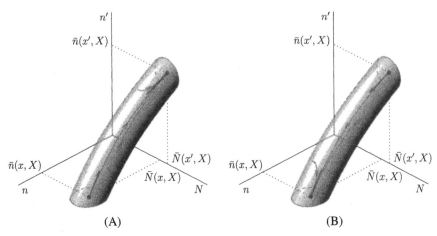

Figure 3.3 Invasion implies substitution.

toward equilibrium (3.17). By contrast, if (3.23) is negative (case B), by simply exchanging the roles of resident and mutant populations, the theorem guarantees that all trajectories starting in the tube converge to equilibrium (3.15).

3.5 THE AD CANONICAL EQUATION

We derive in this section the AD canonical equation, which describes the evolution of each trait x characterizing the community through the ODE

$$\dot{x} = \frac{1}{2}\mu(x)\sigma^2(x)\bar{n}(x, X) \left. \frac{\partial}{\partial x'}\lambda(x, x', X)\right|_{x'=x}, \qquad (3.24)$$

where time spans the evolutionary timescale, $(x, X) \in \mathcal{X}$, and $\mu(x)$ and $\sigma(x)$ are respectively proportional to the probability that a given newborn in the resident population is a mutant and to the standard deviation of the mutational step $(x' - x)$. If equation (3.24) is written for each trait x_i, $i = 1, 2, \ldots$, the result is a set of nonlinear autonomous ODEs

$$\dot{x}_i = k_i(x_i)\bar{n}_i(x_1, x_2, \ldots) \left. \frac{\partial}{\partial x_i'}\lambda_i(x_1, x_2, \ldots, x_i')\right|_{x_i'=x_i}, \quad i = 1, 2, \ldots, \qquad (3.25)$$

where

$$k_i(x_i) = \frac{1}{2}\mu_i(x_i)\sigma_i^2(x_i),$$

often called the *mutational rate*, summarizes the statistics of the mutation process and $\bar{n}_i(x_1, x_2, \ldots) = \bar{n}_j(x_1, x_2, \ldots)$ if x_i and x_j characterize the same population.

The evolution of each trait x is the result of a sequence of advantageous mutations after which the mutant population replaces the resident population and becomes itself the new resident. As we will see, the AD approach turns this step by step evolution into the smooth process (3.24) by considering the limit case of extremely small and rare mutations. Thus, x' can be imagined as close to x as we desire and the resident populations are at the demographic equilibrium (3.6) when the mutation occurs, so that the initial state of the resident-mutant model (3.2) just after the mutation is certainly in the tube of Figure 3.3. From the previous section it therefore follows that if

$$\frac{\partial}{\partial x'}\lambda(x, x', X)\Big|_{x'=x} \qquad (3.26)$$

is positive [negative], then mutant populations characterized by $x' > x$ [$x' < x$] replace the resident population, while mutations characterized by $x' < x$ [$x' > x$] leave no trace in the community.

The quantity (3.26) measures the selection pressure acting on the trait x, selecting for larger trait values if positive and for smaller values if negative, and this is why it is called the *selection derivative* in the AD jargon. In other words, $\dot{x}$ and (3.26) have the same sign, i.e., $\dot{x}_i$ has the sign of

$$\frac{\partial}{\partial x_i'}\lambda_i(x_1, x_2, \ldots, x_i')\Big|_{x_i'=x_i}, \qquad (3.27)$$

which is the selection derivative relative to trait x_i.

As discussed in Sections 1.3 and 1.4 and later in Section 2.3, evolutionary change is a stochastic process with basically two sources of stochasticity: the process of mutation and demographic stochasticity, namely the fact that even advantageous mutations (i.e., mutations with a favorable sign of the selection derivative (3.26)) may fail to invade due to accidental extinction of the mutant population. Thus, a deterministic description of the rate of evolutionary change $\dot{x}$ can only be interpreted as the average evolutionary change among all possible realizations of the processes of mutation and demographic stochasticity, i.e.,

$$\dot{x} = \lim_{dt \to 0}\frac{E[x(t + dt) - x(t)]}{dt}, \qquad (3.28)$$

where $E[\cdot]$ is the standard expected value operator of probability theory and t spans the evolutionary timescale.

Denoting by

$$P(x, x', X, dt)dx' \qquad (3.29)$$

the probability that a community with traits (x, X) at time t will be characterized by traits between (x', X) and $(x' + dx', X)$ at time $t + dt$, (3.28) becomes

$$\dot{x} = \lim_{dt \to 0}\frac{1}{dt}\int_{-\infty}^{+\infty}(x' - x)P(x, x', X, dt)dx'. \qquad (3.30)$$

To explicitly compute the integral in (3.30), one must first determine the probability

(3.29), which is the product of three probabilities:

- the probability P_m that a mutation occurs in the time interval $[t, t + dt]$;

- the probability $P'dx'$ that the mutant trait is between x' and $x' + dx'$;

- the probability P_s that the mutant substitutes the resident.

In the following, the three probabilities are computed separately and then the AD canonical equation (3.24) is derived by performing the integral in (3.30). We closely follow Dieckmann and Law (1996), with emphasis on the logical construction rather than on the mathematical formalism (for a more detailed treatment see Champagnat et al., 2006). To simplify the discussion, we first assume that (3.26) does not vanish at the current evolutionary state (x, X) and discuss the particular case in which (3.26) vanishes at the end of the section.

The Probability P_m

To compute the probability that a mutation occurs in the time interval $[t, t + dt]$, we must distinguish between demographic and evolutionary timescales, since birth events, ultimately responsible of mutations, occur on the demographic timescale. For the moment, we do not consider the extreme case of completely separated timescales, but we introduce a small (positive) timescaling factor ϵ, which will be later used to separate the two timescales by considering the limit $\epsilon \to 0$. Thus, a small amount dt of evolutionary time corresponds to a large amount dt/ϵ of demographic time, which can be subdivided into $1/\epsilon$ small intervals of length dt, where dt is here read on the demographic timescale.

In a small amount dt of demographic time there is space for just a single birth in the resident population. Technically, the probability of more than one birth is $O(dt^2)$ and gives no contribution to the limit (3.30). In fact, the probability of a single birth is

$$b(\bar{n}(x, X), 0, \bar{N}(x, X), x, \cdot, X)\bar{n}(x, X)dt + O(dt^2),$$

where the per-capita birth rate of the resident population is evaluated at the demographic equilibrium (3.15), at which the community is settled before the mutation. This is formally justified by stochastic arguments of the kind we have discussed in Section 2.3 in the context of individual-based models. Stochastic demographic modeling typically assumes birth and death processes as independent Markov processes, where each individual of a population is involved in birth and lethal events that occur with average frequencies (average number of events per individual per unit of demographic time) determined by the present state of the community. Thus, the probability that an event occurs to a given individual in a small amount dt of demographic time is given by the corresponding average frequency scaled by dt plus $O(dt^2)$, while the probability of more than one event in the same time amount is $O(dt^2)$. Going from a stochastic to a deterministic description that captures the average demographic dynamics, as the resident-mutant model (3.2), then reduces to interpreting the average frequencies of birth and lethal events as the per-capita birth and death rates.

The probability that a given newborn in the resident population is characterized by a mutation in the trait x is small and possibly affected by the trait value. It is denoted in the following by $\epsilon\mu(x)$, where ϵ takes into account that mutations become extremely rare on the demographic timescale as ϵ vanishes. Since the traits characterizing the population mutate independently, the probability that a given newborn is a mutant with respect to two traits is $O(\epsilon^2)$ and can be neglected in view of the limit $\epsilon \to 0$.

We can now compute the probability that a mutation occurs in the trait x in the evolutionary time interval $[t, t + dt]$. In each of the $1/\epsilon$ small demographic time intervals composing the evolutionary time interval $[t, t+dt]$ a mutation in x occurs with probability

$$\epsilon\mu(x)b(\bar{n}(x, X), 0, \bar{N}(x, X), x, \cdot, X)\bar{n}(x, X)dt + O(dt^2),$$

and mutations in different intervals occur independently. Thus, in the evolutionary time interval $[t, t+dt]$ there is space for just a single mutation, since the probability of more than one mutation is $O(dt^2)$ and gives no contribution to the limit (3.30). Moreover, since the mutation can occur in any of the $1/\epsilon$ demographic time intervals, the probability of a single mutation in the evolutionary time interval $[t, t+dt]$ is obtained by summing $1/\epsilon$ times the term

$$\left(\epsilon\mu(x)b(\bar{n}(x, X), 0, \bar{N}(x, X), x, \cdot, X)\bar{n}(x, X)dt + O(dt^2)\right)$$
$$\times \left(1 - \epsilon\mu(x)b(\bar{n}(x, X), 0, \bar{N}(x, X), x, \cdot, X)\bar{n}(x, X)dt + O(dt^2)\right)^{(1/\epsilon-1)},$$

which gives the probability that a mutation occurs in one of the demographic time intervals and not in all others. However, since only the terms which are linear in dt do not vanish in the limit (3.30), we can simply write the probability of a mutation in the trait x in the evolutionary time interval $[t, t + dt]$ as

$$P_m(x, X, dt) = \mu(x)b(\bar{n}(x, X), 0, \bar{N}(x, X), x, \cdot, X)\bar{n}(x, X)dt + O(dt^2). \quad (3.31)$$

The Probability $P'dx'$

The mutational step $(x' - x)$ is described as a random variable through a suitable probability distribution P'. Since the distribution may be affected by the trait x, we need to consider a family of distributions parameterized by x. Moreover, positive and negative mutational steps are considered equally likely, so that the distribution of $(x' - x)$ is symmetric around zero in each element of the family and $E[x'] = x$. We denote the distribution family by $P'(x, x' - x)$, where the first argument parameterizes the family and $P'(x, x' - x) = P'(x, x - x')$. Since we later want to consider the limit case of extremely small mutations, the standard deviation of $P'(x, x' - x)$ must vanish with our timescaling parameter ϵ and is denoted in the following by $\epsilon\sigma(x)$. The family $P'(x, x' - x)$ can be easily constructed by suitably scaling a given probability distribution $D(x, x' - x)$ with standard deviation $\sigma(x)$ as follows

$$P'(x, x' - x) = \frac{1}{\epsilon}D\left(x, \frac{x' - x}{\epsilon}\right), \quad (3.32)$$

so that the variance of the family is given by

$$E[(x'-x)^2] = \int_{-\infty}^{+\infty} (x'-x)^2 P'(x,x'-x)dx'$$

$$= \int_{-\infty}^{+\infty} (x'-x)^2 \frac{1}{\epsilon} D\left(x, \frac{x'-x}{\epsilon}\right) dx'$$

$$= \epsilon^2 \int_{-\infty}^{+\infty} \left(\frac{x'-x}{\epsilon}\right)^2 D\left(x, \frac{x'-x}{\epsilon}\right) d\left(\frac{x'-x}{\epsilon}\right) = \epsilon^2 \sigma^2(x).$$

The Probability P_s

Finally, we need to address the probability P_s according to which the mutant population substitutes the resident population at the end of the large amount dt/ϵ of demographic time. Since we will take the limit $\epsilon \to 0$ and we have assumed that the selection derivative (3.26) does not vanish at the current evolutionary state (x,X), by virtue of the invasion implies substitution theorem one might imagine that the probability of substitution is given by

$$\begin{cases} 1 & \text{if } \left. \frac{\partial}{\partial x'}\lambda(x,x',X)\right|_{x'=x} (x'-x) > 0 \\ 0 & \text{otherwise.} \end{cases}$$

However, this would neglect the possible accidental extinction of the mutant population, so that the probability of substitution for a mutation at advantage is not 1, but rather the probability of escaping accidental extinction.

As discussed in Section 3.1, only a stochastic description of birth and death processes can account for accidental extinction risks, and only if the mutant population reaches a large number of individuals can we then map such a number into a low abundance in the resident-mutant model (3.2) and use the invasion implies substitution theorem. Thus, the probability that the mutant population escapes accidental extinction is given by the probability that the population reaches a large number of individuals starting from a single individual.

To compute such a probability, we need a stochastic description of the demography of the mutant population that is representative of what is happening close to the equilibrium (3.15) of the resident-mutant model (3.2). Along the growth of the mutant population, the probabilities that a given mutant individual gives rise to a birth or dies in a small amount dt of demographic time are well approximated by the per-capita birth and death rates of the mutant population evaluated at equilibrium (3.15) and scaled by dt, i.e.,

$$b(0, \bar{n}(x,X), \bar{N}(x,X), x', x, X)dt,$$

$$d(0, \bar{n}(x,X), \bar{N}(x,X), x', x, X)dt.$$

Such probabilities are independent of the actual number of mutant individuals and will be abbreviated in the following by $\lambda_b(x,x',X)dt$ and $\lambda_d(x,x',X)dt$, where the fitness function (3.19) has been split into

$$\lambda(x,x',X) = \lambda_b(x,x',X) - \lambda_d(x,x',X) \qquad (3.33)$$

in accordance with (3.1).

   In Appendix C, it is shown that the probability that the mutant population grows
from 1 to I individuals is given by

$$\frac{1 - \dfrac{\lambda_d(x, x', X)}{\lambda_b(x, x', X)}}{1 - \left(\dfrac{\lambda_d(x, x', X)}{\lambda_b(x, x', X)}\right)^I},$$ (3.34)

where $\lambda_d(x, x', X) < \lambda_b(x, x', X)$, i.e., $\lambda(x, x', X) > 0$, for a mutation at advantage (remember that x' is close to x). Thus, recalling (3.33) and taking the limit
$I \to \infty$, the probability of avoiding accidental extinction is

$$P_s(x, x', X) = \begin{cases} \dfrac{\lambda(x, x', X)}{\lambda_b(x, x', X)} & \text{if } \left.\dfrac{\partial}{\partial x'}\lambda(x, x', X)\right|_{x'=x} (x' - x) > 0 \\ 0 & \text{otherwise.} \end{cases}$$ (3.35)

Derivation of (3.24)

Multiplying the three probabilities P_m, $P'dx'$, and P_s given by (3.31), (3.32), and
(3.35) one obtains

$$\begin{aligned} P(x, x', X, dt)dx' &= P_m(x, X, dt)P'(x, x' - x)P_s(x, x', X)dx' \\ &= \mu(x)\lambda_b(x, x, X)\bar{n}(x, X)P'(x, x' - x)P_s(x, x', X)dx'dt, \end{aligned}$$ (3.36)

where we took into account that properties (3.3) implies

$$\begin{aligned} b(\bar{n}(x, X), 0, \bar{N}(x, X), x, \cdot, X) &= b(\bar{n}(x, X), 0, \bar{N}(x, X), x, x, X) \\ &= b(0, \bar{n}(x, X), \bar{N}(x, X), x, x, X) \\ &= \lambda_b(x, x, X). \end{aligned}$$

Substituting (3.36) into (3.30) we lose all $O(dt^2)$ terms and obtain

$$\dot{x} = \mu(x)\lambda_b(x, x, X)\bar{n}(x, X) \int_{-\infty}^{+\infty} (x' - x)P'(x, x' - x)P_s(x, x', X)dx',$$

where, in view of (3.35), the integral goes from x $[-\infty]$ to $+\infty$ $[x]$ if the selection
derivative (3.26) is positive [negative], i.e.,

$$\dot{x} = \mu(x)\lambda_b(x, x, X)\bar{n}(x, X) \int_{x\,[-\infty]}^{+\infty\,[x]} (x' - x)P'(x, x' - x)\frac{\lambda(x, x', X)}{\lambda_b(x, x', X)}dx'.$$

Now, exploiting the first-order expansion

$$\frac{\lambda(x, x', X)}{\lambda_b(x, x', X)} = \frac{1}{\lambda_b(x, x, X)} \left.\frac{\partial}{\partial x'}\lambda(x, x', X)\right|_{x'=x} (x' - x) + O(|x' - x|^2)$$ (3.37)

(recall that $\lambda(x, x, X) = 0$), which is justified in the limit $\epsilon \to 0$ where $(x' - x)$
becomes infinitesimal, we get

$$\dot{x} = \mu(x)\bar{n}(x, X) \left.\frac{\partial}{\partial x'}\lambda(x, x', X)\right|_{x'=x} \int_{x\,[-\infty]}^{+\infty\,[x]} (x' - x)^2 P'(x, x' - x)dx'.$$

But, due to the symmetry of P', the integral is nothing but half of the variance of the probability distribution $P'(x, x' - x)$, i.e.,

$$\dot{x} = \epsilon^2 \frac{1}{2}\mu(x)\sigma^2(x)\bar{n}(x, X) \left. \frac{\partial}{\partial x'}\lambda(x, x', X)\right|_{x'=x}. \qquad (3.38)$$

Notice that equation (3.38) does not depend on the initial per-capita birth rate $\lambda_b(x, x', X)$ of the mutant population, but only on its initial per-capita growth rate $\lambda(x, x', X)$, so that only the difference (3.1) between per-capita birth and death rates of the resident population matters (as anticipated in Section 3.2). Moreover, equation (3.38) is formally correct only in the limit $\epsilon \to 0$, which completely separates the demographic and evolutionary timescales. Indeed, the two timescales are kept separated by both assumptions of rare and small mutations, which gave the ϵ^2 factor in (3.38). Rare mutations allowed us to consider one mutation at a time and to lump the resident-mutant competition in an infinitesimal evolutionary time, while small mutations smoothed the average evolutionary path described by (3.28) into the solution of an ODE. Technically, one should now define a new evolutionary time variable given by $\epsilon^2 t$ and redefine the evolutionary timescale as the characteristic timescale of this new variable. However, we continue to use the symbol t for time and simply remove the ϵ^2 factor in (3.38), thus obtaining

$$\dot{x} = \frac{1}{2}\mu(x)\sigma^2(x)\bar{n}(x, X) \left. \frac{\partial}{\partial x'}\lambda(x, x', X)\right|_{x'=x},$$

which is the AD canonical equation (3.24) anticipated at the beginning of the section.

The Case of Vanishing Selection Derivative

The AD canonical equation (3.24) has been derived under the assumption that the selection derivative (3.26) does not vanish, i.e., $\dot{x} \neq 0$. If the selection derivative (3.27) relative to trait x_i vanishes at the current evolutionary state $(x_1, x_2, \ldots)$, then mutations of trait x_i are much more likely to accidentally go extinct (see (3.35) and (3.37)) than mutations of traits with nonzero selection derivative. Thus, the possibility that a mutation of trait x_i occurs at the evolutionary state that annihilates the selection derivative (3.27) can be neglected. In other words, the evolutionary rate of change $(\dot{x}_1, \dot{x}_2, \ldots)$ is dominated by the nonvanishing components which "immediately" bring (on the evolutionary timescale) the evolutionary state to points where $\dot{x}_i$ is not vanishing anymore. In fact, the selection derivative (3.27) generically vanishes on sets of zero measure in trait space, which are crossed by the trajectories of the canonical equation (3.25). Thus, as long as the evolutionary rate of change $(\dot{x}_1, \dot{x}_2, \ldots)$ is nonzero, evolution proceeds by pointing in the direction given by (3.25), i.e., along an *evolutionary trajectory* in the evolution set $\mathcal{X}$.

A Schematic Summary

Figure 3.4 schematically summarizes the AD approach, and highlights the relationships between the three models we have introduced: the resident-mutant model (3.2), the resident model (3.5), and the AD canonical equation (3.25). As explicitly

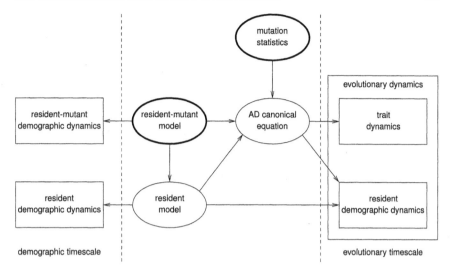

Figure 3.4 Schematic view of the relationships among the resident-mutant model (3.2), the
resident model (3.5), the mutation statistics, the AD canonical equation (3.25),
and the corresponding demographic and evolutionary dynamics.

pointed out in the figure, the resident-mutant model (3.2) and the mutation statistics are the ultimate sources of information from which the AD canonical equation
(3.25) and the dynamics of the community can be formally derived.

Given an ancestral evolutionary condition $(x_1(0), x_2(0), \ldots)$ in the evolution set
$\mathcal{X}$, the AD canonical equation (3.25) describes the evolutionary trajectory $(x_1(t),$
$x_2(t), \ldots), t > 0$, followed by the traits characterizing the community. Meanwhile,
the demographic equilibrium (3.6) of the resident model (3.5) is entrained on the
evolutionary timescale by the variation of the traits.

3.6 EVOLUTIONARY STATE PORTRAITS

The AD canonical equation (3.25) is a continuous-time dynamical system whose
trajectories are defined in the evolution set $\mathcal{X}$. Moreover, the canonical equation
is typically nonlinear, so that evolutionary dynamics can be expected to have the
typical complexity of nonlinear dynamics. The attractors of the canonical equation (3.25), called *evolutionary attractors*, can therefore be *evolutionary equilibria*,
as well as *evolutionary cycles*, *tori*, and *strange attractors*, provided the number
of traits characterizing the community is large enough (see Appendix A). Consistently, evolutionary trajectories will tend toward a point in trait space, or toward
a periodic or aperiodic evolutionary regime (so-called Red Queen dynamics, see
Section 1.6). Evolutionary equilibria, cycles, tori, and strange invariant sets can
also be unstable (saddles or repellors), and evolutionary dynamics can be characterized by several of such invariant sets, e.g., by multiple attractors, pointing out
that long-term evolutionary implications can depend upon the evolutionary paths

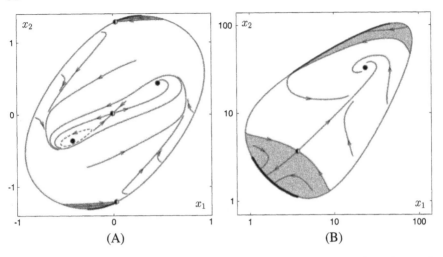

Figure 3.5 Examples of two-dimensional evolutionary state portraits taken from Chapter 5
(A) and Chapter 6 (B). Filled circles: stable evolutionary equilibria; half-filled
circles: evolutionary saddles; solid closed trajectory: stable evolutionary cycle;
dashed closed trajectory: unstable evolutionary cycle; white regions: viable sets;
gray regions: unviable sets; thick segments: extinction segments.

followed in the past, i.e., upon the ancestral evolutionary condition.

The set of all trajectories of the canonical equation (3.25), one for each ancestral condition in the evolution set $\mathcal{X}$, gives the so-called *evolutionary state portrait*
(see Appendix A). Obviously, evolutionary state portraits are useful graphical rep-
resentations only when the trait space is at most three dimensional (as in all our
applications) and are most effective in two dimensions, where they are typically
represented by drawing the boundary of the evolution set $\mathcal{X}$ and a few trajectories
in $\mathcal{X}$ (or finite segments of them) from which all other trajectories can be intuitively
inferred.

Figure 3.5 shows two examples of two-dimensional evolutionary state portraits.
The first is concerned with a resource-consumer community described in Chapter 5,
where x_1 and x_2 are resource and consumer traits, respectively, while the second
is concerned with a mutualistic community described in Chapter 6, where x_1 and
x_2 measure the altruism of two partner populations. Case A points out five equilibria (two stable and a saddle in $\mathcal{X}$ and two saddles on the boundary of $\mathcal{X}$) and two
limit cycles (one stable and one unstable), for a total of three evolutionary attractors, while in case B there are only two equilibria (one stable and a saddle), i.e.,
a unique evolutionary attractor which, however, is not reached from all ancestral
conditions in $\mathcal{X}$. In both cases the evolution set $\mathcal{X}$ is bounded and this makes sense
since extreme trait values are typically unsuitable for the coexistence of resident
populations.

Since $\mu(x)$, $\sigma(x)$, and $\bar{n}(x, X)$ are positive for $(x, X) \in \mathcal{X}$, evolutionary equilibria are points $(\bar{x}, \bar{X})$ in trait space where all selection derivatives (3.27) vanish,

i.e.,

$$\frac{\partial}{\partial x'}\lambda(\bar{x},x',\bar{X})\bigg|_{x'=\bar{x}} = 0, \tag{3.39}$$

for each trait x characterizing the community. At such points, however, the AD canonical equation is not justified anymore, since the invasion implies substitution theorem does not guarantee that invading mutant populations escaping accidental extinction replace the corresponding resident population. Thus, in the vicinity of an evolutionary equilibrium, a deeper analysis of the resident-mutant model (3.2) is required (see next section).

The stability of evolutionary equilibria can be discussed through linearization. In particular, if the community is characterized by a single trait x, the eigenvalue associated with an evolutionary equilibrium $\bar{x}$ in $\mathcal{X}$ is given by (recall (3.39))

$$\frac{\partial}{\partial x}\left(\frac{1}{2}\mu(x)\sigma^2(x)\bar{n}(x)\frac{\partial}{\partial x'}\lambda(x,x')\bigg|_{x'=x}\right)\bigg|_{x=\bar{x}}$$

$$= \frac{1}{2}\mu(\bar{x})\sigma^2(\bar{x})\bar{n}(\bar{x})\left(\frac{\partial^2}{\partial x\partial x'}\lambda(x,x')\bigg|_{x=x'=\bar{x}} + \frac{\partial^2}{\partial x'^2}\lambda(\bar{x},x')\bigg|_{x'=\bar{x}}\right),$$

so that $\bar{x}$ is stable [unstable] if

$$\frac{\partial^2}{\partial x\partial x'}\lambda(x,x')\bigg|_{x=x'=\bar{x}} + \frac{\partial^2}{\partial x'^2}\lambda(\bar{x},x')\bigg|_{x'=\bar{x}} \tag{3.40}$$

is negative [positive]. However, if the community is characterized by two or more traits, then the fitness functions λ_i, $i = 1, 2, \ldots$, do not determine the stability of evolutionary equilibria, since $\mu_i(\bar{x}_i)$, $\sigma_i(\bar{x}_i)$, and $\bar{n}_i(\bar{x}_1, \bar{x}_2, \ldots)$ depend in general upon i and thus critically affect the eigenvalues of the Jacobian matrix associated with an evolutionary equilibrium. For example, we will show in Chapter 5 that a stable evolutionary equilibrium may turn unstable by simply varying the probability of mutation or the standard deviation of the mutational step in one of the resident traits.

As pointed out in Figure 3.5A, evolutionary equilibria can also lie on the boundary of the evolution set $\mathcal{X}$. In fact, as discussed in Section 3.2, the equilibrium abundance of one of the resident populations can vanish on the boundary of $\mathcal{X}$. Thus, boundary equilibria are points on the boundary of $\mathcal{X}$ at which only the selection derivatives relative to the traits of the nonvanishing resident populations annihilate.

The evolutionary state portraits of Figure 3.5 also show that evolutionary trajectories may tend toward the boundary of the evolution set $\mathcal{X}$ (see trajectories starting in the gray regions). In case A, on the boundary of $\mathcal{X}$ the equilibrium abundance $\bar{n}_2(x_1, x_2)$ of the consumer population vanishes and, in fact, the evolutionary trajectories approach the boundary of $\mathcal{X}$ horizontally, since $\dot{x}_2$ vanishes with $\bar{n}_2(x_1, x_2)$ (see (3.25)). By contrast, in case B, the resident demographic equilibrium $(\bar{n}_1(x_1, x_2), \bar{n}_2(x_1, x_2))$ exists together with a strictly positive saddle in the demographic state space (n_1, n_2) and the two equilibria collide on the boundary of $\mathcal{X}$.

In case A, the equilibrium abundance of one of the resident populations gradually vanishes along an evolutionary trajectory approaching the boundary of $\mathcal{X}$.

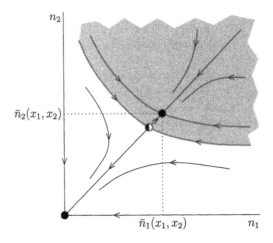

Figure 3.6 The basin of attraction (gray region) of the resident demographic equilibrium $(\bar{n}_1(x_1, x_2), \bar{n}_2(x_1, x_2))$ when (x_1, x_2) is close to the boundary of the evolution set $\mathcal{X}$ in Figure 3.5B.

Therefore, the evolutionary rate of change $\dot{x}$ of each trait x of such a population also vanishes by approaching the boundary of $\mathcal{X}$. In other words, evolution slows down in the vanishing population and the boundary is actually reached (in finite time) because of the evolution of traits characterizing some other resident populations. When the boundary is reached, the vanishing population goes extinct, so that the other resident populations play the role of murderers (so-called evolutionary murder, see Section 1.8). By contrast, if there were no murderers the extinction would be obtained asymptotically (so-called evolutionary runaway).

In case B, the equilibrium abundances of the resident populations do not vanish when the evolutionary trajectory approaches the boundary of $\mathcal{X}$ (since the stable equilibrium (3.6) of the resident model (3.5) approaches a strictly positive saddle), so that the boundary is reached in finite time even in the absence of murderers. This is indeed a case of so-called evolutionary suicide (Section 1.8). In fact, when the evolutionary state (x, X) is close to the boundary, the stable equilibrium (3.6) is close to the saddle and, hence, to the boundary of its basin of attraction given by the stable manifold of the saddle (see Figure 3.6, where the gray region is the basin of attraction of the stable equilibrium $(\bar{n}_1(x_1, x_2), \bar{n}_2(x_1, x_2))$). Small fluctuations of the resident population abundances can then result in a demographic state (n, N) outside the basin of attraction of the stable equilibrium (3.6) (which is, by assumption, the only strictly positive attractor of the resident model (3.5)). Thus, a demographic transient will eventually take place and end on one of the invariant faces of the demographic state space, where one or more resident populations are extinct (the trivial equilibrium where both populations are extinct in the case of Figure 3.6).

We have therefore seen that in both cases reaching the boundary of the evolution set marks the so-called evolutionary extinction of one or more resident populations. However, the two cases are radically different from a physical point of view. In fact, in case A the danger of extinction can be perceived on the evolutionary timescale, because the equilibrium abundance $\bar{n}(x, X)$ of a resident population smoothly declines to zero, while in case B some of the resident populations suddenly collapse without any sign of impending danger. Mutations (x', X) closer to the boundary of $\mathcal{X}$ are at advantage, even if in case A they have a lower resident abundance, $\bar{n}(x', X) < \bar{n}(x, X)$, while in case B they are closer to an unpredictable collapse. After an evolutionary extinction, if the remaining resident populations coexist at a nontrivial demographic equilibrium, the evolutionary dynamics of the community proceed in accordance with a reduced-order canonical equation, since all traits characterizing the extinct populations have been lost.

Finally, as also discussed in Section A.7, the evolution set $\mathcal{X}$ can be divided into the set of ancestral conditions giving rise to trajectories that remain forever in $\mathcal{X}$, called the *viable set* (white regions in Figure 3.5), and the set of ancestral conditions giving rise to evolutionary extinction, called the *unviable set* (gray regions in the figure). A few geometric properties of the viable and unviable sets are worth repeating here, with reference to two-dimensional trait spaces. First of all, the parts of the boundary of the unviable set on which extinction occurs are called *extinction segments* (thick segments in Figure 3.5). At each point of an extinction segment the vector $(\dot{x}_1, \dot{x}_2)$ tangent to the evolutionary trajectory points outside $\mathcal{X}$, so that extinction segments are delimited by points at which the vector $(\dot{x}_1, \dot{x}_2)$ either is tangent to the boundary of $\mathcal{X}$, so-called *tangent points* (see Figure 3.5B), or vanishes at boundary evolutionary equilibria (see Figure 3.5A, where one of the two points delimiting the extinction segment is a boundary equilibrium, actually a saddle). A second interesting geometric feature is that the boundaries separating the viable from the unviable set are either evolutionary trajectories touching a tangent point (see Figure 3.5B) or stable manifolds of saddles which are at the boundary of $\mathcal{X}$ in Figure 3.5A or inside $\mathcal{X}$ in Figure 3.5B. As we will see in Section 3.8, and, in particular, in Chapter 6, these and similar geometric features are crucial for classifying and detecting structural changes in the extinction mechanisms of the evolving community.

3.7 EVOLUTIONARY BRANCHING

In this section, we focus on the dynamics of the traits in the vicinity of evolutionary equilibria $(\bar{x}, \bar{X})$, i.e., close to points in the evolution set $\mathcal{X}$ where all selection derivatives (3.27) vanish. As we will see, in the vicinity of such points a mutant population may invade and coexist at equilibrium with all resident populations, thus giving rise to an extra resident population with traits very similar to those of the population in which the mutation has occurred. Then, the evolutionary dynamics of the community are described by a new canonical equation with one extra ODE for each trait of the new resident population. If the two similar resident populations diverge in trait values, we have a so-called evolutionary branching (see Section 1.7).

In the following, we first show that resident-mutant coexistence is possible for

$X = \bar{X}$ and x, x' close to $\bar{x}$ if

$$\left.\frac{\partial^2}{\partial x \partial x'}\lambda(x, x', \bar{X})\right|_{x=x'=\bar{x}} < 0. \tag{3.41a}$$

Then, we show that two coexisting resident and mutant populations diverge in trait values according to the higher-dimensional canonical equation if

$$\left.\frac{\partial^2}{\partial x'^2}\lambda(\bar{x}, x', \bar{X})\right|_{x'=\bar{x}} > 0. \tag{3.41b}$$

Finally, we partition stable evolutionary equilibria $(\bar{x}, \bar{X})$ into *branching points* (those satisfying conditions (3.41) for at least one trait x) and *terminal points*. Although branching points are technically evolutionary equilibria of the canonical equation, notice that they are not equilibria of the evolutionary process. Conditions (3.41) are called in the following *branching conditions*.

The Coexistence Condition (3.41a)

We now consider the resident-mutant model (3.2) for trait values $X = \bar{X}$ and x, x' close to $\bar{x}$. As done in Section 3.4, we study the sign of the fitness function λ (given $X = \bar{X}$). Recalling that $\lambda(x, x, \bar{X})$ is identically equal to zero, we have

$$\frac{\partial}{\partial x}\lambda(x, x, \bar{X}) = \left.\frac{\partial}{\partial x}\lambda(x, x', \bar{X})\right|_{x'=x} + \left.\frac{\partial}{\partial x'}\lambda(x, x', \bar{X})\right|_{x'=x} = 0,$$

so that, in view of (3.39), we obtain

$$\left.\frac{\partial}{\partial x}\lambda(x, \bar{x}, \bar{X})\right|_{x=\bar{x}} = 0.$$

Thus, the sign of λ for (x, x') in a small neighborhood of $(\bar{x}, \bar{x})$ is generically determined by the second-order terms of the Taylor expansion

$$\begin{aligned}\lambda(x, x', \bar{X}) = &\left.\frac{1}{2}\frac{\partial^2}{\partial x^2}\lambda(x, \bar{x}, \bar{X})\right|_{x=\bar{x}}(x-\bar{x})^2 \\ &+ \left.\frac{\partial^2}{\partial x \partial x'}\lambda(x, x', \bar{X})\right|_{x=x'=\bar{x}}(x-\bar{x})(x'-\bar{x}) \\ &+ \left.\frac{1}{2}\frac{\partial^2}{\partial x'^2}\lambda(\bar{x}, x', \bar{X})\right|_{x'=\bar{x}}(x'-\bar{x})^2 + O(\|(x-\bar{x}, x'-\bar{x})\|^3),\end{aligned} \tag{3.42}$$

or, in matrix notation,

$$\begin{aligned}\lambda(x, x', \bar{X}) = &\frac{1}{2}(x-\bar{x}, x'-\bar{x})\Lambda(\bar{x}, \bar{X})(x-\bar{x}, x'-\bar{x})^T \\ &+ O(\|(x-\bar{x}, x'-\bar{x})\|^3),\end{aligned} \tag{3.43}$$

where the T superscript denotes transposition and $\Lambda(\bar{x}, \bar{X})$ is the so-called Hessian matrix

$$\Lambda(\bar{x}, \bar{X}) = \begin{bmatrix} \left.\dfrac{\partial^2}{\partial x^2}\lambda(x, \bar{x}, \bar{X})\right|_{x=\bar{x}} & \left.\dfrac{\partial^2}{\partial x \partial x'}\lambda(x, x', \bar{X})\right|_{x=x'=\bar{x}} \\[2ex] \left.\dfrac{\partial^2}{\partial x \partial x'}\lambda(x, x', \bar{X})\right|_{x=x'=\bar{x}} & \left.\dfrac{\partial^2}{\partial x'^2}\lambda(\bar{x}, x', \bar{X})\right|_{x'=\bar{x}} \end{bmatrix}.$$

If the determinant of $\Lambda(\bar{x}, \bar{X})$ were positive, then the quadratic form at the right-hand side of (3.43) would be definite in sign, i.e., $\lambda(x, x', \bar{X})$ would have the same sign at all points (x, x') close to $(\bar{x}, \bar{x})$. But this is not possible, because the invasion eigenvalues $\lambda(x, x', X)$ and $\lambda(x', x, X)$ have opposite sign if x and x' are different from $\bar{x}$ and sufficiently close (see Section 3.4). Thus, the determinant of $\Lambda(\bar{x}, \bar{X})$ must be negative (the case of null determinant is nongeneric and therefore omitted).

To obtain the boundaries of the regions of the (x, x') plane in which $\lambda(x, x', \bar{X})$ is positive or negative, let us rewrite (3.42) by introducing the polar coordinates $x = \bar{x} + \rho \cos \theta$ and $x' = \bar{x} + \rho \sin \theta$, thus obtaining

$$
\lambda(x, x', \bar{X}) = \left(\frac{\partial^2}{\partial x^2} \lambda(x, \bar{x}, \bar{X}) \Big|_{x=\bar{x}} \cos^2 \theta \right.
$$

$$
+ 2 \frac{\partial^2}{\partial x \partial x'} \lambda(x, x', \bar{X}) \Big|_{x=x'=\bar{x}} \cos \theta \sin \theta
$$

$$
\left. + \frac{\partial^2}{\partial x'^2} \lambda(\bar{x}, x', \bar{X}) \Big|_{x'=\bar{x}} \sin^2 \theta \right) \frac{\rho^2}{2} + O(\rho^3). \qquad (3.44)
$$

Then, taking into account that

$$
0 = \frac{\partial^2}{\partial x^2} \lambda(x, x, \bar{X}) = \frac{\partial^2}{\partial x^2} \lambda(x, x', \bar{X}) \Big|_{x'=x}
$$

$$
+ 2 \frac{\partial^2}{\partial x \partial x'} \lambda(x, x', \bar{X}) \Big|_{x'=x}
$$

$$
+ \frac{\partial^2}{\partial x'^2} \lambda(x, x', \bar{X}) \Big|_{x'=x}
$$

($\lambda(x, x, \bar{X})$ is identically equal to zero), i.e.,

$$
2 \frac{\partial^2}{\partial x \partial x'} \lambda(x, x', \bar{X}) \Big|_{x'=x} = - \frac{\partial^2}{\partial x^2} \lambda(x, x', \bar{X}) \Big|_{x'=x} - \frac{\partial^2}{\partial x'^2} \lambda(x, x', \bar{X}) \Big|_{x'=x},
$$
$$
(3.45)
$$

we can write (3.44) in the new form

$$
\lambda(x, x', \bar{X}) = \left(\frac{\partial^2}{\partial x^2} \lambda(x, \bar{x}, \bar{X}) \Big|_{x=\bar{x}} \cos \theta - \frac{\partial^2}{\partial x'^2} \lambda(\bar{x}, x', \bar{X}) \Big|_{x'=\bar{x}} \sin \theta \right)
$$

$$
\times (\cos \theta - \sin \theta) \frac{\rho^2}{2} + O(\rho^3). \qquad (3.46)
$$

Thus, varying θ from 0 to 2π with ρ sufficiently small, $\lambda(x, x', \bar{X})$ changes sign roughly when $\theta = \pi/4$ (and $3\pi/4$), and when $\theta = \theta_c$ (and $\theta_c + \pi$), where

$$
\tan \theta_c = \frac{\frac{\partial^2}{\partial x^2} \lambda(x, \bar{x}, \bar{X}) \Big|_{x=\bar{x}}}{\frac{\partial^2}{\partial x'^2} \lambda(\bar{x}, x', \bar{X}) \Big|_{x'=\bar{x}}}.
$$

Notice that, generically, $\theta_c \neq \pi/4$ (and $3\pi/4$), since otherwise the determinant of $\Lambda(\bar{x}, \bar{X})$ would vanish. There are therefore two curves in the (x, x') plane passing through point $(\bar{x}, \bar{x})$ with slope $\pi/4$ and θ_c, along which $\lambda(x, x', \bar{X}) = 0$.

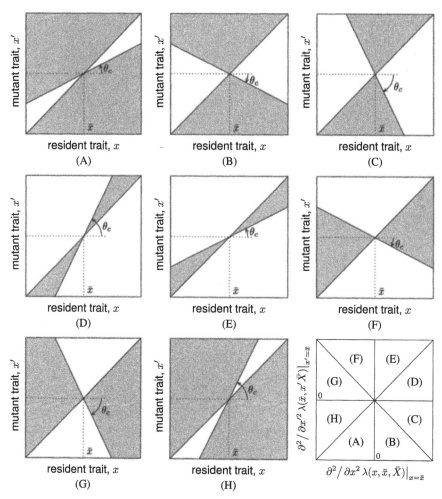

Figure 3.7 Classification of evolutionary equilibria $(\bar{x}, \bar{X}) \in \mathcal{X}$. Panels A–H show the
sign of the fitness function $\lambda(x, x', \bar{X})$ (positive: white regions; negative: gray
regions) in a small neighborhood of point $(\bar{x}, \bar{x})$ of the (x, x') plane and correspond to cases A–H of the bottom-right panel.

These two curves, one of which is obviously the diagonal $x' = x$, partition the
neighborhood of $(\bar{x}, \bar{x})$ into four regions: in two of them $\lambda(x, x', \bar{X})$ is positive
and in the other two it is negative.

As depicted in Figure 3.7, we can distinguish between eight qualitatively different cases, identified by different inequalities among the second derivatives of
the fitness function $\lambda(x, x', \bar{X})$ (see the bottom-right panel). Panels A–H picture
the sign of $\lambda(x, x', \bar{X})$ in a small neighborhood of $(\bar{x}, \bar{x})$, by reporting in white
[gray] the regions of the (x, x') plane in which $\lambda(x, x', \bar{X})$ is positive [negative],
while $\lambda(x, x', \bar{X}) = 0$ on the boundaries of such regions (the boundary passing
through $(\bar{x}, \bar{x})$ with slope θ_c is approximated by a straight line). Moreover, in the

particular case of single-trait communities (i.e., when X is empty), by combining (3.40), (3.45), and the bottom-right panel of Figure 3.7, it is easy to check that the evolutionary equilibrium $\bar{x}$ is stable in cases A–D and unstable in cases E–H.

In AD jargon, the diagrams A–H are called *pairwise invasibility plots* (PIP) (Metz et al., 1996). Clearly, a PIP gives information on the possible invasion of the mutant population: if the point (x, x') lies in a white [gray] region, then the invasion eigenvalue $\lambda(x, x', \bar{X})$ is positive [negative] (i.e., the equilibrium (3.15) of the resident-mutant model (3.2) is unstable [stable], as discussed in Section 3.4), and the mutant population invades [goes extinct]. The term "pairwise" refers to the fact that if the symmetric point (x', x) lies in a white [gray] region, then the invasion eigenvalue $\lambda(x', x, \bar{X})$ is positive [negative] (i.e., the equilibrium (3.17) is unstable [stable]).

The superposition of the PIP with its mirror image along the diagonal contains three types of regions, as shown in Figure 3.8: white-white regions (white regions ① in panels C–F), in which both equilibria (3.15) and (3.17) are unstable; white-gray regions (light gray regions ② and ③ in all panels), in which one of the two equilibria is stable while the other is unstable; and gray-gray regions (dark gray regions ④ in panels A, B, G, and H), in which both equilibria are stable. Moreover, the invasion implies substitution theorem guarantees that in region ② the mutant population replaces the resident population and, by the symmetry of resident and mutant roles, that in region ③ the resident population replaces the mutant population (see the demographic state portraits ② and ③ in Figure 3.9). Finally, as shown in Appendix D, in the two other regions a unique strictly positive equilibrium of the resident-mutant model (3.2) exists in the tube of Figure 3.3 and is stable in region ① and unstable (saddle) in region ④ (see state portraits ① and ④ in Figure 3.9).

In conclusion, Figure 3.9 shows all state portraits of the resident-mutant model (3.2) that can be obtained for (x, x') in a small neighborhood of $(\bar{x}, \bar{x})$. Due to the symmetry of resident and mutant roles the state portraits corresponding to two symmetric points (x, x') and (x', x) are obtained one from the other by exchanging x and n with x' and n'. In other words, if point (x, x') lies in region ①, ②, ③, ④, then point (x', x) lies in region ①, ③, ②, ④, respectively.

PIPs and their mirror images along the diagonal are graphical tools commonly used in the AD context to picture the sign of the fitness function $\lambda(x, x', \bar{X})$ even far from evolutionary equilibria. The curves crossing the diagonal at evolutionary equilibria can be approximated by straight lines only locally, as in Figures 3.7 and 3.8, but need to be numerically produced in the large (see Section 3.8). An example, concerning a single-trait community described in Chapter 7, is shown in Figure 3.10A and points out three evolutionary equilibria. Notice, however, that far from small neighborhoods around evolutionary equilibria there can also be resident-mutant behaviors different from those of Figure 3.9, as pointed out by region ⑤ in Figure 3.10B (shown only for $x' > x$ due to the symmetry of resident and mutant roles).

As anticipated at the beginning of the section, the analysis of the resident-mutant model (3.2) in the parameter plane (x, x'), summarized in Figure 3.8, shows that mutant and resident populations can coexist in the vicinity of an evolutionary equilibrium (see, for example, the two evolutionary equilibria on the boundary of region

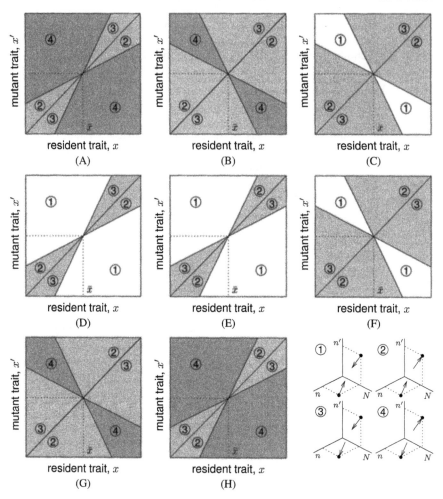

Figure 3.8 Classification of evolutionary equilibria $(\bar{x}, \bar{X}) \in \mathcal{X}$. Panels A–H show the sign of the invasion eigenvalues $\lambda(x, x', \bar{X})$ (positive in regions ① and ②; negative in regions ③ and ④) and $\lambda(x', x, \bar{X})$ (positive in regions ① and ③; negative in regions ② and ④) in a small neighborhood of point $(\bar{x}, \bar{x})$ of the (x, x') plane and correspond to cases A–H of Figure 3.7. As sketched in the bottom-right panel, the equilibria (3.15) and (3.17) of the resident-mutant model (3.2) are both unstable [stable] in region ① (white) [④ (dark gray)], while (3.15) is unstable [stable] and (3.17) is stable [unstable] in region ② [③] (light gray).

① in Figure 3.10B). By inspection of Figure 3.8, resident-mutant coexistence occurs in cases C–F, where region ① is present. Looking at the bottom-right panel of Figure 3.7, cases C–F are identified by the condition

$$
\left.\frac{\partial^2}{\partial x^2}\lambda(x, \bar{x}, \bar{X})\right|_{x=\bar{x}} + \left.\frac{\partial^2}{\partial x'^2}\lambda(\bar{x}, x', \bar{X})\right|_{x'=\bar{x}} > 0, \qquad (3.47)
$$

which, in view of (3.45), is nothing but (3.41a).

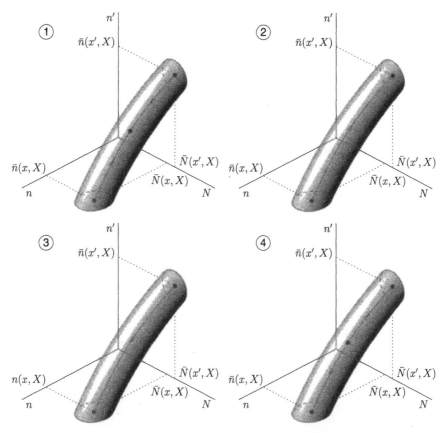

Figure 3.9 State portraits of the resident-mutant model (3.2), restricted to the invariant tube connecting equilibria (3.15) and (3.17) and corresponding to regions ①–④ of Figure 3.8.

The Divergence Condition (3.41b)

Once resident and mutant populations coexist at a strictly positive equilibrium in the demographic state space (n, n', N), the mutant population becomes, by definition, a resident population, so that there are two similar resident populations with abundances n and n', characterized by the same trait values except one, which is x in one population and x' in the other (both close to $\bar{x}$). In the following, we rename x and n with x_1 and n_1, x' and n' with x_2 and n_2, and pack all other traits characterizing populations 1 and 2 into vectors $X^{(1)}$ and $X^{(2)}$. Similarly, we pack all traits characterizing the other resident populations and their abundances into vectors $X^{(\mathrm{r})}$ and $N^{(\mathrm{r})}$.

We can hence denote by

$$n_1 = \bar{n}_1(x_1, x_2, X^{(1)}, X^{(2)}, X^{(\mathrm{r})}), \tag{3.48a}$$

$$n_2 = \bar{n}_2(x_1, x_2, X^{(1)}, X^{(2)}, X^{(\mathrm{r})}), \tag{3.48b}$$

$$N^{(\mathrm{r})} = \bar{N}^{(\mathrm{r})}(x_1, x_2, X^{(1)}, X^{(2)}, X^{(\mathrm{r})}) \tag{3.48c}$$

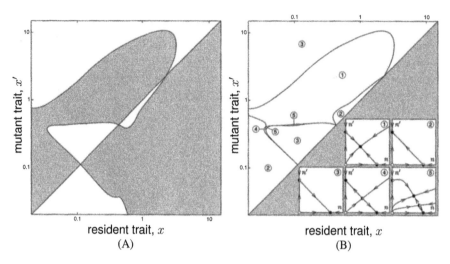

mutant trait, x'

resident trait, x

(A)

mutant trait, x'

resident trait, x

(B)

Figure 3.10 (A) PIP concerning a single-trait community described in Chapter 7. (B) Analysis of the (two-dimensional) resident-mutant model in the parameter plane (x, x') (shown only for $x' > x$) and corresponding demographic state portraits (filled circles: stable equilibria; half-filled circle: saddles; empty circles: repellors).

the new resident demographic equilibrium and derive the new canonical equation describing the evolution of the traits x_1 and x_2, i.e.,

$$\dot{x}_1 = \frac{1}{2}\mu(x_1)\sigma^2(x_1)\bar{n}_1(x_1, x_2, X^{(1)}, X^{(2)}, X^{(r)})$$

$$\times \frac{\partial}{\partial x_1'}\lambda_1(x_1, x_2, X^{(1)}, X^{(2)}, X^{(r)}, x_1')\Big|_{x_1' = x_1}, \tag{3.49a}$$

$$\dot{x}_2 = \frac{1}{2}\mu(x_2)\sigma^2(x_2)\bar{n}_2(x_1, x_2, X^{(1)}, X^{(2)}, X^{(r)})$$

$$\times \frac{\partial}{\partial x_2'}\lambda_2(x_1, x_2, X^{(1)}, X^{(2)}, X^{(r)}, x_2')\Big|_{x_2' = x_2} \tag{3.49b}$$

(and similarly for $X^{(1)}$, $X^{(2)}$, $X^{(r)}$), where the functions μ and σ are taken from (3.24). Compared with the canonical equation (3.25), the new canonical equation (3.49) has one extra ODE for each trait characterizing the new resident population, one of which is x_2, while the others are components of $X^{(2)}$. As for the new evolution set $\mathcal{X}^{(1,2,r)}$, we know that close to $(\bar{x}, \bar{x})$ its intersection with the plane (x_1, x_2), defined by

$$X^{(r)} = \bar{X}^{(r)}, \quad X^{(1)} = X^{(2)} = \bar{X}^{(1)}, \tag{3.50}$$

coincides with the white regions of Figure 3.8. The structure of $\mathcal{X}^{(1,2,r)}$ far from point $(\bar{x}, \bar{x}, \bar{X}^{(1)}, \bar{X}^{(1)}, \bar{X}^{(r)})$ can be obtained by analyzing the equilibrium (3.48) of the new resident model in the parameter space $(x_1, x_2, X^{(1)}, X^{(2)}, X^{(r)})$; the analysis, however, can be restricted to $x_2 > x_1$, since any one of the two similar resident populations can be considered as population 1 or 2.

Notice that the trait space $(x_1, x_2, X^{(1)}, X^{(2)}, X^{(r)})$ of the new resident model has higher dimension than the trait space (x, x', X) of the resident-mutant model (3.2) whenever the resident population 1 is characterized by more than one trait. However, in all our applications, we consider communities where each resident population is characterized by a single trait (see Table 3.1), so that the new resident model coincides with the resident-mutant model (3.2). For example, the analysis of the two-dimensional resident-mutant model shown in Figure 3.10B reveals that the evolution set $\mathcal{X}^{(1,2)}$ of the two-dimensional canonical equation describing the evolution of the traits x_1 and x_2 (x and x' in Figure 3.10B) is composed of regions ① and ⑤, where the equilibrium $(\bar{n}_1(x_1, x_2), \bar{n}_2(x_1, x_2))$ is stable and strictly positive.

The initial conditions $x_1(0)$, $x_2(0)$, $X^{(1)}(0)$, $X^{(2)}(0)$, $X^{(r)}(0)$ of the canonical equation (3.49) are close to point

$$\left(\bar{x}, \bar{x}, \bar{X}^{(1)}, \bar{X}^{(1)}, \bar{X}^{(r)} \right), \tag{3.51}$$

and characterized by $X^{(1)}(0) = X^{(2)}(0)$, since populations 1 and 2 initially differ only in trait x, which takes values $x_1(0)$ and $x_2(0)$, respectively. A local analysis carried out in Appendix D shows that point (3.51) is an equilibrium of the canonical equation (3.49) whose stability cannot be studied through linearization. However, we will show that evolutionary trajectories with initial conditions close to point (3.51) give rise to evolutionary rates of change $\dot{x}_1$ and $\dot{x}_2$ of opposite sign and that x_1 and x_2 diverge [converge] if

$$\left. \frac{\partial^2}{\partial x'^2} \lambda(\bar{x}, x', \bar{X}) \right|_{x'=\bar{x}} \tag{3.52}$$

is positive [negative]. Thus, if we look at the intersection of $\mathcal{X}^{(1,2,r)}$ with the plane (x_1, x_2) defined by (3.50) (see white regions of cases C–F of Figure 3.8), the evolutionary trajectories described by the canonical equation (3.49) are as in Figure 3.11, where horizontal and vertical arrows take into account that one of the two resident equilibrium abundances $\bar{n}_1$ and $\bar{n}_2$, and hence $\dot{x}_1$ or $\dot{x}_2$, vanishes on the boundaries of white regions.

In particular, if (3.52) is positive (cases D–F), then x_1 and x_2 initially diverge and a nonlocal analysis of the canonical equation (3.49) is required to investigate the evolutionary fate of the community. By contrast, if (3.52) is negative (case C), then x_1 and x_2 converge and the evolutionary trajectory tends to the boundary of the evolution set $\mathcal{X}^{(1,2,r)}$, where one of the two similar resident populations is driven to extinction by the evolution of the other (evolutionary murder). Notice that there is a particular evolutionary trajectory reaching the evolutionary equilibrium (3.51), along which the two similar resident populations do not vanish and merge in trait value (the dual phenomenon of evolutionary branching that we call *evolutionary merging*). Thus, case C explains why merging of two different populations into a single one, though theoretically possible, must be considered unlikely (see Section 1.7).

It is also interesting to note that positive [negative] values of (3.52) mean that the resident trait $\bar{x}$ is a local minimum [maximum] of the fitness landscape $\lambda(\bar{x}, x', \bar{X})$

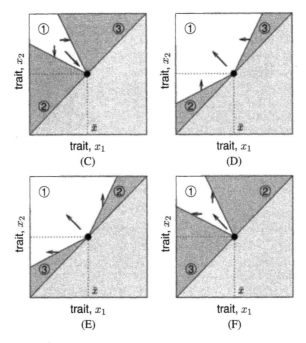

Figure 3.11 Evolutionary branching (D–F) and evolutionary merging (C).

experienced by the mutation x' at the evolutionary equilibrium $(\bar{x}, \bar{X})$. At fitness minima [maxima], points just above and below point $(\bar{x}, \bar{x})$ in the panels of Figure 3.7 lie in white [gray] regions, i.e., cases D–G [A–C, H]. Thus, small mutations x' either larger or smaller than $\bar{x}$ invade at fitness minima, while no small mutations invade at fitness maxima. Moreover, since the stability of evolutionary equilibria is independent of the sign of (3.52), we conclude that evolution on adaptive fitness landscapes (see Section 2.7) may drive communities toward fitness maxima, as traditionally expected, as well as toward fitness minima (Abrams et al., 1993b).

Branching and Terminal Points

We are now in the position of classifying evolutionary equilibria $(\bar{x}, \bar{X})$ of the canonical equation (3.25) according to their stability, to the possibility of resident-mutant coexistence in their vicinity, and to the initial behavior of the higher-dimensional canonical equation if resident-mutant coexistence is possible.

In evolutionary game theory (discussed in Section 2.5) an *evolutionarily stable strategy* (ESS) is defined as a point in trait space (x, X) at which no mutant population can invade. Consequently, ESSs are evolutionary equilibria $(\bar{x}, \bar{X})$, since we have seen in Section 3.4 that when the selection derivative (3.26) does not vanish mutant traits either larger or smaller than the resident trait invade. As first noticed by Eshel and Motro (1981) (see also Eshel, 1983; Nowak, 1990), ESSs, despite of their names, can also be unstable. This means that even if by an extraordinary fluke the resident populations settle down at an unstable ESS, the slightest perturbation

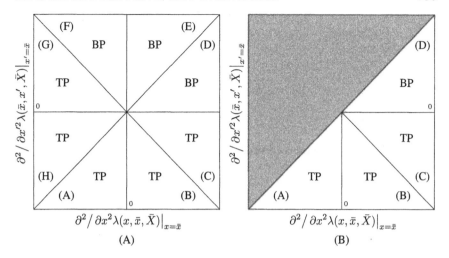

Figure 3.12 Classification of evolutionary equilibria with respect to trait x. (A) Multitrait communities: TP and BP labels apply only if the evolutionary equilibrium $(\bar{x}, \bar{X})$ is stable. (B) Single-trait communities: unshaded (A–D) [shaded (E–H)] sectors correspond to stable [unstable] evolutionary equilibria, so that evolutionary branching occurs only in case D.

of any parameter affecting $(\bar{x}, \bar{X})$ (such as the demographic and environmental parameters of the resident-mutant model) would leave the evolutionary state out of equilibrium and beget an evolutionary transient toward another evolutionary attractor (this is why unstable ESSs are called "Gardens of Eden").

To avoid any confusion and contradiction, we do not follow here the classical definition of evolutionary game theory and limit our analysis to stable evolutionary equilibria. More precisely, as anticipated at the beginning of the section, we call branching points (BP) all stable evolutionary equilibria at which at least one of the traits can branch, i.e., at which both branching conditions (3.41a) and (3.41b) hold with respect to at least one trait x. By contrast, all other stable evolutionary equilibria (i.e., nonbranching points) are called terminal points (TP), because evolution has a halt at these points.

As shown in Figure 3.12A, the branching conditions (3.41) are satisfied in the three sectors D–F (already introduced in Figure 3.7). Thus, an evolutionary equilibrium is a TP if it is stable and the second derivatives of its invasion eigenvalues $\lambda(x, x', \bar{X})$ are in one of the five remaining sectors A–C, G, and H for each trait x characterizing the community.

Notice that at a TP $(\bar{x}, \bar{X})$ corresponding to sectors A–C and H, small mutations x' cannot invade (fitness maxima, see Figure 3.7). This means that these TPs somehow correspond to the ESSs of evolutionary game theory. But this is not so in case G (fitness minimum), where a small mutation x' leads to invasion against the resident $\bar{x}$. Since in this case resident and mutant populations cannot coexist for (x, x') in a small neighborhood of $(\bar{x}, \bar{x})$ (see Figure 3.8), each invasion leads to the substitution of the resident population, i.e., to an evolutionary state (x, X)

different from the equilibrium $(\bar{x}, \bar{X})$ but very close to it. However, the stability of the equilibrium guarantees that the evolutionary state (x, X) tends again toward $(\bar{x}, \bar{X})$. Another interesting case is that of TPs corresponding to sector C, where resident and mutant populations can coexist in the vicinity of the evolutionary equilibrium. However, as shown in Figure 3.11C, the two similar resident populations evolve one toward the other until one of the two goes extinct. Thus, TPs can be characterized by temporary invasion, a mechanism that is absolutely not present in the notion of ESS.

If the branching conditions (3.41) hold at a stable evolutionary equilibrium with respect to several traits characterizing different resident populations, it is a matter of chance which population will branch first. However, after a branching has occurred, the evolution of the community is governed by the corresponding higher-dimensional canonical equation (3.49), which will possibly lead to another BP. Notice that the timing of branching events is not at all predicted by the canonical equation, since it is intrinsically related to the stochastic process of mutation. The canonical equation guarantees only that in the vicinity of a BP one of the resident populations allowed to branch will sooner or later do so. At each branching, the species to which the two initially similar resident populations belong increases its polymorphism by gaining one morph. BPs are therefore sources of diversity for the community: they increase species polymorphism and, in particular, turn monomorphic species into polymorphic ones.

Finally, it is worth noticing that in the particular but important case of single-trait communities (i.e., when X is empty), the nature of evolutionary equilibria can be more easily identified. In fact, as already noticed, the evolutionary equilibria $\bar{x}$ are stable in sectors A–D and unstable in sectors E–H. Moreover, if $\bar{x}$ is stable, then (3.40) is negative, so that condition (3.41b) implies (3.41a). Thus, condition (3.41b) alone guarantees that a stable evolutionary equilibrium $\bar{x}$ is a BP, as shown in Figure 3.12B.

3.8 THE ROLE OF BIFURCATION ANALYSIS

The dependence of evolutionary dynamics upon demographic and environmental parameters is a problem of great concern and is actually the core of all applications considered in this book. In this section we show that the numerical analysis of the so-called *bifurcations* of the AD canonical equation is an effective tool for facing this problem. Indeed, a bifurcation of a dynamical system is a particular parameter combination at which the state portrait of the system undergoes a structural change (see Appendix A).

The AD canonical equation (3.25) typically depends upon many demographic and environmental parameters that are not explicitly pointed out in (3.25). In particular, we can subdivide such parameters into three classes, each identified by a parameter vector: $p^{(r)}$ characterizing the resident model (3.5), $p^{(rm)}$ characterizing the resident-mutant model (3.2) but not the resident model (3.5), and $p^{(m)}$ characterizing the statistics of the mutation process. Moreover, the parameter vector $p^{(r)}$ [$p^{(rm)}$] can be further split into two vectors of first- and second-level parameters, say $p^{(\mathrm{r},1)}$ and $p^{(\mathrm{r},2)}$ [$p^{(\mathrm{rm},1)}$ and $p^{(\mathrm{rm},2)}$], which respectively identify trait-independent demographic and environmental quantities and parameters that control the trait dependence of trait-dependent quantities. The distinction between first- and second-level parameters is not needed for studying the parameter dependence of evolutionary dynamics, but is conceptually very important. In fact, while first-level parameters typically refer to measurable physical quantities, like birth and death rates, resource carrying capacities, and handling times, second-level parameters describe how such quantities are affected by the traits of interacting populations. Quantitative measures of second-level parameters therefore require a deep understanding of the underlying biology, where genetic details might play a major role. Thus, most of the time, only a few demographic and environmental parameters are taken to be trait-dependent and their trait dependence is postulated on the ground of qualitative and phenomenological arguments through simple mathematical functions, whose shape is controlled by parameters $p^{(\mathrm{r},2)}$ [$p^{(\mathrm{rm},2)}$].

Once all parameters characterizing the evolving community are explicitly pointed out, the resident-mutant model (3.2) becomes

$$\dot{n} = nf(n, n', N, x, x', X, p^{(\mathrm{r})}, p^{(\mathrm{rm})}), \tag{3.53a}$$

$$\dot{n}' = n'f(n', n, N, x', x, X, p^{(\mathrm{r})}, p^{(\mathrm{rm})}), \tag{3.53b}$$

$$\dot{N} = F(n, n', N, x, x', X, p^{(\mathrm{r})}, p^{(\mathrm{rm})}), \tag{3.53c}$$

while the resident model (3.5) takes the form

$$\dot{n} = nf(n, 0, N, x, \cdot, X, p^{(\mathrm{r})}, \cdot), \tag{3.54a}$$

$$\dot{N} = F(n, 0, N, x, \cdot, X, p^{(\mathrm{r})}, \cdot), \tag{3.54b}$$

and the AD canonical equation (3.24) can be written as

$$\dot{x} = k(x, X, p^{(\mathrm{m})})\bar{n}(x, X, p^{(\mathrm{r})}) \left. \frac{\partial}{\partial x'} \lambda(x, x', X, p^{(\mathrm{r})}, p^{(\mathrm{rm})}) \right|_{x'=x}, \tag{3.55}$$

where

$$k(x, X, p^{(\mathrm{m})}) = \frac{1}{2}\mu(x, p^{(\mathrm{m})})\sigma^2(x, p^{(\mathrm{m})})$$

is the parametric mutational rate and

$$\lambda(x, x', X, p^{(\mathrm{r})}, p^{(\mathrm{rm})})$$
$$= f(0, \bar{n}(x, X, p^{(\mathrm{r})}), \bar{N}(x, X, p^{(\mathrm{r})}), x', x, X, p^{(\mathrm{r})}, p^{(\mathrm{rm})}) \tag{3.56}$$

is the parametric fitness function (recall (3.19)).

Notice that the equilibrium abundances $\bar{n}$ and $\bar{N}$ of the resident populations and the evolution set $\mathcal{X}$ depend only on vector $p^{(\mathrm{r})}$ characterizing the resident model (3.5). Thus, for each $p^{(\mathrm{r})}$ there is a different evolution set where the evolutionary dynamics of the community are defined, and this set is invariant under perturbations of $p^{(\mathrm{rm})}$ and $p^{(\mathrm{m})}$. This means that the dependence of the evolutionary dynamics on $p^{(\mathrm{rm})}$ and $p^{(\mathrm{m})}$ is relatively easier to investigate, since perturbed evolutionary dynamics remain defined in the same evolution set, which does not need to be recomputed for each parameter setting. Moreover, since the fitness function (3.56) depends only on $p^{(\mathrm{r})}$ and $p^{(\mathrm{rm})}$, the location of evolutionary equilibria $(\bar{x}, \bar{X})$ in

$\mathcal{X}(p^{(\mathrm{r})})$ is not affected by $p^{(\mathrm{m})}$. However, as already discussed in Section 3.6, $p^{(\mathrm{m})}$ affects the stability of evolutionary equilibria, with the relevant exception of single-trait communities.

In the analysis of dynamical systems affected by parameters, the study of the impact of parametric perturbations on state portraits is usually restricted to the impact on the asymptotic behavior of trajectories, namely the attractors, saddles, and repellors of the system. If a parameter is slightly perturbed, by continuity the position and the form of attractors, saddles, and repellors in state space will vary smoothly (e.g., a cycle might become slightly bigger and faster) but all trajectories will remain topologically the same (e.g., an attracting cycle will remain attractive). As discussed in more detail in Appendix A, the above continuity argument fails at so-called bifurcation points in parameter space, which correspond to collisions of attractors, saddles (with associated stable and unstable manifolds), and repellors in state space. At bifurcation points, small variations of the parameters entail significant changes in model behavior. For example, an equilibrium can be stable (i.e., attract all nearby trajectories) for a given parameter setting but lose its stability if a parameter is varied even of an infinitesimal amount. If this is the case, for the new parameter value the state of the system will tend not toward the equilibrium but toward another attractor. In two-dimensional parameter spaces, bifurcation points identify bifurcation curves which partition the parameter plane into regions. All systems corresponding to the same region have the same qualitative behavior because they have topologically equivalent state portraits, while the comparison of two state portraits, one for each side of a bifurcation curve, points out a structural change in the dynamics of the system. Thus, to produce a complete catalog of the modes of behavior of a parameterized system, we must perform a complete bifurcation analysis, namely determine all its bifurcation curves.

As we will see in Chapters 5, 6, and 10, bifurcation curves can be many (even infinite!) in small windows of the relevant parameter ranges. Since systems are often crude and qualitative descriptions of reality, the identification of bifurcation curves in tiny regions of parameter space seems not very relevant for the understanding of the behavior of the system. However, if the bifurcation curves have not been detected in all regions, one cannot be sure to have identified all key features of the system. Most of the time, in fact, the conclusions one can extract from a complete bifurcation analysis are hardly obtainable from a partial analysis.

In this chapter, we have introduced three dynamical systems: the resident-mutant model (3.53), the resident model (3.54), and the AD canonical equation (3.55), the latter constrained within the evolution set $\mathcal{X}(p^{(\mathrm{r})})$. As seen in Section 3.2, the computation of $\mathcal{X}(p^{(\mathrm{r})})$ for a given $p^{(\mathrm{r})}$ reduces to the analysis of the sign and stability of the equilibrium

$$(n, N) = \left(\bar{n}(x, X, p^{(\mathrm{r})}), \bar{N}(x, X, p^{(\mathrm{r})}) \right) \tag{3.57}$$

of the resident model (3.54) with respect to the traits (x, X), i.e., to the bifurcation analysis of model (3.54) with respect to the traits (x, X) considered as parameters. In fact, if point (x, X) crosses the boundary of $\mathcal{X}(p^{(\mathrm{r})})$, the equilibrium (3.57) either disappears by colliding with a strictly positive saddle (so-called saddle-node bifurcation), or loses stability and gives rise to a small stable limit cycle (so-called

Hopf bifurcation, see Chapter 9), or loses positivity (and stability) by colliding with a saddle lying on one of the faces of the demographic state space (n, N) (so-called transcritical bifurcation).

Notice that a bifurcation diagram of a two-dimensional resident model has already been presented in Figure 3.10B (x and x' must be read as x_1 and x_2 in the figure), where the nature of the bifurcation curves separating different regions can be easily identified by comparing the corresponding state portraits. But examples of bifurcation curves have been encountered even before in Figures 3.7 and 3.8, where the boundaries separating the different regions of PIPs and their mirror images are nothing but transcritical bifurcation curves of the resident-mutant model (3.2) in the parameter plane (x, x') (see Appendix D for more details).

Once the evolution set $\mathcal{X}(p^{(r)})$ has been computed for all parameters $p^{(r)}$ in a given domain, each parameter setting $(p^{(r)}, p^{(rm)}, p^{(m)})$ corresponds to one canonical equation, defined in the corresponding evolution set $\mathcal{X}(p^{(r)})$ and, hence, to one specific evolutionary state portrait. In the bifurcation analysis of the AD canonical equation we can distinguish between three categories of bifurcations. The bifurcations that generically occur in dynamical systems in which the state is unconstrained are call *standard bifurcations*, to distinguish them from those that critically involve the boundary of the evolution set $\mathcal{X}$, which we call *extinction bifurcations* (see Section A.7). The transitions between branching and terminal points are called *branching bifurcations*.

Standard bifurcations correspond to collisions of evolutionary attractors, saddles (with associated stable and unstable manifolds), and repellors in trait space. For example, in Figure 3.5A, the unstable limit cycle (dashed line) may shrink, as a parameter is varied, and collide with the stable equilibrium inside the cycle (filled circle) at a (subcritical) Hopf bifurcation, so that an unstable equilibrium remains as the parameter is further varied. Analogously, in Figure 3.5B, the stable equilibrium (filled circle) and the saddle (half-filled circle) may collide at a saddle-node bifurcation and disappear for further parameter variation. In both cases, however, the topology of evolutionary trajectories close to the boundary of the evolution set $\mathcal{X}$ is not affected by the bifurcation, since only invariant sets far from the boundary of $\mathcal{X}$ are involved.

By contrast, extinction bifurcations involve, by definition, structural changes of the evolutionary state portrait at the boundary of the evolution set $\mathcal{X}$. More precisely (see Section A.7), they correspond to the appearance/disappearance of an extinction segment, through the collision of two tangent points, or to the collision of a tangent point with an evolutionary attractor, repellor, saddle (or its associated stable or unstable manifolds), or with the trajectory emanating from a different tangent point. For example, we can imagine that, varying a parameter, the two tangent points delimiting the extinction segment in the top part of Figure 3.5B (thick segment) come closer one to the other until they collide, causing the disappearance of the extinction segment (as well as that of the associated part of the unviable set), i.e., a structural change of the evolutionary state portrait involving the boundary of the evolution set $\mathcal{X}$. Alternatively, again in Figure 3.5B, the stable equilibrium can move toward the boundary of $\mathcal{X}$ in the direction of the extinction segment and finally collide with the tangent point at the boundary of the viable set. Before that collision, however, the unstable manifold of the saddle necessarily has to come in contact with the tangent point, thus changing its endpoint from the stable equilibrium to a point on the extinction segment. All together, the last two bifurcations cause the annihilation of the viable set.

Finally, branching bifurcations are transitions from branching to nonbranching (i.e., terminal) evolutionary equilibria. Thus, they involve the change of sign of one of the left-hand sides of the branching conditions (3.41), i.e.,

$$\frac{\partial^2}{\partial x \partial x'} \lambda(x, x', \bar{X}, p^{(\mathrm{r})}, p^{(\mathrm{rm})}) \bigg|_{x=x'=\bar{x}} = 0, \qquad (3.58\mathrm{a})$$

$$\frac{\partial^2}{\partial x'^2} \lambda(\bar{x}, x', \bar{X}, p^{(\mathrm{r})}, p^{(\mathrm{rm})}) \bigg|_{x'=\bar{x}} = 0. \qquad (3.58\mathrm{b})$$

Any region of a two-dimensional parameter space characterized by a stable evolutionary equilibrium can then be partitioned into subregions delimited by curves on which either (3.58a) or (3.58b) holds. Then, each subregion can be accordingly labeled TP or BP, with respect to a particular trait x or to all traits of the community, and the boundaries separating TP from BP subregions are branching bifurcation curves. Notice that the evolutionary state portraits corresponding to parameter settings on opposite sides of a branching bifurcation curve are topologically equivalent. Thus, strictly speaking, branching bifurcations are not bifurcations in the classical sense. However, the transition of an evolutionary equilibrium from the TP class to the BP class marks a structural change of the evolutionary dynamics of the community. In fact, a stable evolutionary equilibrium is an endpoint of evolutionary dynamics if it is a TP, while the community continues to evolve in a higher-dimensional trait space in the BP case.

We have shown in this section that the systematic analysis of the bifurcations of the resident-mutant model (3.53), of the resident model (3.54), and of the AD canonical equation (3.55) can answer all questions regarding the impact of strategic parameters on the evolution of the community. As already discussed in Section 2.9, one might hope to detect bifurcations by simulating the above models for various parameter settings and initial conditions, and by identifying for each simulation the characteristics of the obtained attractor. However, the same operation must be performed for repellors and saddles. While repellors can be obtained through backward simulation, saddles, which together with their stable and unstable manifolds play a fundamental role in bifurcation analysis, cannot be obtained. Moreover the "brute force" simulation approach is never effective and accurate when the bifurcation is related to an attractor losing stability. In fact, the length of the simulation needs to be dramatically increased while approaching the bifurcation in parameter space, and it is very difficult, if not impossible, to understand if the attractor has changed with respect to previous simulations. Thus, as explained in Appendix A, numerical bifurcation analysis cannot be performed through simulations, but through so-called *continuation* techniques (Kuznetsov, 2004, Chapter 10), which produce a bifurcation curve by continuing in parameter space the conditions characterizing the bifurcation.

We now close the section with a technical note for readers interested in performing numerical analyses of the AD canonical equation. So far we have implicitly

assumed that the resident demographic equilibrium (3.57) is known in closed form
and its expression has in fact been used in the evaluation of the fitness function
(3.56), in the canonical equation (3.55), and in the branching conditions (3.58).
However, there are cases (e.g., Chapter 7) in which the equilibrium (3.57) is not
known analytically. In such situations, the canonical equation (3.55) is actually a
differential-algebraic system of the form

$$\dot{x} = k(x, X, p^{(m)}) \, n \, \frac{\partial}{\partial x'} f(0, n, N, x', x, X, p^{(r)}, p^{(rm)}) \Big|_{x'=x}, \qquad (3.59a)$$

$$0 = nf(n, 0, N, x, \cdot, X, p^{(r)}, \cdot), \qquad (3.59b)$$

$$0 = F(n, 0, N, x, \cdot, X, p^{(r)}, \cdot) \qquad (3.59c)$$

(and similarly for X). System (3.59) can be numerically integrated in the evolution set $\mathcal{X}(p^{(r)})$ by means of standard algorithms for differential-algebraic systems, which keep track of the solution (3.57) of the algebraic equations (3.59b) and
(3.59c) at each step of the integration.

Alternatively, solutions of (3.59) can be approximated by integrating the *slow-
fast system*

$$\dot{x} = k(x, X, p^{(m)}) \, n \, \frac{\partial}{\partial x'} f(0, n, N, x', x, X, p^{(r)}, p^{(rm)}) \Big|_{x'=x}, \qquad (3.60a)$$

$$\epsilon \dot{n} = nf(n, 0, N, x, \cdot, X, p^{(r)}, \cdot), \qquad (3.60b)$$

$$\epsilon \dot{N} = F(n, 0, N, x, \cdot, X, p^{(r)}, \cdot), \qquad (3.60c)$$

where the small positive timescaling factor ϵ forces the demographic state (n, N) to
quickly converge (on the demographic timescale) to the equilibrium (3.57), while
the slow evolutionary state (x, X) remains practically constant. In other words,
during the integration of model (3.60), any deviation of (n, N) from the equilibrium
(3.57) is quickly damped, so that (n, N) remains as close to (3.57) as ϵ is small.
The discussion of the literature concerning slow-fast systems and the numerical
methods for their analysis is not within the scope of this book. A few details,
however, are given in Chapter 9 and in Appendix B.

Evolutionary state portraits can therefore be easily generated through (3.59) or
approximated through (3.60) for any given parameter setting $(p^{(r)}, p^{(rm)}, p^{(m)})$,
even if the resident demographic equilibrium (3.57) is not known analytically.
Models (3.59) and (3.60) can also be used to perform numerical bifurcation analysis of evolutionary equilibria. In particular, the branching condition (3.58a) involves the derivative of the equilibrium (3.57) with respect to trait x, which can be
obtained from the definition of equilibrium (3.57), i.e., from the equations

$$f(\bar{n}(x, X, p^{(r)}), 0, \bar{N}(x, X, p^{(r)}), x, \cdot, X, p^{(r)}, \cdot) = 0, \qquad (3.61a)$$

$$F(\bar{n}(x, X, p^{(r)}), 0, \bar{N}(x, X, p^{(r)}), x, \cdot, X, p^{(r)}, \cdot) = 0, \qquad (3.61b)$$

which hold for all $(x, X) \in \mathcal{X}(p^{(r)})$. In fact, by differentiating (3.61) with respect
to x and X, one gets a set of linear algebraic equations in the unknowns

$$\frac{\partial}{\partial x} \bar{n}(x, X, p^{(r)}), \quad \frac{\partial}{\partial X} \bar{n}(x, X, p^{(r)}), \quad \frac{\partial}{\partial x} \bar{N}(x, X, p^{(r)}), \quad \frac{\partial}{\partial X} \bar{N}(x, X, p^{(r)}),$$

which generically has a unique solution for $(x, X) \in \mathcal{X}(p^{(\mathrm{r})})$. However, it is worth noticing that the numerical analysis of bifurcations involving nonstationary trajectories of models (3.59) and (3.60) (e.g., limit cycles) is problematic, since effective continuation algorithms for differential-algebraic systems are still not available (see Ascher and Spiteri, 1994) and standard algorithms are not suited for slow-fast systems.

3.9 WHAT SHOULD WE EXPECT FROM THE AD CANONICAL EQUATION

Before applying all the machinery developed in this chapter to various problems, we want to briefly review the assumptions made through the derivation of the AD canonical equation (3.25), to comment on its applicability. The assumptions of rare and small mutations and the omission of any detail on genetics were extensively discussed in Chapter 2 with the conclusion that, at least for qualitative purposes, the AD canonical equation can be used to describe also situations where the above simplifying conditions are most likely not matched.

Analogously, the exclusion of spatial heterogeneity and physiological details of coevolving populations and the restriction to stationary demographic regimes and to mutations affecting only a single trait (see conditions a, b, and c″ of Section 2.8, where references to extensions of the AD canonical equation are given), are not severe limitations if one is interested in qualitative and not in quantitative aspects of evolutionary dynamics.

On the other hand, quantitative aspects are hard to grasp and would require a deep understanding of the underlying biology. For example, as one can see in all applications considered in Chapters 4–10, the dependence of the demographic parameters upon the traits is specified by rather simple mathematical functions with typically no empirical underpinning. The shape of such functions is controlled by the second-level parameters of the canonical equation (3.55) ($p^{(\mathrm{r},2)}$ and $p^{(\mathrm{rm},2)}$ in previous section), and the physical interpretation of such parameters is sometimes obscure. This is certainly one of the weakest points of the AD approach. The same weakness, however, affects all phenotypic approaches to long-term evolutionary dynamics (see Chapter 2), where the relation between adaptive traits and invasion fitnesses results from assumptions made at some stage on qualitative and phenomenological ground. The difference between AD and other approaches is that AD makes such assumptions at a lower level. Instead of postulating a direct relation between adaptive traits and invasion fitnesses (or payoffs), AD postulates a relation between adaptive traits and demographic parameters and derives invasion fitnesses from the underlying resident-mutant demographic dynamics. Although the biological relevance of modeling assumptions can be criticized in both cases, at the level of demographic parameters there is often more intuition on the qualitative effects of adaptive traits, so that the resulting evolutionary model may be more credible.

In Section 3.2, we also assumed that the resident demographic equilibrium (3.6) is the only strictly positive attractor of the resident model (3.5). However, this assumption is not necessary. In fact, if model (3.5) has several stable and strictly positive equilibria, for each one of them there is a corresponding canonical equation. Of course, the selection pressure summarized by the selection derivative (3.26) may be different if evaluated at different resident demographic equilibria. For example, at a low abundance equilibrium, an adaptive trait positively correlated to competitive abilities at the expense of fertility may decrease, since encounters are rare at low abundances and reproduction is the best investment. By contrast, at a high abundance equilibrium, contests are so frequent that it is an advantage for individuals to be better competitors even at the expense of reduced fertility. Thus, starting at given trait values, different evolutionary trajectories may develop for communities settled at different resident demographic equilibria.

The evolution sets associated with different resident demographic equilibria are also different. As a consequence, if an evolutionary trajectory reaches the boundary of its associated evolution set and if such a boundary is characterized by a saddle-node bifurcation of the corresponding resident demographic equilibrium, then the extinction of one or more resident populations is not guaranteed. In fact, if the resident model (3.5) has other strictly positive attractors, all resident populations may survive the disappearance of the equilibrium at which they have been settled during the last evolutionary transient, by being rescued (on the demographic timescale) on another attractor. Thus, reaching the boundary of the evolution set might induce the switch to a new resident demographic attractor and therefore trigger a discontinuity in the selection pressure on some of the traits characterizing the community. Of course, in the above discussion, we have implicitly assumed that multiple demographic attractors are not too close one to the other; otherwise small demographic fluctuations may let the resident populations to randomly visit both attractors, thus averaging the corresponding selection pressures.

Another assumption that we can somehow relax is that of small mutations. The evolutionary consequence of a large mutation cannot be investigated by means of the analysis carried out in Sections 3.4–3.7, where the resident and mutant traits x and x' are always assumed to be very close. However, the resident-mutant model (3.2) contains all the information needed to study the fate of any mutation. As we have seen in the example of Figure 3.10B, the bifurcation analysis of the resident-mutant model (3.2) allows one to subdivide the (x, x') plane into regions associated with particular outcomes of the resident-mutant dynamics. The catalog of possible behaviors is, however, wider than in the case of small mutations. Indeed, large mutations can trigger nonstationary resident-mutant coexistence, extinction of other resident populations, and switches between resident attractors, even if the current evolutionary state is far from the boundary of the evolution set.

Genetic mutations are not the only way to inject a new morph in an evolving community (see Section 1.3), since similar or radically different individuals, even belonging to species not present in the community, can enter the community through immigration. The immigration of a new population can be taken into account by adding the corresponding population to the resident model (3.5). If, after a large mutation or an immigration event, the community finds a rest at a demographic equilibrium, one can derive the corresponding AD canonical equation and use it for describing the next phase of evolution induced by small mutations. Thus, in the most general case, evolutionary dynamics are described by smooth trajectories of suitable AD canonical equations punctuated from time to time by large mutation or immigration events. While it is rather obvious that the injection of radically different populations can increase the diversity of the community, the AD approach interestingly shows that diversity can also emerge through small mutational steps.

Also the assumption of stationary demographic coexistence of the resident populations can, in principle, be relaxed. In Dieckmann and Law (1996), the AD canonical equation is extended to the case of nonstationary resident demographic coexistence, by averaging the expected evolutionary change (3.28) on the resident demographic attractor and by assuming that each invading mutant is able to replace the corresponding resident. Unfortunately, nonstationary attractors are never known in closed form, so that the resulting canonical equation can be mainly used for simulations, while its bifurcation analysis may be problematic. A significant exception to this predicament is that of populations with contrasting demographic timescales. In fact, as shown in Chapter 9, in such a case nonstationary demographic attractors can be explicitly derived by exploiting the slow-fast nature of the resident model (3.5).

Despite the numerous and often rather crude modeling assumptions required by the AD approach, the numerical bifurcation analysis of the AD canonical equation is a powerful tool for deriving qualitative answers to many interesting phenomenological questions. To support this statement, in the remaining chapters of this book, we answer the following questions:

- Can intraspecific resident-mutant competition promote diversity in evolving communities?

- Which demographic or environmental factors may enhance Red Queen scenarios?

- What are the evolutionary consequences of catastrophic bifurcations of evolutionary attractors?

- Can evolutionary branching and extinction events concatenate forever?

- Can demographic bistability lead to Red Queen dynamics in which recurrent switches between resident attractors trigger evolutionary reversals?

- Can evolution promote nonstationary demographic regimes?

- Does evolution in a constant abiotic environment have the power to generate never-ending chaotic Red Queen dynamics?

Chapter Four

Evolutionary Branching and the Origin of Diversity

In this chapter we show how continuous marginal innovations subject to severe competition may give rise to increasing diversity in evolving systems. The analysis is performed by pointing out that the AD canonical equations describing the evolution of a family of systems with increasing number of adaptive traits always lead to a branching point. The application that has motivated this study comes from economics, where the emergence of technological variety arising from market interaction and technological innovation has been ascertained. Existing products in the market compete with innovative ones, resulting in a slow and continuous evolution of the underlying technological characteristics of successful products. When technological evolution reaches an equilibrium, it can either be an evolutionary terminal point, where marginal innovations are not able to bring new technological advancement in the market, or a branching point, where new products coexist along with established ones and diversify from them through further innovations. Thus, technological branching can explain product variety without requiring exogenous major breakthroughs. The limitations of the AD approach to economics, as well as some promising further applications, are briefly discussed in the concluding section. Most of this chapter is taken from Dercole et al. (2008).

4.1 INTRODUCTION

The process of evolution not only is fundamental to our understanding of nature, but it underpins many of the phenomena of self-organization encountered in several fields of science: "The essential point to grasp," wrote Schumpeter (1942), "is that in dealing with capitalism we are dealing with an evolutionary process." The evolutionary driver meant by Schumpeter was innovation, not only technological, but organizational as well: "The opening up of new markets, foreign or domestic, and the organizational developments... illustrate the same process of industrial mutation — if I may use that biological term — that incessantly revolutionizes the economic structure from within, incessantly destroying the old one, incessantly creating a new one."

Technological change is a major driver of economic development (Burda and Wyplosz, 1997; Harberger, 1998). New growth theory has claimed the understanding of the implications of technological advancement for economic policy making mainly focusing on efficiency gains (see, for instance, Romer, 1990; Grossman and Helpman, 1991; Aghion and Howitt, 1992; Kortum, 1997; Peretto, 1998; Segerstrom, 1998; Young, 1998). One of the fundamental empirical trends in economic development is the trend toward growing variety. Although some, like Schumpeter (1912), early realized that variety in consumer goods is "one of the fundamental impulses that set and keep the capitalist engine in motion," relatively little attention has traditionally been devoted to the systematic exploration of the nature of diversity in economics.

Diversity is variously argued to be a major factor in the fostering of innovation and growth, an important strategy for hedging against intractable uncertainty and ignorance, the principal means to mitigate the effects of "lock-in" under increasing returns, and a potentially effective response to some fundamental problems of social choice. Grübler (1998) argues that technological diversity is both a means and a result of economic development. Saviotti (1996), a relevant contribution on the subject, establishes two explicit hypotheses linking variety to economic development: (1) the growth in variety is a necessary requirement for long-term economic development; (2) variety growth, leading to new sectors, and productivity growth in preexisting sectors, are complementary and not independent aspects of economic development. Stirling (1998), who provides an excellent literature review on diversity in the economy, concludes that the concept of diversity (and especially technological diversity) is of considerable general significance in economics.

The aim of this chapter is to propose a formal model which describes the interaction of technology with its social and physical environment and explains the emergence of technological diversity. Viewed through the lenses of AD, technological change is mainly based on a large number of small (marginal) intentional or spontaneous innovations, recombinations, and rearrangements of technological and economic characteristic traits. Firms compete in terms of the efficiency with which they produce or by changing products and processes. Efficiency gains, as well as changes in products or processes, are measured by "characteristic traits." When a new technological variant enters the market, it is subjected to severe selection by customers and other agents such as banks, courts of appeal, democratic vote, and so on. Under these circumstances, the AD modeling approach predicts the following series of facts that one can often observe in real economies, at least at a stylized level:

- *The AD canonical equation.* Technological innovations are either rejected or win the competition with established products, thus becoming the new predominant type. A small variation of the technological characteristic traits is associated with each invasion and substitution event. The result is a slow and smooth evolution of the traits described by the so-called AD canonical equation.

- *Long-term scenarios.* Evolution can slow down and approach an equilibrium, but it can also tend toward a cyclic or chaotic regime (Red Queen dynamics, see Section 1.6). Moreover, it is not said that all evolutionary paths tend toward the same attractor: in other words, the long-term implications of the innovation-competition process can strongly depend upon the innovation paths followed in the past. Finally, technological change can also transform particular products that in the past were predominant types into obsolete products that are swept off the market.

- *Emergence of diversity.* Evolutionary equilibria can be terminal points of technological change and, in particular, they can correspond to the well-known ESS, where no marginal innovation can penetrate the market (Nash, 1950, 1951; Hamilton, 1967; Maynard Smith and Price, 1973; Maynard Smith, 1974, 1982). However, evolutionary equilibria can also be branching points, where new variant can penetrate, coexist, and then diversify with respect to previously established products. Marginal innovations can therefore induce the emergence of technological variety through technological branching. Indeed, repeated branchings can give rise to rich clusters of different products coexisting in the market.

- *Exogenous factors.* The above processes of disappearance and emergence of specific technologies are largely influenced, if not dominated, by consumer behavior and other exogenous market conditions, which act as the economic filter for innovations and either pull or suppress the diffusion of new technologies (see, e.g., Brooks, 1980; Hodgson, 1997; Kelm, 1997).

In the next section we review the general AD framework, by adapting it to the modeling of technological change. This is a summary of the previous chapter with the exclusion, however, of all technical details. Then, we present the first application of AD to a specific problem of technological change. The problem we discuss is intentionally very simple, in order to obtain the AD canonical equation in closed form and easily point out the stylized properties mentioned above. Finally, we discuss the limitations and the advantages of the AD approach to economics and give a short overview of the wide scope of evolutionary phenomena that AD could potentially explain in economics.

4.2 A MARKET MODEL AND ITS AD CANONICAL EQUATION

In this section we present the general framework of AD by focusing our attention on a market with a given number of coexisting products (e.g., goods, artifacts, services, or more general economic entities), hereafter called *established products.*
We base our model on four technical assumptions:

- Each product is identified by a *characteristic trait* (simply *trait* in the following) quantifying its features by a positive real number. We assume that products with a higher trait value are technologically more advanced. However, this does not imply that more advanced products are necessarily preferred by consumers, since elasticities of the products as well as budgetary constraints are also important. Simple examples of characteristic traits are the waterproof characteristic for watches, Internet capabilities for mobile phones, and the features of the graphical user interface for software.

- In the absence of innovations, product densities reach a market equilibrium. The characteristic timescale of the dynamics of product densities is called the *market timescale.*

- Innovation events are rare on the market timescale, i.e., they occur on a longer timescale, which we call the *evolutionary timescale*. In other words, we assume that market clearing occurs instantaneously on the evolutionary timescale. The separation between the market and evolutionary timescales allows one to assume that the established products are at market equilibrium, when an innovative product appears, and that the market is challenged by one innovation at a time.

- Innovations are small, i.e., the innovative trait differs only slightly from the trait of one of the established products. In other words, we consider the case of "marginal" innovations, which generate new but similar versions of the existing products.

The starting point of AD is the description of the dynamics of the product densities in the market (e.g., the number of items owned by 1000 persons) through an ODE system. Denote by $n_1, ..., n_P$ and $x_1, ..., x_P$ the densities and traits of P established products. On the market timescale (fast *market dynamics*), the traits are constant while the densities vary in accordance with P ODEs of the form

$$\dot{n}_j = n_j f_j^{(P)}(n_1, ..., n_P, x_1, ..., x_P), \quad j = 1, ..., P, \qquad (4.1)$$

where $f_j^{(P)}$ is the relative diffusion rate of the jth product. For example, if $P = 1$, there is a single product in the market and its diffusion can be described through the classical logistic growth equation (see, e.g., Fisher and Pry, 1971):

$$\dot{n}_1 = r(x_1)n_1 \left(1 - \frac{n_1}{K(x_1)}\right),$$

where $r(x_1)$ is the maximum diffusion rate and $K(x_1)$ is the market equilibrium density.

In the following, model (4.1) is assumed to have a stable and strictly positive equilibrium $\bar{n}_j^{(P)}(x_1, ..., x_P)$, $j = 1, ..., P$, called *market equilibrium*, for each $(x_1, ..., x_P)$ belonging to an open set $\mathcal{X}^{(P)}$ of the trait space, called *market evolution set*. Outside $\mathcal{X}^{(P)}$, the market has no strictly positive attractors, so that some of the products are forced to exit the market.

The AD Canonical Equation

The dynamics of the traits, hereafter called *evolutionary dynamics*, should reflect the characteristics of the innovation and competition processes acting on the market, which, however, are not included in model (4.1). To describe the competition between established and innovative products, we split the ith product into two sub-products (established and innovative, respectively) with densities n_i and n_i' and traits x_i and x_i' (close to x_i), so that the market model becomes

$$\dot{n}_j = n_j f_j^{(P+1)}(n_1, ..., n_P, n_i', x_1, ..., x_P, x_i'), \quad j = 1, ..., P, \qquad (4.2a)$$

$$\dot{n}_i' = n_i' f_i^{(P+1)}(n_1, ..., n_{i-1}, n_i', n_{i+1}, ..., n_P, n_i, x_1, ..., x_{i-1}, x_i', x_{i+1}, ..., x_P, x_i),$$

$$\qquad (4.2b)$$

where the relative diffusion rate of the innovative product is obtained from that of the ith established product by exchanging density n_i and trait x_i with n_i' and x_i', respectively.

Obviously, model (4.2) contains more information than model (4.1). Indeed, model (4.1) can be immediately derived from (4.2) by disregarding the equation of the innovative product and setting $n_i' = 0$, thus obtaining

$$f_j^{(P)}(n_1,...,n_P,x_1,...,x_P) = f_j^{(P+1)}(n_1,...,n_P,0,x_1,...,x_P,\cdot),$$

$j = 1,...,P$, where the dot ($\cdot$) stands for any value of x_i'.

We can now derive the evolutionary dynamics of the market. Since model (4.1) is, by assumption, at equilibrium when an innovation occurs, the initial condition in model (4.2) is $(\bar{n}_1^{(P)},...,\bar{n}_P^{(P)},n_i'(0))$ (the dependence of $\bar{n}_j^{(P)}$ upon $(x_1,...,x_P)$ is omitted for simplicity and $t = 0$ is the time at which the innovation occurs) with $n_i'(0)$ small, since the innovative product is initially present in a few items. Thus, the innovative product penetrates the market, i.e., $\dot{n}_i'(0) > 0$ [disappears, i.e., $\dot{n}_i'(0) < 0$], if the so-called *invasion fitness* of the innovation

$$\lambda_i^{(P)}(x_1,...,x_P,x_i')$$
$$= f_i^{(P+1)}(\bar{n}_1^{(P)},...,\bar{n}_{i-1}^{(P)},0,\bar{n}_{i+1}^{(P)},...,\bar{n}_P^{(P)},\bar{n}_i^{(P)},x_1,...,x_{i-1},x_i',x_{i+1},...,x_P,x_i)$$
$$(4.3)$$

(see definition (3.19) in Chapter 3) is positive [negative].

The invasion fitness $\lambda_i^{(P)}$ is the relative diffusion rate of a few innovative items in the market set by the P established products. Innovations are imagined to originate by chance but their fate depends on their competitiveness, i.e., on their capacity to penetrate into the market. Competitiveness is therefore a concept relevant on the market timescale, which necessarily depends on the innovative trait x_i', as well as on the current market conditions defined by the traits $x_1,...,x_P$ of the established products. In other words, the invasion fitness of the innovative product provides a summary of the underlying market selection process. As we have seen in the previous chapter, such a summary and a proper stochastic description of the innovation process are necessary and sufficient to make the step to macroevolutionary considerations on the evolutionary timescale.

If $\lambda_i^{(P)}(x_1,...,x_P,x_i') < 0$, it follows from model (4.2) that just after the innovation $\dot{n}_i' < 0$, i.e., the innovative product does not penetrate and actually exits the market. Thus, the final result is still a set of P established products with traits $x_1,...,x_P$ and densities $\bar{n}_1^{(P)},...,\bar{n}_P^{(P)}$. By contrast, if $\lambda_i^{(P)}(x_1,...,x_P,x_i') > 0$, the innovative product initially penetrates and, under very general conditions, the ith established product exits the market and is replaced by its new version. Thus, in this case, the trajectory of model (4.2) originating at $(\bar{n}_1^{(P)},...,\bar{n}_P^{(P)},n_i'(0))$ ends at the stable equilibrium

$$\left(\bar{n}_1^{(P)},...,\bar{n}_{i-1}^{(P)},0,\bar{n}_{i+1}^{(P)},...,\bar{n}_P^{(P)},\bar{n}_i^{(P)}\right)$$

(where $\bar{n}_j^{(P)}$, $j = 1,...,P$, are here evaluated at $(x_1,...,x_{i-1},x_i',x_{i+1},...,x_P) \in \mathcal{X}^{(P)}$), i.e., the final result is a new set of P established products where the ith

product has slightly changed its technological content. In other words, each innovation brings a new trait into the market, but competition between established and
innovative products selects the winner, namely the trait that remains in the market.

As discussed in the previous chapter (see, in particular, Section 3.4), the condition under which the innovative product replaces the established one, known as the
invasion implies substitution theorem, requires

$$\frac{\partial}{\partial x_i'}\lambda_i^{(P)}(x_1,...,x_P,x_i')\bigg|_{x_i'=x_i} (x_i'-x_i) > 0. \qquad (4.4)$$

By expanding $\lambda_i^{(P)}$ in Taylor series with respect to x_i', and recalling that innovations
are marginal and appear in the market only in a few items (i.e., x_i' is close to x_i and
$n_i'(0)$ is very small), one obtains

$$\dot{n}_i'(0) \simeq n_i'(0)\frac{\partial}{\partial x_i'}\lambda_i^{(P)}(x_1,...,x_P,x_i')\bigg|_{x_i'=x_i} (x_i'-x_i), \qquad (4.5)$$

so that condition (4.4) implies the initial penetration of the innovative product.

The quantity

$$\frac{\partial}{\partial x_i'}\lambda_i^{(P)}(x_1,...,x_P,x_i')\bigg|_{x_i'=x_i} \qquad (4.6)$$

is called the *selection derivative*, the vector with components (4.6), $i = 1,...,P$,
is called the *selection gradient*. Thus, as long as the selection gradient does not
vanish, the evolutionary dynamics are characterized by

$$\dot{x}_i \begin{cases} > 0 & \text{if} \quad \frac{\partial}{\partial x_i'}\lambda_i^{(P)}(x_1,...,x_P,x_i')\bigg|_{x_i'=x_i} > 0 \\ < 0 & \text{if} \quad \frac{\partial}{\partial x_i'}\lambda_i^{(P)}(x_1,...,x_P,x_i')\bigg|_{x_i'=x_i} < 0, \end{cases} \quad i = 1,...,P.$$

The selection gradient gives the direction of technological change and describes a
continuous feedback between the innovation and competition processes. In fact,
technological change, $\dot{x}_i$, depends on consumption patterns that develop on the
market timescale in accordance with model (4.2) and are summarized by the invasion fitness (4.3). In turn, consumption patterns are affected by the current market
conditions condensed in the traits $x_1,...,x_P$ of the established products.

The process of technological change can be further specified by making suitable
assumptions on the frequency and distribution of innovations. The rate of technological change $\dot{x}_i$ is influenced by three primary factors: how often an innovation
occurs in product i; how large is the trait change entailed by the innovation; and
how likely it is that the initially scarce group of innovative products penetrates the
market. As we have shown in the previous chapter, by suitably modeling these three
factors, the evolutionary process proceeds by a large number of subsequent pene-
trations and substitutions and can be approximated, in the limit case of marginal
innovations, by the following ODE system:

$$\dot{x}_i = \frac{1}{2}\mu(x_i)\sigma^2(x_i)\bar{n}_i^{(P)}(x_1,...,x_P)\frac{\partial}{\partial x_i'}\lambda_i^{(P)}(x_1,...,x_P,x_i')\bigg|_{x_i'=x_i}, \qquad (4.7)$$

$i = 1, ..., P$, called the AD *canonical equation*. With reference to the ith product, $\mu(x_i)$ is proportional to the probability of an innovation in the production of new items, $\mu(x_i)\bar{n}_i^{(P)}(x_1, ..., x_P)$ is hence proportional to the number of innovations that are put on the market per unit of evolutionary time, and $\sigma^2(x_i)$ is the variance of the technological impact of an innovation (with zero expected impact). The probability of penetration consists of two factors. First, if the selection derivative (4.6) does not vanish, only innovations with trait value either larger or smaller than that of the established product can penetrate; in other words, half of the innovations are at selective disadvantage. This leads to the factor $1/2$ in the canonical equation. Second, innovations at selective advantage may be accidentally lost in the initial phase of invasion when they are present only in a few items. The probability of success in the initial phase of invasion is proportional to the selective advantage of the innovation, as measured by the selection derivative (4.6).

In conclusion, model (4.7) describes the technological coevolution of P products coexisting in the market under the assumption of rare innovations of small effect. At any time on the innovation timescale, the products coexist at the market equilibrium $\bar{n}^{(P)}(x_1, ..., x_P)$ corresponding to the current technologies $x_1, ..., x_P$.

Long-term Scenarios

In contrast to prevailing economic theories that focus on the properties of the equilibrium, the AD approach is based on a dynamical framework that accounts for the full dynamics of technological change and its concomitant changes in the market, including, for instance, the description of the evolutionary transient. Notice that the evolutionary model (4.7) is an autonomous system of ODEs. Thus, economic systems perpetually reshape themselves, thereby changing their own basis in terms of technologies in use and market conditions, which are both condensed in the traits $x_1, ..., x_P$.

It is important to note that the AD canonical equation models a coevolutionary context where innovation in one product leads to coevolutionary changes in all other related products in the market under consideration. The importance of this mutual interaction is best described by Ziman (2000), who says "material artifacts cannot be considered in isolation from their cognitive and social correlates... as the artifact changes, so does the cloud of ideas and social activities that surround it."

Moreover, model (4.7) is, in general, nonlinear, which means that the interactions between technology and its market are capable of giving rise to a rich set of scenarios. In the simplest case one can imagine (Figure 4.1A), technological change converges to a particular combination of the traits, no matter what the initial conditions are. This is the typical situation where a set of products or services have reached a so high standard to become practically unbeatable, like the Chinese, French, and Italian cuisine. A wilder evolutionary scenario is that of never-ending ups and downs of the traits, like those repeatedly recorded in the fashion market (see, e.g., the skirt length of women's formal evening dresses reconstructed by Lowe and Lowe, 1990, from the analysis of fashion magazines over two centuries). In this case the traits evolve either toward a limit cycle (Figure 4.1B) or toward a complex aperiodic attractor (Red Queen dynamics). Another scenario of interest

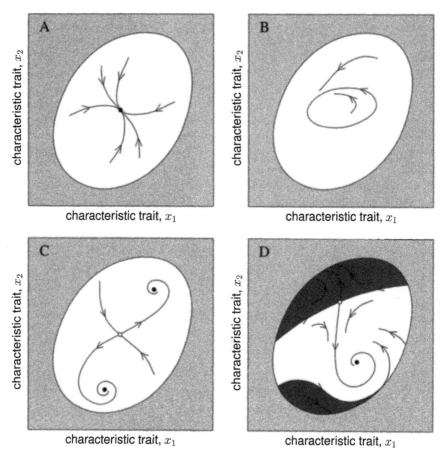

Figure 4.1 Possible evolutionary scenarios: convergence toward an equilibrium (A) or a limit cycle (B) from any initial conditions; alternative equilibria (C); evolutionary extinction of a product (D).

(Figure 4.1C) is that of alternative equilibria (or attractors). This means that the long-term implications of the innovation-competition process can depend upon the innovation paths followed in the past. Such path dependence could for example explain divergence phenomena discussed in development economics, where some developing countries seem to fall into a technological and economic underdevelopment trap, while, industrialized countries converge to a high technological level. Finally, it can also happen (Figure 4.1D) that some evolutionary trajectories reach the boundary of the market evolution set $\mathcal{X}^{(P)}$, where one or more products cannot be sustained in the market (evolutionary extinction, or dismission, of a product; on the boundary of $\mathcal{X}^{(P)}$ the market equilibrium $\bar{n}_j^{(P)}(x_1,...,x_P)$, $j = 1,...,P$, loses one of its properties, namely existence, stability, and positivity). This is, for example, what happened to telex technology and what is expected to happen in the near future to fax technology.

Emergence of Diversity

In accordance with the jargon introduced in the previous chapter, we call *evolutionary equilibrium* a constant solution of the canonical equation (4.7), i.e., a set of traits $(\bar{x}_1, ..., \bar{x}_P)$ at which all selection derivatives (4.6) vanish:

$$\frac{\partial}{\partial x_i'} \lambda_i^{(P)}(\bar{x}_1, ..., \bar{x}_P, x_i') \bigg|_{x_i' = x_i} = 0, \quad i = 1, ..., P.$$

When the evolutionary dynamics have found a halt at a stable evolutionary equilibrium $(\bar{x}_1, ..., \bar{x}_P)$, the first-order term in the expansion (4.5) vanishes, so that to establish if an innovation is initially successful or not one must consider the second-order term of the Taylor expansion of the fitness function $\lambda_i^{(P)}$, thus obtaining

$$\dot{n}_i'(0) \simeq \frac{1}{2} n_i'(0) \frac{\partial^2}{\partial x_i'^2} \lambda_i^{(P)}(\bar{x}_1, ..., \bar{x}_P, x_i') \bigg|_{x_i' = \bar{x}_i} (x_i' - x_i)^2.$$

The result is that the innovation initially penetrates if

$$\frac{\partial^2}{\partial x_i'^2} \lambda_i^{(P)}(\bar{x}_1, ..., \bar{x}_P, x_i') \bigg|_{x_i' = \bar{x}_i} > 0 \qquad (4.8)$$

no matter if the trait value x_i' is larger or smaller than the established trait $\bar{x}_i$. If condition (4.8) holds with the opposite inequality sign for all $i = 1, ..., P$, then the evolutionary equilibrium is protected against invasion and therefore corresponds to a so-called *evolutionarily stable strategy* (ESS) as defined in evolutionary game theory (see Section 2.5).

Understanding the consequences of an innovation that occurs when the market is in the vicinity of the equilibrium conditions $(\bar{x}_1, ..., \bar{x}_P)$ is not an easy problem, since we cannot rely on the invasion implies substitution theorem, whose requirement (4.4) does not hold at an evolutionary equilibrium. However, as shown in detail in the previous chapter, coexistence between the innovative and the established products is possible for $(x_1, ..., x_P)$ close to $(\bar{x}_1, ..., \bar{x}_P)$ (and x_i' close to x_i) only if

$$\frac{\partial^2}{\partial x_i \partial x_i'} \lambda_i^{(P)}(\bar{x}_1, ..., \bar{x}_{i-1}, x_i, \bar{x}_{i+1}, ..., \bar{x}_P, x_i') \bigg|_{x_i = x_i' = \bar{x}_i} < 0. \qquad (4.9)$$

Moreover, once the innovative product is established, with trait x_{P+1} and density n_{P+1} at a stable and strictly positive equilibrium $\bar{n}_j^{(P+1)}(x_1, ..., x_{P+1})$, $j = 1, ..., P+1$, of the market model (4.2), the traits x_i and x_{P+1} initially diversify, in accordance with the $(P+1)$-dimensional canonical equation

$$\dot{x}_i = \frac{1}{2} \mu(x_i) \sigma^2(x_i) \bar{n}_i^{(P+1)}(x_1, ..., x_{P+1}) \frac{\partial}{\partial x_i'} \lambda_i^{(P+1)}(x_1, ..., x_{P+1}, x_i') \bigg|_{x_i' = x_i}, \qquad (4.10)$$

if condition (4.8) is satisfied.

Thus, if one of the conditions (4.8) and (4.9) holds with the opposite inequality sign for all $i = 1, ..., P$, then the stable evolutionary equilibrium $(\bar{x}_1, ..., \bar{x}_P)$ is an evolutionary terminal point. Technological evolution by means of small innovations can therefore drive an economic system to a terminal point of the evolutionary process, a trap from which the system may escape only if radically different products are exogenously injected into the market.

By contrast, if conditions (4.8) and (4.9) both hold for some i, the evolutionary equilibrium is a so-called branching point; namely $(\bar{x}_1, ..., \bar{x}_P)$ identify the market conditions where a new technological variety can be established, through the branching of product i into two initially similar products. In other words, technological branching occurs when the selective forces acting on the market first allow the coexistence of two slightly different products (condition (4.9)) and then become repulsive, therefore favoring the diversification of two technologies originating from the same trait (condition (4.8)). Think, for example, to mobile and fixed phones: the first mobile phones were heavy car phones, different from fixed phones only in the presence of an antenna instead of a wire.

If branching can occur in several products, the AD canonical equation does not tell which of the products will branch first. However, after a branching, the $(P+1)$-dimensional canonical equation (4.10) can be used to investigate further possible branching events. Since no limit exists on the number of possible repeated branchings there is room for the formation of rich clusters of products. Long sequences of technological branchings are empirically evident in almost every market segment. Consumers worldwide can witness that an increasing number of products that match their expectations are available on the market (see, e.g., Grübler, 1998; Saviotti, 2001). For example, Ausubel (1990) showed that the average number of items on sale in a typical large U.S. supermarket has increased from 2000 in 1950 to 18,000 items in the 1990s.

Exogenous Factors

The market competition model (4.2) and the statistics of the innovation process certainly depend upon many exogenous factors, such as consumer preferences, social and political structures, international relationships, and availability of natural resources. To simplify the analysis, these factors can be left out of the model, but they can also be included and measured through strategic parameters explicitly appearing in the AD canonical equation (4.7). In the absence of relevant exogenous fluctuations, such parameters can be considered constant, so that the role played by exogenous factors on technological change can be identified by studying the canonical equation for all possible parameter settings. This naturally calls for bifurcation analysis (see Section 3.8 and Appendix A).

Marginal Innovations vs. Major Breakthroughs

Before presenting an explicit application of AD in economics, it is worth stressing that the analysis described until now applies only to marginal innovations. In the case of a radical innovation (e.g., a major invention or the establishment of new import/export protocols), namely when the innovative product trait is remarkably different from all other traits present in the market, the outcome of the competition must be established by means of model (4.2). For example, radically innovative products can penetrate the market without substituting any previous established product, thus increasing product diversity. Once the new market equilibrium has been determined, the method of analysis discussed in this section can be used again to detect the consequences of new marginal innovations. The process of technological change is therefore described as a continuous process due to marginal innovations punctuated by major breakthroughs. The AD approach interestingly shows that the emergence of new products can be attributed to both marginal and radical innovation events. While it is rather obvious, as noted already by Schumpeter (1912), that major breakthroughs can generate product variety, it is much less obvious that product diversification can emerge through marginal innovational steps.

4.3 A SIMPLE EXAMPLE OF TECHNOLOGICAL BRANCHING

We now present the first formal application of AD in economics. The problem we consider is intentionally simple in order to obtain the AD canonical equation in closed form and clearly identify the stylized properties mentioned in the Introduction. Although the model is far from being empirically testable, it provides some insights on the market conditions that favor technological branching and the emergence of product clusters.

We assume that P different products, each characterized by a single technological trait x taking values $x_1, ..., x_P$, compete in the market according to the following model:

$$\dot{n}_j = r(x_j)n_j \left(1 - \frac{1}{K(x_j)} \sum_{l=1}^{P} \alpha(x_j, x_l)n_l \right), \quad j = 1, ..., P, \qquad (4.11)$$

where the functions $r(x_j)$, $K(x_j)$, and $\alpha(x_j, x_l)$ describe the market environment and have the following interpretation. The function $r(x_j)$ is the maximum diffusion rate of the jth product, which is realized only when the product is present in the market in small quantities (n_j very small) and there are no competitors ($n_l = 0$ for all $l \neq j$). The function $r(x_j)$ therefore measures the penetration power of the product in an empty market. Similarly, $K(x_j)$ is the equilibrium density reached by the jth product after penetration into an empty market, under the scaling property $\alpha(x_j, x_j) = 1$. Thus, $K(x_j)$ is a measure of the product density absorbable by the market and will be hereafter called *market capacity*. Finally, the *competition function* $\alpha(x_j, x_l)$ measures the reduction of the rate of diffusion of the jth product due to the presence in the market of the lth competitor.

Model (4.11) describes a purely competitive market. However, the analysis performed in the following can be extended to a wide spectrum of market interactions, ranging from competition to cooperation, and exploitation. Various notions of *Homo reciprocans* and *Homo economicus* could be modeled, reflecting certain social mechanisms and institutions that punish antisocial behavior (Boyd and Richerson, 1992; Fehr and Gächter, 1998) but also reward image scoring (Nowak and Sigmund, 1998).

Model (4.11) is the simplest type of competition model (Hofbauer and Sigmund, 1998), where the relative diffusion rate $\dot{n}_j/n_j$ of the jth product is a linear combination of all product densities (so-called bilinear competition model). It generically has a unique strictly positive equilibrium satisfying the following system of P linear algebraic equations:

$$\sum_{l=1}^{P} \alpha(x_j, x_l)\bar{n}_l^{(P)}(x_1,...,x_P) = K(x_j), \quad j = 1,...,P. \tag{4.12}$$

Let us assume that r is independent upon the trait and that the market capacity and competition functions are lognormal, i.e., given by

$$K(x_j) = K_0 \exp\left(-\frac{1}{2K_2^2}\left(\ln\left(\frac{x_j}{K_1}\right)\right)^2\right) \tag{4.13}$$

and

$$\alpha(x_j, x_l) = \exp\left(\frac{\ln^2 \alpha_1}{2\alpha_2^2}\right)\exp\left(-\frac{1}{2\alpha_2^2}\left(\ln\left(\frac{\alpha_1 x_j}{x_l}\right)\right)^2\right), \tag{4.14}$$

where K_0, K_1, K_2, α_1, and α_2 are constant positive parameters.

Three parameters, K_0, K_1, and K_2, characterize the market capacity function (4.13), which is bell shaped and peaks at some intermediate trait, given by K_1. The economic interpretation is that in a single-product market K_1 is the technology that is most absorbable by the market, while the equilibrium density of a technologically very poor or very sophisticated product vanishes with a sensitivity controlled by K_2. High or low sensitivity (small and large values of K_2) respectively represent market structures where products concentrate around the technological characteristic trait K_1 or where consumers are to a large degree indifferent to different products satisfying a specific need. Graphically (see Figure 4.2A), K_2 is a measure of the wideness of the bell shape representing the market capacity function.

The competition function (4.14) depends only upon the ratio x_j/x_l of its arguments and tends to zero when such a ratio tends to either zero or infinity, reflecting the fact that very diversified products compete only weakly (e.g., Ferrari and Ford in the car market). Two parameters, namely α_1 and α_2, control the shape of the competition function. For $\alpha_1 = 1$, competition is symmetric, i.e., $\alpha(x_j, x_l) = \alpha(x_l, x_j)$, and $\alpha(x_j, x_l)$ is maximum (and equal to one) for $x_j = x_l$. For $\alpha_1 \neq 1$, competition is asymmetric and $\alpha(x_j, x_l)$ is maximum for $x_l = \alpha_1 x_j$ (see (4.14) and Figure 4.2B). This implies that for $\alpha_1 > 1$ products with higher technological content tend to have a competitive advantage (the rate of diffusion of the jth product is maximally depleted by competitors with trait $x_l = \alpha_1 x_j > x_j$, which, in turn, mildly suffer from the jth product, since $\alpha(\alpha_1 x_j, x_j) < 1 < \alpha(x_j, \alpha_1 x_j)$). Analogously, for $\alpha_1 < 1$ products with less technological content are better competitors. The parameter α_2 controls the sensitivity of competition with respect to the ratio of competing traits. High sensitivity (i.e., small α_2) means that only very similar products compete, while for large α_2 competition remains high even between quite different products. Graphically (see Figure 4.2B), α_2 is a measure of the wideness of the bell shape representing the competition function with respect to the competing trait x_l.

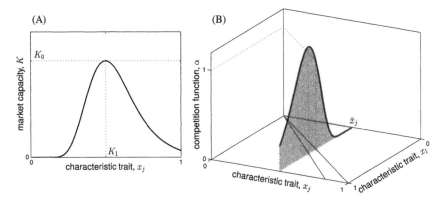

Figure 4.2 (A) The market capacity function; parameter values $K_0 = 1000$, $K_1 = 0.5$, $K_2 = 0.3$. (B) The strength of competition exerted by the lth product on the rate of diffusion of the jth product with trait $\bar{x}_j$, as a slice of the competition function $\alpha(x_j, x_l)$ along the plane $x_j = \bar{x}_j$; straight lines on the (x_j, x_l) plane: $x_l = x_j$ on which $\alpha(x_j, x_l) = 1$; $x_l = \alpha_1 x_j$ on which $\alpha(x_j, x_l)$ is maximum; parameter values $\alpha_1 = 1.2$, $\alpha_2 = 0.3$, $\bar{x}_j = 0.5$.

Consider now a market with a single established product (density n and trait x) and denote by n' and x' the density and trait of the innovative product. From (4.11) and (4.12), it follows that the market equilibrium density $\bar{n}(x)$, the invasion fitness $\lambda(x, x')$, and the selection derivative $\partial\lambda/\partial x'|_{x'=x}$ (see (4.3) and (4.6)) are given by

$$\bar{n}(x) = K(x),$$

$$\lambda(x, x') = r\left(1 - \alpha(x', x)\frac{K(x)}{K(x')}\right),$$

$$\frac{\partial}{\partial x'}\lambda(x, x')\Big|_{x'=x} = \frac{r}{x}\left(\frac{1}{\alpha_2^2}\ln(\alpha_1) - \frac{1}{K_2^2}\ln\left(\frac{x}{K_1}\right)\right).$$

Therefore, the AD canonical equation (see (4.7)) is

$$\dot{x} = \frac{1}{2}\mu\sigma^2\bar{n}(x)\frac{\partial}{\partial x'}\lambda(x, x')\Big|_{x'=x}, \tag{4.15}$$

where μ and σ are assumed to be trait-independent.

Equation (4.15) admits a unique evolutionary equilibrium

$$\bar{x} = K_1\alpha_1^{\left(\frac{K_2}{\alpha_2}\right)^2}, \tag{4.16}$$

which is always an attractor, since its associated eigenvalue has the same sign as

$$\frac{d}{dx}\left(\frac{\partial}{\partial x'}\lambda(x, x')\Big|_{x'=x}\right)\Big|_{x=\bar{x}} = -\frac{r}{(\bar{x}K_2)^2} < 0. \tag{4.17}$$

Thus, in a market with a single product, repeated innovations and replacements of old variants with new ones drive the technological trait x toward the equilibrium value $\bar{x}$. At $\bar{x}$ two selective forces acting on the market balance: the desire to produce a better competitor by being distinct in technological content (under asymmetric competition) and the tendency to cultivate the median consumer to maximize the number of product items in the market (battle for market share). If, for example, higher technological traits are favored ($\alpha_1 > 1$), the economic intuition for reaching the evolutionary equilibrium is that there are cognitive, informational, or physical limitations of consumers to absorb high technology, or simply budget constraints. Such limitations are modeled by the ratio K_2/α_2 (see (4.16)). Notice that if $\alpha_1 > 1$ [$\alpha_1 < 1$] the equilibrium level of technology is larger [smaller] than K_1, so that the product density is not maximized at equilibrium. In particular, when x is slightly larger [smaller] than K_1, a penetrating innovative product conquers the market, even if this implies a loss in product density.

To assess if the evolutionary equilibrium $\bar{x}$ marks the end of technological change or is a branching point, we can use the branching condition (4.8), i.e.,

$$\frac{\partial^2}{\partial x'^2}\lambda(\bar{x}, x')\bigg|_{x'=\bar{x}} = \frac{r}{\bar{x}^2}\left(\frac{1}{\alpha_2^2} - \frac{1}{K_2^2}\right) > 0, \qquad (4.18)$$

and notice that condition (4.9), i.e.,

$$\frac{\partial^2}{\partial x \partial x'}\lambda(x, x')\bigg|_{x=x'=\bar{x}} < 0,$$

is implied by (4.17) and (4.18), since

$$\frac{\partial^2}{\partial x \partial x'}\lambda(x, x')\bigg|_{x=x'=\bar{x}} = \frac{d}{dx}\left(\frac{\partial}{\partial x'}\lambda(x, x')\bigg|_{x'=x}\right)\bigg|_{x=\bar{x}} - \frac{\partial^2}{\partial x'^2}\lambda(\bar{x}, x')\bigg|_{x'=\bar{x}}.$$

Thus, the evolutionary equilibrium $\bar{x}$ is a branching point if the sensitivity K_2 of the market capacity function is larger than the sensitivity α_2 of the competition function. Of course, in the opposite case, $\bar{x}$ is an evolutionary terminal point and the critical condition $K_2 = \alpha_2$ marks a so-called branching bifurcation (see the classification of bifurcations introduced in Section 3.8). Although the two sensitivities govern the dynamics of product densities on the market timescale (they are parameters of the market competition model (4.11)), they ultimately manifest themselves on the evolutionary timescale. The difference $K_2 - \alpha_2$ is a measure of the strength of diversification through technological change. Taking into account the geometric characteristics of the competition and market capacity functions, we can say that our simple model suggests that technological branching occurs when the market capacity is flatter than the competition function (see Figure 4.2). In other words, technological branching is expected in markets which are capable to absorb a rich variety of single technologies and/or easily offer competition-safe niches even to rather similar products.

A relatively flat market capacity would arise in situations when consumers are, to a large extent, indifferent to products satisfying a specific need. A typical example in the food market would be when consumers are indifferent to various sources of protein, be it red meat, white meat, or meat imitations like soya products. The market capacity could also be interpreted as an aggregated utility function. In this case, the curvature, i.e., the second derivative of the utility function, is a measure for risk aversion. Hence, a relatively flat market capacity could be interpreted as

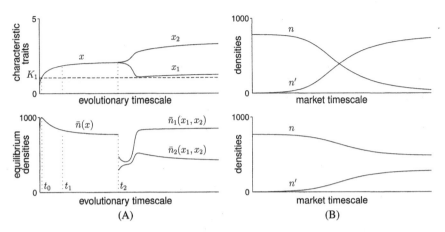

Figure 4.3 Evolutionary dynamics under asymmetric competition. (A) Characteristic traits
(top) and equilibrium densities (bottom) obtained through simulation of models
(4.15) and (4.19) with initial condition $x(0) = 0.5 < K_1$ (see dashed line). (B)
Two examples of market dynamics obtained through simulation of model (4.11):
product substitution (top, $x = 2$, $x' = x * 1.01$, $n(0) = \bar{n}(x) = 786.45$,
and $n'(0) = 1$); branching (bottom, $x = \bar{x}=2.0736$, $x' = \bar{x} * 1.01$, $n(0) =$
$\bar{n}(x) = 766.49$, and $n'(0) = 1$); on the evolutionary timescale, these examples
correspond to times t_1 and t_2 in (A). Parameter values: $r = 1$, $\alpha_1 = 1.2$,
$\alpha_2 = 0.5$, $K_0 = 1000$, $K_1 = 1$, $K_2 = 1$, $\mu = 1$, $\sigma = 1$.

a more risk-taking representative agent (in this case consumer). Given such an
interpretation, our model says that less risk aversion leads to product diversification.

On the other hand, competition functions are narrow when, despite of a relatively small difference in the characteristic traits, the respective products weakly
suffer from each other. For example, the competition between Rolex and Swatch
watches, or between stocks within the NASDAQ index, could be modeled by narrow competition functions.

A specific example of evolutionary dynamics under asymmetric competition
($\alpha_1 > 1$) is shown in Figure 4.3 for a particular parameter setting (specified in
the caption) for which the equilibrium $\bar{x}$ is a branching point. Starting with a single product with trait x smaller than K_1, the trait first increases toward $\bar{x}$ (which
is larger than K_1 since $\alpha_1 > 1$, see (4.16)), as shown in Figure 4.3A (top panel,
$0 < t < t_2$). On the evolutionary timescale, the equilibrium density $\bar{n}(x)$ (bottom
panel) first increases, as long as $x < K_1$ ($0 < t < t_0$), and then declines when
$x > K_1$ ($t_0 < t < t_2$). Figure 4.3B shows the invasion transients on the market
timescale of two particular successful innovations. The first one (top panel) corresponds to the market conditions holding at time t_1 in Figure 4.3A: the density n'
of the innovative product is initially very small, but then grows toward an equilibrium, while the density n of the established product declines to zero, thus revealing
the complete product substitution. In the second transient (bottom panel), corresponding to the branching occurring at time t_2, the innovative product penetrates
the market but does not replace the established product, as shown by the graph of

n, which declines but does not vanish.

After the branching has occurred, the evolutionary dynamics are the result of the competition between a challenging innovative product and two distinct established products characterized by densities n_1 and n_2 and traits x_1 and x_2, respectively. The analysis of the two-product market can be performed by studying the corresponding two-dimensional canonical equation. Denoting by n_i' and x_i' the density and trait of the innovative product, $i = 1, 2$, and recalling (4.11) and (4.12), the two-product market equilibrium $\bar{n}_i(x_1, x_2)$, the invasion fitness $\lambda_i(x_1, x_2, x_i')$, and the selection gradient are given by

$$
\left[\begin{array}{c} \bar{n}_1(x_1, x_2) \\ \bar{n}_2(x_1, x_2) \end{array} \right] = \frac{1}{1 - \alpha(x_1, x_2)\alpha(x_2, x_1)} \left[\begin{array}{c} K(x_1) - \alpha(x_1, x_2)K(x_2) \\ -\alpha(x_2, x_1)K(x_1) + K(x_2) \end{array} \right],
$$

$$
\lambda_i(x_1, x_2, x_i') = r\left(1 - \frac{1}{K(x_i')}\left(\alpha(x_i', x_1)\bar{n}_1(x_1, x_2) + \alpha(x_i', x_2)\bar{n}_2(x_1, x_2)\right)\right),
$$

$$
\left. \frac{\partial}{\partial x_i'}\lambda_i(x_1, x_2, x_i')\right|_{x_i'=x_i} = -\frac{r}{K(x_i)}
$$

$$
\times \left(\left. \frac{\partial}{\partial x_i'}\alpha(x_i', x_1)\right|_{x_i'=x_i} \bar{n}_1(x_1, x_2) + \left. \frac{\partial}{\partial x_i'}\alpha(x_i', x_2)\right|_{x_i'=x_i} \bar{n}_2(x_1, x_2)\right)
$$

$$
+ \frac{r}{K(x_i)^2}\frac{d}{dx_i}K(x_i)\left(\alpha(x_i, x_1)\bar{n}_1(x_1, x_2) + \alpha(x_i, x_2)\bar{n}_2(x_1, x_2)\right),
$$

$i = 1, 2$, provided that $x_1 \neq x_2$. Thus, the second-order AD canonical equation is given by

$$
\dot{x}_i = \frac{1}{2}\mu\sigma^2\bar{n}_i(x_1, x_2) \left. \frac{\partial}{\partial x_i'}\lambda_i(x_1, x_2, x_i')\right|_{x_i'=x_i}, \quad i = 1, 2. \tag{4.19}
$$

The relevant trajectory of model (4.19) is that originating from point $(\bar{x}, \bar{x} + \epsilon)$ (ϵ very small), corresponding to the market condition at time t_2 just after the coexistence between products 1 and 2 has been established (i.e., just after the invasion transient depicted in the lower panel of Figure 4.3B). The evolution of the traits x_1 and x_2 and that of the corresponding product densities $\bar{n}_1(x_1, x_2)$ and $\bar{n}_2(x_1, x_2)$ along such a trajectory are shown in Figure 4.3A ($t > t_2$). The trait in one branch permanently increases, while in the other it initially decreases. This was expected because at a branching point the old and the new version of the product coexist under opposite selection pressures. Notice that the product associated with the upper branch (away from K_1) has a lower density, i.e., a lower market share, but a competitive advantage with respect to the product in the lower branch, which, by contrast, resists competition being close to the trait (K_1) that matches the median consumer. Finally, Figure 4.3A shows that the evolutionary dynamics drive the traits x_1 and x_2 of the two coexisting products toward a stable evolutionary equilibrium $(\bar{x}_1, \bar{x}_2)$. At this equilibrium the branching conditions (4.8) and (4.9) have been numerically tested by varying all the parameters of the model and the result is that again conditions (4.8) and (4.9) hold for both products if $K_2 > \alpha_2$.

Of course, to understand the evolution of the system after the second branching, the analysis should be repeated, by deriving the new (three-dimensional) canonical equation. By means of a systematic bifurcation analysis of three- and higher-dimensional canonical equations, we checked that the traits of P coexisting products always converge toward a unique stable evolutionary equilibrium at which the branching conditions (4.8) and (4.9) hold for all products if $K_2 > \alpha_2$ (more weakly as P increases). This numerical analysis has been performed for wide ranges of all parameters and always leads to the same conclusion, namely that $K_2 > \alpha_2$ implies the formation of rich clusters of products through a long sequence of technological branchings.

4.4 DISCUSSION AND CONCLUSIONS

The purpose of this chapter was to show how AD could be used to explain the emergence of technological variety in economic systems. Since some of the assumptions underlying AD are rather extreme, one must be careful in applying it to real situations. For example, technological change is an economic phenomenon taking place at different levels of temporal aggregation of the economy, involving individual consumers, businesses, markets, science, technology, formal and informal institutions, and culture at wider levels (Hayek, 1967; Nelson, 1995; North, 1997). In real economic systems the market and evolutionary timescales are sometimes comparable, while AD requires that they are fully separated. However, quite frequently competition and technological change occur on contrasting timescales. Technological change slowly proceeds by means of continual replacement of established entities by novel ones on the microlevel, i.e., as a result of the fast interaction between economic actors on the market timescale. In consequence, AD provides a reasonable approximation of the process of technological change with the major promise of elucidating the long-term effects of the interplay between the single entities on a micro level and the evolutionary fate of the system on a macro level.

Perhaps, the most relevant advantage of AD with respect to other approaches is the possibility of clearly explaining the emergence of technological diversity and the formation of rich clusters of products. Indeed, it is empirically evident that technological diversity is a natural characteristic of industries undergoing technological change (see, e.g., Metcalfe, 1988; Bernard et al., 1994). However, there is little room for technological diversity in classical economic models: when the best practice is common knowledge, it is instantly adopted and diversity has no theoretical justification (Jonard and Yildizoglu, 1999). Yet diversity is the basis for consumer choice and a prerequisite for competition. With the advent of evolutionary approaches and institutional approaches, the role of variety, as Schumpeter called it, became a renascent topic (see Stirling, 1998, and various papers by Saviotti for a review of the literature on the economics of diversity). There is detailed empirical and theoretical work in areas like consumer characteristics, production processes and organizational forms, research strategies, competences and learning processes, technologies and modes of innovation, investor expectations and customer choice, and competition.

Some of the evolutionary approaches to industrial dynamics explain the emergence of diversity by including uncertainty in the diffusion process, bounded rationality, imperfect information, demand slacks, and endogenously determined market structures (Nelson and Winter, 1982; Dalle, 1998; Saviotti, 2001; Witt, 2001). However, in all models the selective pressures and the mechanical nature of diffusion lead to just one dominant technology. De Palma et al. (1998) show that, in the presence of network externalities, diversity prevails as long as the effect of consumer heterogeneity overrules the effect of network externalities. These models, however, have to assume diversified markets from the beginning and are not explicit on the emergence of diversity. Another popular hypothesis in economic theory is that local interaction (e.g., localization of imitation and localization of network externalities) is a condition for aggregate diversity (see, e.g., Nelson and Winter, 1982; Jonard and Yildizoglu, 1998a,b). In these models diversity is explained through geographically disjoint technological path dependencies leading to localized positive feedback economies such as agglomeration economies (Engländer, 1926; Ritschl, 1927; Palander, 1935; Arthur, 1990; Porter, 1990; Matsuyama, 1995). Earlier models in spatial economics, which can be associated with the names of von Thünen (1826), Weber (1909), Christaller (1933), and Loesch (1941), see locational patterns as independent of history, inevitable, and thus leading to a unique equilibrium determined by, among other factors, geographical endowments, infrastructures and firms' needs. Geographically disjoint technological development is, however, in conflict with the empirical observations of spatial clusters, which consist of a complex of competing and complementary firms (or even branches within firms) involved in producing similar goods and services (see, e.g., Marshall, 1920; Dunning, 2000).

In contrast with all the above attempts, AD endogenously allows the emergence of diversity through the mechanism of technological branching, and does not require geographically disjoint market environments. The coexistence between marginally innovative and already established technologies is not possible as long as the penetration power of innovations is sufficiently strong (i.e., as long as the selection gradient does not vanish, see (4.6)). Only when technological change slows down at an evolutionary equilibrium the market opens up the possibility for more product diversity. Whether this is the case or not can be tested through the branching conditions (4.8) and (4.9), which only involve elements of the market model (4.1). The branching conditions fully characterize the market at the evolutionary equilibrium, whether it allows the coexistence of similar technologies (condition (4.9)), and whether its selective pressure diversifies them or not (condition (4.8)). Evolutionary equilibria are then classified as terminal points of technological change if branching is not possible (among which we find the ESS of evolutionary game theory), or as branching points. Repeated branching therefore represents a route toward product variety, and the identification of the market conditions which favor such a route is one of the question AD might help to elucidate. In our simple application, for example, we found that repeated branching is expected in market sectors which are not targeted on a specific technology and offer competition-safe niches even to rather similar products.

As this chapter represents a first attempt to develop a dynamical model of technological change consistent with AD and compatible with several results from existing economic models, we have also shown that AD carries the potential to lead to new insights in the analysis of the metabolism and development of traits of economic systems. Many are, in fact, the evolutionary phenomena that might be tackled by means of the AD approach. For example, apart from the increase in numbers of products through technological branching, we also see an increase in product complexity. Illustrative for increasing complexity is the fact that the 1885 Rover safety bicycle consisted of about 500 parts, a modern car involves as many as 30,000 components, and a Boeing 747 has roughly 3.5 million (Ayres, 1988). Thus, product complexity could be modeled as a trait of a suitable AD model, to investigate the economic conditions that lead to increasing complexity and their consequence on technological change and economic development.

Another scenario that can be interpreted by the AD approach is the convergence toward an underdevelopment trap, an evolutionary terminal point that can be broken only by a radical innovation. Modeling aggregate traits such as the level of technological development, one could use AD to show that developing countries are often destined to reach an evolutionary terminal point corresponding to a low technological development, from which they can hardly escape. In fact, for many developing countries, the relevant technological traits are defined more by epigenetic codes such as formal institutions and tacit social norms, which are more difficult to change radically as they acquire more and longer-lasting information than individual agents. On these lines Greif (1994) argues that "the capacity of societal organization to change is a function of history, since institutions are combined of organizations and cultural beliefs... and past organizations and beliefs influence historically subsequent games, organizations and equilibria." The work of Hayek (1967) is more inspired by the idea of spontaneous evolution of conventions and institutions (Vromen, 1995) explaining radical changes of epigenetic codes.

Chapter Five

Multiple Attractors and Cyclic
Evolutionary Regimes

We show in this chapter that resource-consumer communities can be characterized by multiple evolutionary attractors and cyclic evolutionary regimes. We consider a standard resource-consumer demographic model with one adaptive trait for each population and derive the corresponding AD canonical equation, namely an evolutionary model composed of two ODEs, one for the resource trait and one for the consumer trait. Then, we perform a complete bifurcation analysis of the evolutionary model with respect to various demographic and environmental parameters and show that up to three evolutionary attractors can be present. Moreover, the evolutionary dynamics can easily promote resource diversity, as well as the evolutionary extinction of the consumer. Interesting biological properties can be extracted from these findings, which are in agreement with the existing literature. Most of the chapter is taken from Dercole et al. (2003).

5.1 INTRODUCTION

The theoretical work developed so far in the literature has shown that evolutionary dynamics can be as complex as one can imagine. For example, nonstationary evolutionary regimes (the so-called Red Queen dynamics, see Section 1.6) are possible (Dieckmann et al., 1995; Dercole et al., 2003, 2006), as well as evolutionary extinction (Matsuda and Abrams, 1994a; Ferrière, 2000; Dercole, 2005; Dercole et al., 2006) and branching (Geritz et al., 1999; Kisdi, 1999; Doebeli and Dieckmann, 2000; Dercole and Rinaldi, 2002; Dercole, 2003). Moreover, an evolving community can also have alternative evolutionary attractors, in which case the fate of the system is determined by its ancestral conditions (Dieckmann et al., 1995; Dercole et al., 2003, 2006).

As discussed in Section 3.8, once an evolutionary model is available in terms of ODEs, the powerful machinery of numerical bifurcation analysis can be applied to it. This is mandatory if the aim is to detect the impact of some strategic parameters on the evolution of the community. Systematic bifurcation analysis with respect to key demographic and environmental parameters could, for example, explain why ecosystems differ at various latitudes, altitudes, and depths. The few bifurcation studies of evolutionary models performed before Dercole et al. (2003) (see, e.g., Marrow et al., 1992; Matsuda and Abrams, 1994b; Dieckmann et al., 1995) are far from satisfactory: they are inaccurate because they have been carried out mainly through simulation and they are incomplete because they refer to nongeneric cases or point out only some aspects of the full bifurcation diagram. Here we report the bifurcation study performed in Dercole et al. (2003), which is the first complete bifurcation analysis of a typical evolutionary model.

The problem we tackle is the coevolution of resource and consumer adaptive traits, a subject that has received a great deal of attention in the last decade (see Abrams, 2000, for a review). We consider two populations (resource and consumer), and two adaptive traits (one for each population), and the bifurcation analysis of the evolutionary model (the AD canonical equation) is performed with respect to pairs of biologically relevant demographic and environmental parameters.

The chapter is organized as follows. In the next section we focus on the well-known Rosenzweig-MacArthur resource-consumer model (Rosenzweig and Mac-Arthur, 1963) and derive the corresponding AD canonical equation. Then, we present the bifurcation analysis of the evolutionary model, and show how interesting biological conclusions can be extracted from it. Some comments and comparisons with the literature close the chapter.

5.2 A MODEL OF RESOURCE-CONSUMER COEVOLUTION

We consider a two-population assembly composed of resource and consumer, respectively described by their population abundances n_1 and n_2 and adaptive traits x_1 and x_2. The traits are real variables obtained from the actual phenotypic traits through suitable nonlinear scalings that map the positive interval of the actual trait into the real axis. In view of the importance of individual body size in determining interactions between resources and consumers (Cohen et al., 1993; Dieckmann et al., 1995), we imagine x_1 and x_2 as scaled body sizes of adult resource and consumer individuals.

The model that has most often been used in the last few decades to predict resource and consumer abundances on the demographic timescale is the well-known Rosenzweig-MacArthur model:

$$\dot{n}_1 = rn_1 - cn_1^2 - \frac{an_1}{1 + a\tau n_1}n_2, \tag{5.1a}$$

$$\dot{n}_2 = e\frac{an_1}{1 + a\tau n_1}n_2 - dn_2, \tag{5.1b}$$

where r and c are resource net growth rate and intraspecific competition, while a, τ, e, and d are consumer attack rate, handling time, efficiency, and net death rate (see Section 3.3 for a more detailed description of model parameters). In the following, we will fix the parameters in such a way that $e > d\tau$, because, otherwise, the consumer population cannot grow even in the presence of an infinitely abundant resource population. Moreover, we will also impose that $a\tau r/c < 1$, because under this condition model (5.1) cannot have limit cycles (see the Example: Rosenzweig-MacArthur Resource-Consumer Model in Appendix A). Under these assumptions, model (5.1) has a globally stable equilibrium, namely the trivial equilibrium $(r/c, 0)$ (where $K = r/c$ is the so-called resource carrying capacity), if $e < d(\tau + c/(ra))$, and the strictly positive equilibrium

$$(\bar{n}_1, \bar{n}_2) = \left(\frac{d}{a(e - d\tau)}, \frac{e\,(ar(e - d\tau) - cd)}{a^2(e - d\tau)^2} \right), \tag{5.2}$$

if

$$e > d\left(\tau + \frac{c}{ra}\right). \tag{5.3}$$

The transition between the two cases is a transcritical bifurcation (see Appendix A) at which the strictly positive equilibrium (5.2) collides and exchanges stability with the trivial equilibrium $(r/c, 0)$, thus marking the extinction of the consumer population.

To study the evolution of resource and consumer body sizes x_1 and x_2, one must first specify how the parameters of the resident model (5.1) depend upon the traits. The number of possibilities is practically unlimited, because even for well-identified resource and consumer species there are many meaningful options. Thus, at this level of abstraction, it is reasonable to limit the number of parameters sensitive to the traits and avoid trait dependencies that could give rise to biologically unrealistic evolutionary dynamics, like the unlimited growth of a trait (the so-called evolutionary runaway). Our choice has been to assume that r, e, and d are constant parameters, while c, a, and τ are trait-dependent. The particular functional forms specifying how c, a, and τ depend upon the traits are classic forms used in this context (see, e.g., Dieckmann et al., 1995). Since x_1 and x_2 are unbounded real variables obtained from the actual phenotypic traits through suitable nonlinear scalings, the maximum and minimum feasible body sizes of resource [consumer] adult individuals correspond to the limit values $+\infty$ and $-\infty$ of x_1 [x_2]. This allows us to consider an unbounded trait space and to specify c, a, and τ as bounded functions of the traits.

Resource intraspecific competition c is assumed to be mild at intermediate body sizes through the function

$$c(x_1) = \frac{c_1 + c_2 (x_1 - c_0)^2}{1 + c_3 \left(c_1 + c_2 (x_1 - c_0)^2\right)}, \tag{5.4a}$$

where the parameter c_0 (which can be either positive or negative) is the value of x_1 (called *optimum resource trait* in the following) at which intraspecific competition is minimum (and equal to $c_1/(1 + c_1 c_3)$, see Figure 5.1A). For x_1 far from c_0, intraspecific competition saturates at $1/c_3$, so that intermediate resource sizes are favored in the absence of consumers.

The consumer attack rate a is taken to be the following bell-shaped function:

$$a(x_1, x_2) = a_0 + a_4 \exp\left(-\left(\frac{x_1}{a_1}\right)^2 + 2a_3 \frac{x_1 x_2}{a_1 a_2} - \left(\frac{x_2}{a_2}\right)^2 \right), \tag{5.4b}$$

with $a_3 < 1$. This reflects that consumers show some degree of specialization in the size of the harvested resource relative to their own size. If resource and consumer sizes are "tuned," i.e., if $x_1 = x_2 = 0$, the consumer attack rate is maximum (and equal to $a_0 + a_4$, see Figure 5.1B). When resource and consumer traits are far from being tuned, the consumer attack rate drops to a_0.

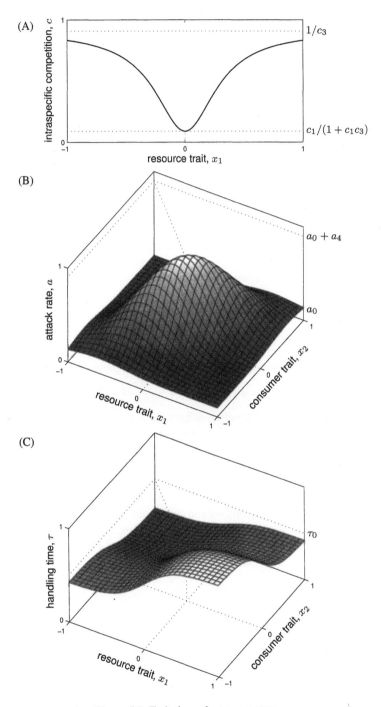

Figure 5.1 Trait-dependent parameters.

The consumer handling time τ is the product of an increasing sigmoidal function of the resource trait x_1 and of a decreasing sigmoidal function of the consumer trait x_2, i.e.,

$$\tau(x_1, x_2) = \tau_0\left(1 + \tau_1 - \frac{2\tau_1}{1 + \exp(\tau_3 x_1)}\right)\left(1 + \tau_2 - \frac{2\tau_2}{1 + \exp(-\tau_4 x_2)}\right), \quad (5.4c)$$

which takes into account that handling time typically increases [decreases] with resource [consumer] size. Here τ_0 is the handling time corresponding to the tuned situation ($x_1 = x_2 = 0$, see Figure 5.1C).

In the following, we fix r, d, and most of the parameters shaping functions c, a, and τ (parameter values are indicated in the caption of Figure 5.3). In particular, we set an upper bound on τ and a lower bound on e, so that the inequalities

$$e > d\tau(x_1, x_2), \quad a(x_1, x_2)\tau(x_1, x_2)\frac{r}{c(x_1)} < 1$$

hold for all (x_1, x_2). As already said, these two conditions guarantee that consumers can grow when the resource is abundant and that cyclic coexistence of the two populations is not possible. Thus, the resource and consumer populations can coexist only at the demographic equilibrium

$$\bar{n}_1(x_1, x_2) = \frac{d}{a(x_1, x_2)(e - d\tau(x_1, x_2))}, \quad (5.5a)$$

$$\bar{n}_2(x_1, x_2) = \frac{e\left(a(x_1, x_2)r(e - d\tau(x_1, x_2)) - c(x_1)d\right)}{a(x_1, x_2)^2(e - d\tau(x_1, x_2))^2} \quad (5.5b)$$

(see (5.2)), which is stable and strictly positive for all pairs (x_1, x_2) such that

$$e > d\left(\tau(x_1, x_2) + \frac{c(x_1)}{ra(x_1, x_2)}\right) \quad (5.6)$$

(see (5.3)). Condition (5.6) therefore defines the evolution set $\mathcal{X}$ of the community (bounded ovoid region in Figure 5.2). On the boundary of $\mathcal{X}$, the consumer equilibrium abundance $\bar{n}_2(x_1, x_2)$ vanishes, i.e., the consumer population goes extinct if the traits reach the boundary of the evolution set.

Model (5.1), together with the trait-dependent parameters (5.4), specifies the resident model of the resource-consumer community. If we now imagine that a mutant population is also present, we can enlarge the resident model (5.1) by adding a third ODE, and by slightly modifying the equations of the resident populations, to take the mutant population into account. In the case of a mutation in the resource population, the resident-mutant model is

$$\dot{n}_1 = n_1\left(r - c(x_1)n_1 - \gamma(x_1, x_1')n_1'\right.$$

$$\left. - \frac{a(x_1, x_2)}{1 + a(x_1, x_2)\tau(x_1, x_2)n_1 + a(x_1', x_2)\tau(x_1', x_2)n_1'}n_2\right), \quad (5.7a)$$

$$\dot{n}_1' = n_1'\left(r - \gamma(x_1', x_1)n_1 - c(x_1')n_1'\right.$$

$$\left. - \frac{a(x_1', x_2)}{1 + a(x_1, x_2)\tau(x_1, x_2)n_1 + a(x_1', x_2)\tau(x_1', x_2)n_1'}n_2\right), \quad (5.7b)$$

$$\dot{n}_2 = n_2\left(e\frac{a(x_1, x_2)n_1 + a(x_1', x_2)n_1'}{1 + a(x_1, x_2)\tau(x_1, x_2)n_1 + a(x_1', x_2)\tau(x_1', x_2)n_1'} - d\right), \quad (5.7c)$$

where $\gamma(x_1, x_1')$ gives the competition coefficient characterizing reduced birth rate and/or increased death rate in the resource resident population due to the competition with the resource mutant population (necessarily $\gamma(x_1, x_1) = c(x_1)$). As discussed in Section 3.3, resource intraspecific competition is said to be symmetric if the competition function $\alpha(x_1, x_1') = \gamma(x_1, x_1')/c(x_1)$ is symmetric with respect to the diagonal $x_1' = x_1$. We assume a constant competition function, so that $\gamma(x_1, x_1') = c(x_1)$.

Similarly, in the case of a mutation in the consumer population, the resident-mutant model is

$$\dot{n}_1 = n_1\left(r - c(x_1)n_1 - \frac{a(x_1, x_2)}{1 + a(x_1, x_2)\tau(x_1, x_2)n_1}n_2 \right.$$
$$\left. - \frac{a(x_1, x_2')}{1 + a(x_1, x_2')\tau(x_1, x_2')n_1}n_2' \right), \tag{5.8a}$$

$$\dot{n}_2 = n_2\left(e\frac{a(x_1, x_2)n_1}{1 + a(x_1, x_2)\tau(x_1, x_2)n_1} - d \right), \tag{5.8b}$$

$$\dot{n}_2' = n_2'\left(e\frac{a(x_1, x_2')n_1}{1 + a(x_1, x_2')\tau(x_1, x_2')n_1} - d \right). \tag{5.8c}$$

At this point, the AD canonical equation governing the evolution of the resource and consumer traits x_1 and x_2 in the evolution set $\mathcal{X}$ can be derived. In accordance with the definition (3.19) in Chapter 3, the invasion fitnesses of the resource and consumer mutant populations are given by

$$\lambda_1(x_1, x_2, x_1') = \left(r - c(x_1')\bar{n}_1(x_1, x_2) - \frac{a(x_1', x_2)\bar{n}_2(x_1, x_2)}{1 + a(x_1, x_2)\tau(x_1, x_2)\bar{n}_1(x_1, x_2)} \right),$$
$$\tag{5.9a}$$

$$\lambda_2(x_1, x_2, x_2') = \left(e\frac{a(x_1, x_2')\bar{n}_1(x_1, x_2)}{1 + a(x_1, x_2')\tau(x_1, x_2')\bar{n}_1(x_1, x_2)} - d \right), \tag{5.9b}$$

and the canonical equation reads

$$\dot{x}_i = k_i\bar{n}_i(x_1, x_2)\left.\frac{\partial}{\partial x_i'}\lambda_i(x_1, x_2, x_i')\right|_{x_i' = x_i}, \quad i = 1, 2, \tag{5.10}$$

where $k_i = 1/2\mu_i\sigma_i^2$, $i = 1, 2$, are constant mutational rates, proportional to the frequency and variance of mutations in the resource and consumer populations, respectively. The explicit expressions of the selection derivatives $\partial\lambda_i/\partial x_i'|_{x_i'=x_i}$, $i = 1, 2$, and of the fitness second derivatives needed in the following for testing if evolutionary branching can occur (see (5.11)), are not reported because they are very long. However, they can be generated and handled by means of symbolic computation, as done in Dercole et al. (2003).

Model (5.10) describes the evolution of the resource and consumer traits under the assumption of rare and random mutations of small effects. An example of evolutionary state portrait of model (5.10) is given in Figure 5.2, which points out all relevant invariant sets (equilibria, limit cycles, and saddle separatrices, i.e., the so-called stable and unstable manifolds of saddles). Some evolutionary trajectories (see thin gray regions in Figure 5.2) reach the boundary of the evolution set $\mathcal{X}$, thus

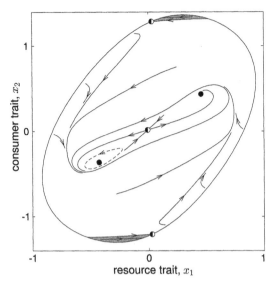

Figure 5.2 Evolutionary state portrait of the AD canonical equation (5.10). Evolutionary trajectories are defined in the evolution set $\mathcal{X}$ (bounded ovoid region). The portrait
is characterized by three equilibria (two stable foci, filled circles, and one saddle,
half-filled circle) and two limit cycles (one stable, solid closed trajectory, and
one unstable, dashed trajectory). Parameter values as in Figure 5.3, region ⑪.

implying the extinction of the consumer population. It is worth noticing that this
form of evolutionary extinction is always an evolutionary murder (see Sections 1.8
and 3.6). In fact, the boundary of the evolution set corresponds to a transcritical bifurcation of the resident model (5.1) at which $\bar{n}_2(x_1, x_2) = 0$, so that the consumer
evolutionary rate $\dot{x}_2$ vanishes while approaching the boundary of $\mathcal{X}$ (see (5.10)).
Thus, when extinction occurs, the consumer trait is constant, while the resource
trait varies, so that it is licit to say that consumers are murdered by the resource.

5.3 THE CATALOG OF EVOLUTIONARY SCENARIOS

We now present the bifurcation analysis of the AD canonical equation (5.10) derived in the previous section, in order to discuss all qualitatively different evolutionary scenarios that are possible for different demographic and environmental
parameter settings. Local and global codimension-1 bifurcations with respect to
various parameters can be numerically detected and continued by means of spe-
cialized software (see Appendix A), so that bifurcation diagrams with respect to
any pair of parameters can be produced.

The first surprising result is that the evolutionary model is much richer than the
resident demographic model. In fact, while model (5.1) is characterized by only
one transcritical bifurcation, many bifurcations have been detected in model (5.10).
Figure 5.3 shows these bifurcations in the parameter space (e, c_0) where e is con-

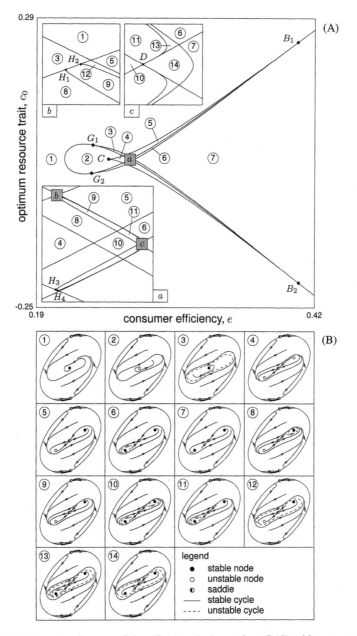

Figure 5.3 Bifurcation diagram of the AD canonical equation (5.10) with respect to consumer efficiency e and optimum resource trait c_0 (A; panels a, b, and c are magnified views, see Table 5.1 for bifurcation curves) and sketches of evolutionary state portraits (B). The bifurcation diagram shows a strong (not exact) horizontal symmetry. The state portraits corresponding to unnumbered regions are equivalent (through a sort of 180-degree rotation) to those in the symmetric regions. Parameter values: $r = 0.5$, $d = 0.05$, $c_1 = 0.5$, $c_2 = 1$, $c_3 = 0.01$, $a_0 = 0.01$, $a_1 = 1$, $a_2 = 1$, $a_3 = 0.6$, $a_4 = 1$, $\tau_0 = 0.9$, $\tau_1 = \tau_2 = 0.5$, $\tau_3 = \tau_4 = 1$, $k_1 = k_2 = 1$.

Table 5.1 Bifurcation curves in Figure 5.3.

transition	bifurcation
①–②	supercritical Hopf (stable cycle)
①–③, ⑤–⑫, ⑥–⑬, ⑦–⑭	tangent of limit cycles
①–⑤, ①–⑦, ②–④, ③–⑧, ③–⑨, ③–⑫	saddle-node
②–③, ④–⑧, ⑤–⑥, ⑧–⑩, ⑨–⑪, ⑫–⑬	subcritical Hopf (unstable cycle)
③–⑨	homoclinic (to saddle-node)
⑥–⑦, ⑧–⑨, ⑨–⑫, ⑩–⑪, ⑪–⑬, ⑬–⑭	homoclinic (unstable cycle)

sumer efficiency and c_0 is the optimum resource trait. In general, both parameters are influenced by environmental factors. For example, the efficiency of a herbivore depends upon the caloric content of the grass, which, in turn, is mainly fixed by humidity, temperature, and soil composition. Figure 5.3 points out that there are 14 regions characterized by qualitatively different evolutionary state portraits. The nature of the bifurcation curve separating region ⓘ from region ⓙ can be easily identified by comparing the two state portraits ⓘ and ⓙ. For example, the transition from ① to ② is a supercritical Hopf bifurcation (a stable equilibrium becomes unstable and surrounded by a stable limit cycle), while the transition from ① to ③ is a tangent bifurcation of limit cycles; and so on (see Table 5.1 for the complete list).

Evolutionary attractors can be equilibria or limit cycles and the existence of alternative attractors is a rather common case. When they exist, attracting cycles surround all equilibria. Actually, there can be up to three alternative attractors (two equilibria and one cycle), as shown by the evolutionary state portraits ⑩, ⑪, ⑬, and ⑭. The evolutionary extinction of the consumer population can occur in all cases, as pointed out by the small gray regions present in all state portraits, but no bifurcations critically involving the boundary of the evolution set $\mathcal{X}$ (so-called extinction bifurcations, see Section 3.8) are detected in the parameter ranges explored in Figure 5.3.

There are 10 codimension-2 bifurcation points (filled circles in Figure 5.3A), at which several bifurcation curves merge. In particular, there are one cusp (C, where three equilibria collide all together), two generalized Hopf (G_1 and G_2, where the Hopf bifurcation turns from supercritical to subcritical and vice versa), two Bogdanov-Takens (B_1 and B_2, where an equilibrium has two vanishing eigenvalues), four noncentral-saddle-node homoclinic loops (H_1, H_2, H_3, and H_4, where a homoclinic bifurcation turns from standard-saddle to saddle-node equilibrium type and vice versa), and one double homoclinic loop (D, where two homoclinic loops involving the same saddle are present) (see Kuznetsov, 2004, for a detailed treatment of codimension-2 bifurcations).

No other bifurcation curves and codimension-2 bifurcation points are present in the two extra bifurcation diagrams presented in Figure 5.4, where the evolutionary state portraits are intentionally not shown in order to stress that they are part of those of Figure 5.3B. The parameter on the horizontal axis of these two bifurcation diagrams is still the efficiency of the consumer, while the parameter on the vertical axis is related to two important characteristics of the mutation and predation processes, namely the ratio k_1/k_2 between the resource and consumer mutational rates and the consumer handling time τ_0 corresponding to the maximum attack rate (i.e., in the tuned situation $x_1 = x_2 = 0$).

Figures 5.3 and 5.4 show that bifurcation curves can be so numerous in small windows of parameter space that several layers of magnification are required to clearly point out the complete bifurcation structure. Detailing the analysis to this extent might seem like nonsense, in particular, when models mimic only qualitatively the real biology. However, the details of the small windows are essential for checking the consistency of the overall bifurcation structure, from which general biological conclusions can be extracted. For example, one could be interested in identifying the demographic and environmental factors favoring Red Queen dynamics (i.e., the possibility of nonstationary evolution of the traits, which here can only take the form of cyclic evolution). For this, one should extract from each bifurcation diagram the regions ②–④, ⑧–⑭, where at least one of the evolutionary attractors is a limit cycle. The result is Figure 5.5, which shows where cyclic evolution is the only possible outcome (light gray regions) and where stationary evolution is also possible (dark gray regions). Figure 5.5 indicates that Red Queen dynamics occur only for intermediate values of consumer efficiency. Thus, slow environmental drifts entraining slow but continuous variations of consumer efficiency can promote the disappearance of Red Queen dynamics. However, if efficiency decreases, Red Queen dynamics disappear smoothly through a supercritical Hopf bifurcation (where the attracting evolutionary cycle shrinks to a point). By contrast, if efficiency increases, Red Queen dynamics disappear discontinuously through a catastrophic bifurcation (tangent bifurcation of limit cycles). Figure 5.5 also indicates other biologically relevant properties, such as the fact that Red Queen dynamics are facilitated by high [low] resource [consumer] mutational rates and by low consumer handling times. This last result suggests that the highest chances for cyclic evolution are obtained when $\tau_0 = 0$, i.e., when the Rosenzweig-MacArthur model degenerates into the Lotka-Volterra model (Lotka, 1920; Volterra, 1926). This brings us to the following rather intriguing conclusion: the Lotka-Volterra assumptions (which do not give rise to demographic cycles) can easily explain evolutionary cycles, while the Rosenzweig-MacArthur assumptions (which easily give rise to demographic cycles for sufficiently high handling time) are less prone to support Red Queen dynamics.

Extra information can be added to the bifurcation diagrams of Figures 5.3 and 5.4 by specifying if the stable evolutionary equilibria $(\bar{x}_1, \bar{x}_2)$ are terminal points (TP) or branching points (BP) of the AD canonical equation (5.10) and, in the latter case, if branching involves the resource, the consumer, or both populations. This can be

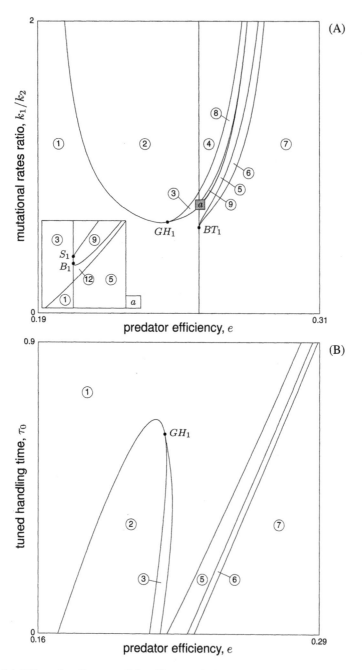

Figure 5.4 Bifurcation diagram of the AD canonical equation (5.10) with respect to consumer efficiency, e, and mutational rates ratio, k_1/k_2 (A), tuned handling time, τ_0 (B). See Table 5.1 for bifurcation curves and Figure 5.3 for evolutionary state portraits and parameter values.

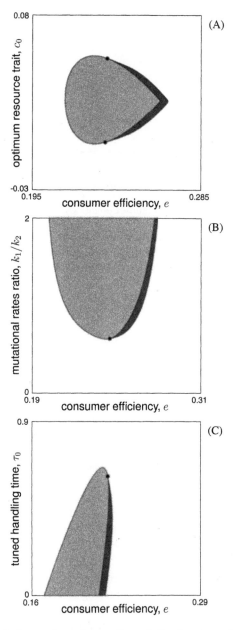

Figure 5.5 Cyclic evolutionary dynamics: cyclic evolution is not possible in white regions, while it is possible in shaded regions; in light gray regions cyclic evolution is the only long-term form of evolution, while in dark gray regions both stationary and cyclic evolution are possible. Panels A, B, and C are extracted from the bifurcation diagrams of Figures 5.3, 5.4A, and 5.4B, respectively.

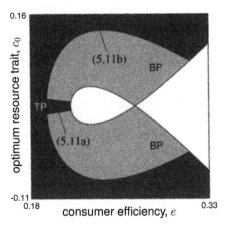

Figure 5.6 In the shaded regions (extracted from Figure 5.3) the AD canonical equation
(5.10) has only one stable evolutionary equilibrium, which is either a branching
point (BP) for the resource population or a terminal point (TP).

easily done by finding out which one of the branching conditions

$$\frac{\partial^2}{\partial x_i \partial x_i'} \lambda_i(x_1, x_2, x_i') \Big|_{\substack{x_1 = \bar{x}_1 \\ x_2 = \bar{x}_2 \\ x_i' = \bar{x}_i}} < 0, \tag{5.11a}$$

$$\frac{\partial^2}{\partial x_i'^2} \lambda_i(\bar{x}_1, \bar{x}_2, x_i') \Big|_{x_i' = \bar{x}_i} > 0 \tag{5.11b}$$

(see (3.41) in Chapter 3), $i = 1, 2$, are satisfied. In all the numerical experiments
we have performed, consumer mutants do not invade, so that there are only two
possibilities: $(\bar{x}_1, \bar{x}_2)$ is a branching point for the resource population or an evolutionarily terminal point. This property is consistent with the well-known principle
of "competitive exclusion" (Hardin, 1960), which states that under general conditions two consumers competing for the same resource cannot coexist at equilibrium. In any case, our findings are perfectly in line with biological observations
and principles, which support the idea that consumers are promoters of resource
species diversity (Brown and Vincent, 1992). Figure 5.6 partly summarizes the result: the shaded areas group regions ①, ③, ⑤, ⑧, ⑨, ⑫ of Figure 5.3 where the
canonical equation has only one stable equilibrium and the lines separating light
(BP) and dark (TP) gray regions are branching bifurcations, where the branching
conditions (5.11) are critically satisfied for the resource (i.e., with $i = 1$). The result is rather interesting if it is complemented with what has already been pointed
out about the disappearance of Red Queen dynamics induced by variations of consumer efficiency. In fact, the overall conclusion is that Red Queen dynamics disappear abruptly if consumer efficiency increases and smoothly if consumer efficiency
decreases. However, in the latter case, as soon as Red Queen dynamics disappear
(transition from region ② to region ① of Figure 5.3), evolutionary branching can
occur in the resource population (transition from the white to the BP region of Figure 5.6). Thus, environmental drifts of any sign can give rise to discontinuities in

the evolutionary dynamics of the resource-consumer community. This observation proves once more that evolution is an astonishingly complex dynamical process.

5.4 DISCUSSION AND CONCLUSIONS

The problem of resource-consumer coevolution has been investigated in this chapter from a purely mathematical point of view. For this, the classic Rosenzweig-MacArthur model (logistic resource and consumer with saturating functional response) has first been transformed into a resident model by considering two adaptive traits, one characterizing resource intraspecific competition and one characterizing consumer attack rate and handling time. Then, two three-dimensional resident-mutant models have been formulated by adding the equation for the resource and consumer mutant populations, respectively, and the AD canonical equation describing the evolutionary dynamics of the two traits has been derived.

The bifurcation analysis of the AD canonical equation has shown that the dynamics of the traits on the evolutionary timescale are much more complex than the dynamics of the populations on the demographic timescale. The numerically produced bifurcation diagrams have proved to be powerful tools for extracting qualitative information on the impact of various factors on coevolution. Conclusions like those we have obtained on the impact of environmental drifts on evolutionary cycles (Red Queen dynamics) cannot be derived without performing a complete bifurcation analysis of the evolutionary model. A general encouraging message emerging from this study is that other important biological problems, such as the evolution of mutualism, cannibalism, and parasitism, could most likely be studied successfully through the bifurcation analysis of the AD canonical equation.

Limiting the discussion to the problem of resource-consumer coevolution, we can say that the results presented in Dercole et al. (2003) (and summarized in this chapter) are by far much more complete than those available before. Indeed, the only comparable result is the bifurcation analysis presented in Dieckmann et al. (1995), where a bifurcation diagram similar to that of Figure 5.4A was obtained through simulation. That bifurcation diagram is incomplete and derived for a quite degenerate case, i.e., for a Lotka-Volterra model ($\tau_0 = 0$ in our model) with a very special parameter combination reducing the number of bifurcation curves to six. However, despite this double degeneracy, the analysis in Dieckmann et al. (1995) points out Red Queen dynamics, multiple evolutionary attractors, and the evolutionary murder of the consumer. A comparison with the nonmathematical literature (see, e.g., Pimentel, 1961, 1968; Abrams, 2000) is also quite favorable, because some of our general results, such as the role of consumers in promoting resource species diversity, are emerging also from empirical studies (Chown and Smith, 1993; van Damme and Pickford, 1995; Walker, 1997; Stone, 1998).

Even if what we have presented in this chapter might seem rather general, there are a number of possible interesting extensions. First, one could investigate the dynamics of more complex population assemblies, composed, for example, of one resident consumer population and two resident resource populations. The outcome of such a study could be that consumer branching is possible, because this is not in conflict with the principle of competitive exclusion. Second, one could be interested in detecting the resource-consumer evolutionary dynamics under the assumption that the two populations can coexist by cycling at ecological timescale. This extension is absolutely not trivial, because the derivation of the evolutionary model is rather difficult in this case. However, the problem is of great interest because its analysis could perhaps help to answer the very intriguing question: does evolution destabilize populations? For this reason, we will return to this problem in Chapter 9. Third, one could be interested in extending the analysis to the coevolution of tritrophic food chains composed of a resource, a consumer, and a predator population. From the results obtained in this chapter (i.e., evolutionary dynamics of ditrophic food chains are much more complex than the corresponding demographic dynamics), one should naturally be inclined to conjecture that chaotic evolutionary dynamics should be possible in tritrophic food chains. The validity of this conjecture is proved in Chapter 10.

Chapter Six

Catastrophes of Evolutionary Regimes

We show in this chapter that the set of unviable evolutionary states, namely the set of all ancestral conditions giving rise to evolutionary trajectories leading to the extinction of one or more populations, depends discontinuously upon demographic and environmental parameters. In particular, we show that the discontinuities of the unviable set are produced by catastrophes of evolutionary attractors, namely by bifurcations at which microscopic parameter perturbations beget transients toward macroscopically different evolutionary regimes. Not all catastrophes produce discontinuities of the unviable set. However, the discontinuity is guaranteed whenever the catastrophic bifurcation involves the critical values of the adaptive traits at which evolutionary extinction takes place, and the basin of attraction of the evolutionary attractor is not vanishing at the bifurcation. All this has important implications in a conservation perspective and is pointed out by studying a two-species model of obligate mutualism and by focusing on the long-standing puzzle posed by cooperative behaviors: their evolutionary persistence in spite of the fact that slight cheats arising by mutations could gradually erode the cooperative interaction. Parts of this chapter are taken from Ferrière et al. (2002) and Dercole (2005).

6.1 INTRODUCTION

Mutually beneficial interactions between members of different species play a central role in all ecosystems (Boucher et al., 1992; Thompson, 1994; Bronstein, 2001b). Despite the widespread occurrence and obvious importance of mutualistic interactions, the theory of mutualistic coevolution is rather limited (see Kiester et al., 1984; Law, 1985; Frank, 1994, 1996; Law and Dieckmann, 1997), by contrast with the well-developed coevolutionary theory of competition, host-parasite, and prey-predator interactions (surveyed in Roughgarden, 1983a, Frank, 1996, and Abrams, 2000, respectively). This lack of theory has prevented resolution of the most basic and long-standing puzzle posed by mutualisms: their persistence in spite of apparent evolutionary instability. Interspecific mutualisms inherently exhibit conflicts of interest between the interacting species in that selection should favor cheating strategies, which are displayed by individuals that reap mutualistic benefits while providing fewer commodities to the partner species (Axelrod and Hamilton, 1981; Soberon Mainero and Martinez del Rio, 1985; Bull and Rice, 1991; Addicott, 1996). Slight cheats arising by mutation could gradually erode the mutualistic interaction, leading to dissolution or reciprocal extinction (Roberts and Sherratt, 1998; Doebeli and Knowlton, 1998). Although cheating has been often assumed to be limited, recent empirical findings indicate that cheating is rampant in most mutualisms (Poulin and Grutter, 1996; Johnson et al., 1997; Foster and De-lay, 1998; Irwin and Brody, 1998; Addicott and Bao, 1999; Currie et al., 1999); in some cases, cheaters have been associated with mutualisms over long spans of evolutionary time (Pellmyr et al., 1996; Machado et al., 1996; Addicott, 1985). Here, we offer a general explanation for the evolutionary origin of cheaters and the unexpected stability of mutualistic associations subject to cheating, by synthesizing two recent contributions on the subject (Ferrière et al., 2002; Dercole, 2005).

The chapter is organized as follows. In the next section we consider a two-species model of obligate mutualism, where both species cannot survive without the partner's support and are characterized by an adaptive trait measuring their mutualistic attitude. In the third section we present the bifurcation analysis of the corresponding AD canonical equation, pointing out various evolutionary scenarios: from the evolutionary suicide of the entire community to stationary and cyclic evolutionary regimes. Part of this analysis is based on the detection and continuation of so-called extinction bifurcations, which critically involve the boundary of the trait space where evolutionary extinction takes place (see Sections 3.8 and A.7). These bifurcations play a fundamental role in the long-term persistence of populations, since their occurrence can be associated with a discontinuous expansion of the set of ancestral conditions giving rise, in the long run, to population extinction. Finally, in Section 6.4 we show that evolutionary branching may also occur and lead to species polymorphism. As discussed in Chapter 3, once the two-species association has converged to an evolutionary equilibrium, two conspecific populations characterized by slightly different mutualistic attitudes can coexist at evolutionary advantage with respect to intermediate trait values. The two coexisting morphs are therefore under opposite selection pressures and further diverge in their mutualistic attitudes (in sexual species, this requires mechanisms of assortative mating, inducing individuals to mate with phenotypically similar partners, thus inhibiting the generation of intermediate trait values due to population interbreeding; see the discussion in Section 1.7). The simulations performed with an individual-based model (Section 2.3) show that the divergence of the two coexisting morphs is permanent: one of the two populations evolves toward good mutualism while the other evolves toward cheating. Moreover, the simulations show that the new three-population association undergoes repeated branchings, which explains the origin of rich spectra of cheaters in complex mutualistic associations.

6.2 A MODEL FOR THE EVOLUTION OF COOPERATION

Ferrière et al. (2002) offered a general explanation for the evolutionary origin of cheaters and the surprising sustainability of mutualistic associations by assuming a competitive premium for "good mutualists" that provide commodities in large amounts. Provided commodities represent a limited resource for the partner species, so that there is intraspecific competition for commodities (Addicott, 1985; Iwasa et al., 1995; Bultman et al., 2000), and competition in nature is often asymmetric (Brooks and Dodson, 1965; Lawton and Hassell, 1981; Karban, 1986;

Callaway and Walker, 1997); i.e., cheaters or good mutualists are better competitors. Clearly, if any competitive asymmetry were to give advantage to cheaters, there would be no way to sustain mutualistic interactions. However, individuals often discriminate among partners according to the quantity of rewards they provide and associate differentially with higher reward producers (Bull and Rice, 1991; Christensen et al., 1991; Mitchell, 1994; Anstett et al., 1998). Thus, a competitive advantage to good mutualists may explain a richer range of evolutionary outcomes.

Ferrière et al. (2002) analyzed the case of a two-species obligate mutualism (i.e., both species cannot survive without the partner's support; see Doebeli and Dieckmann, 2000, for a nonobligate case) and assumed that each species is initially monomorphic (i.e., composed of a single population of identical individuals) and characterized by a single adaptive trait that measures the rate at which commodities are provided to the partner. Thus, low [high] trait values correspond to cheaters [good mutualists]. Provision of commodities is costly in terms of reproduction or survival, and cheaters incur a reduced cost (Boucher et al., 1992; Maynard Smith and Szathmary, 1995; Herre et al., 1999; Bronstein, 2001a).

Demographic Dynamics

The demographic interactions between species 1 (abundance n_1, mutualistic trait x_1) and species 2 (abundance n_2, mutualistic trait x_2) are described by the following two-dimensional resident model:

$$\dot{n}_1 = n_1 \left(-r_1(x_1) - d_1 n_1 + x_2 n_2 \left(1 - c_1 n_1\right)\right), \tag{6.1a}$$

$$\dot{n}_2 = n_2 \left(-r_2(x_2) - d_2 n_2 + x_1 n_1 \left(1 - c_2 n_2\right)\right). \tag{6.1b}$$

The mutualistic traits x_1 and x_2 are measured as per-capita rates of commodities trading; thus, $x_1 n_1$ and $x_2 n_2$ represent the probabilities per unit of time that one partner individual receives benefit from the mutualistic interaction. Intraspecific competition for commodities provided by the partner species is expressed by the linear density-dependent factors $(1 - c_1 n_1)$ and $(1 - c_2 n_2)$ (Wolin, 1985). The terms $-d_1 n_1$ and $-d_2 n_2$ measure the detrimental effect of intraspecific competition for other resources. The mutualism being obligate, the net rates of increase (per capita), $-r_1(x_1)$ and $-r_2(x_2)$, are negative, and $r_1(x_1)$ and $r_2(x_2)$ increase with x_1 and x_2, respectively, to reflect the direct cost of producing commodities. The functions $r_1(x_1) = r_{11}(x_1 + x_1^2)$ and $r_2(x_2) = r_{21}(x_2 + x_2^2)$ are used to perform the numerical analysis, where r_{11} and r_{21} (first derivatives of r_1 and r_2 at $x_1 = 0$ and $x_2 = 0$, respectively), as well as c_1, c_2, d_1, and d_2, are positive parameters.

The bifurcation analysis of the resident model (6.1), reported in Figure 6.1, shows that the extinction equilibrium $(n_1, n_2) = (0, 0)$ is always locally stable (with respect to the positive quadrant of the (n_1, n_2) plane) and that, depending on the trait values x_1 and x_2, there may also exist two positive equilibria, one stable (denoted by $(\bar{n}_1, \bar{n}_2)$ in the following) and one unstable (a saddle). The transition between the two cases (none or two positive equilibria) is a saddle-node bifurcation (see Section A.8) through which the stable equilibrium and the saddle collide and disappear. Straightforward computations give the condition satisfied by the

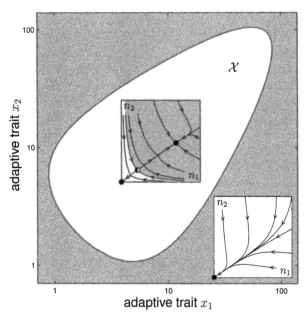

Figure 6.1 Bifurcation analysis of the resident model (6.1) with respect to the mutualistic traits x_1 and x_2 (main panel, logarithmic scale) and corresponding demographic state portraits (subpanels; filled circles: stable equilibria; half-filled circle: saddle; shaded area: basin of attraction of the stable equilibrium $(\bar{n}_1, \bar{n}_2)$). The (saddle-node) bifurcation curve bounds the evolution set $\mathcal{X}$ of the community. For $(x_1, x_2) \notin \mathcal{X}$ the two populations go extinct (bottom-right subpanel). For $(x_1, x_2) \in \mathcal{X}$ the coexistence of the two populations is possible (central subpanel). When the boundary of $\mathcal{X}$ is approached from inside, the stable equilibrium $(\bar{n}_1, \bar{n}_2)$ and the saddle approach each other until they collide. Trait values in subpanels: $x_1 = x_2 = 1$ and $x_1 = x_2 = 1.8$. Parameter values: $c_1 = 2$, $c_2 = 4$, $d_1 = 1$, $d_2 = 2$, $r_{11} = r_{21} = 10^{-3}$.

model parameters at this bifurcation, as well as explicit formulas for $\bar{n}_1(x_1, x_2)$ and $\bar{n}_2(x_1, x_2)$. Specifically, if the stable equilibrium $(\bar{n}_1, \bar{n}_2)$ exists, $\bar{n}_2$ is the larger real solution of

$$An_2^2 + Bn_2 + C = 0, \qquad (6.2)$$

where

$$A = x_1 x_2 c_2 + x_2 c_1 d_2,$$
$$B = -x_1 x_2 - x_1 r_1(x_1) c_2 + x_2 c_1 r_2(x_2) + d_1 d_2,$$
$$C = x_1 r_1(x_1) + d_1 r_2(x_2),$$

i.e. (see (6.1b)),

$$\bar{n}_1(x_1, x_2) = \frac{r_2(x_2) + d_2 \bar{n}_2(x_1, x_2)}{x_1 (1 - c_2 \bar{n}_2(x_1, x_2))}, \qquad (6.3a)$$

$$\bar{n}_2(x_1, x_2) = \frac{-B + \sqrt{B^2 - 4AC}}{2A}, \qquad (6.3b)$$

and the bifurcation condition is the annihilation of the discriminant $B^2 - 4AC$ of
(6.2). The corresponding bifurcation curve in the (x_1, x_2) trait space (the closed
ovoid curve depicted in Figure 6.1) is the boundary of the so-called evolution set $\mathcal{X}$
of the mutualistic community, which is the set of pairs (x_1, x_2) for which the two
partner populations are able to coexist according to model (6.1).

If (x_1, x_2) lies outside $\mathcal{X}$, model (6.1) has no positive equilibria and the mutualistic association goes extinct on the short-term demographic timescale, leaving no room for evolution (see Figure 6.1 bottom-right panel). By contrast, if
$(x_1, x_2) \in \mathcal{X}$, then the two mutualistic partners can coexist at $(\bar{n}_1, \bar{n}_2)$ (see Figure 6.1 central panel), so that the mutation-selection process can drive the evolution
of the adaptive traits. In other words, the evolutionary dynamics of the community
are only defined within the evolution set $\mathcal{X}$. Moreover, if an evolutionary trajectory approaches the boundary of $\mathcal{X}$, the resident demographic equilibrium $(\bar{n}_1, \bar{n}_2)$
becomes closer and closer to the saddle, i.e., closer and closer to the boundary of
its basin of attraction (shaded area in the central subpanel in Figure 6.1). Thus,
when the evolutionary state of the community is close to the boundary of $\mathcal{X}$, ar-
bitrarily small demographic fluctuations result in a demographic state (n_1, n_2) in
the basin of attraction of the extinction equilibrium $(0, 0)$ (white area in the central
subpanel in Figure 6.1), so that the evolutionary suicide of the entire community is
guaranteed (see Sections 1.8 and 3.6).

Evolutionary Dynamics

As we learned in Chapter 3, to construct a mathematical model for the coevolution
of x_1 and x_2, it is assumed that individuals' births, interactions, and deaths described by the resident model (6.1) occur on a short, demographic, timescale over
which the species abundances n_1 and n_2 quickly equilibrate at $(\bar{n}_1, \bar{n}_2)$. Rare and
small mutations in the traits arise on a long, evolutionary, timescale. The evolutionary process comprises a sequence of trait substitutions caused by selection of
successful mutants that win the competition against residents on the demographic
timescale.

To derive the evolutionary dynamics of the traits, one has to extend the resident
model (6.1) by considering the presence of a mutant population, i.e.,

$$\dot{n}_1 = n_1(-r_1(x_1) - d_1(n_1 + n_1') + x_2 n_2 (1 - c_1 n_1 - \gamma_1(x_1' - x_1)n_1')), \quad (6.4\text{a})$$
$$\dot{n}_1' = n_1'(-r_1(x_1') - d_1(n_1 + n_1') + x_2 n_2 (1 - \gamma_1(x_1 - x_1')n_1 - c_1 n_1')), \quad (6.4\text{b})$$
$$\dot{n}_2 = n_2(-r_2(x_2) - d_2 n_2 + (x_1 n_1 + x_1' n_1')(1 - c_2 n_2)), \quad (6.4\text{c})$$

for the case of a mutant trait x_1' with population abundance n_1', and

$$\dot{n}_1 = n_1(-r_1(x_1) - d_1 n_1 + (x_2 n_2 + x_2' n_2')(1 - c_1 n_1)), \quad (6.5\text{a})$$
$$\dot{n}_2 = n_2(-r_2(x_2) - d_2(n_2 + n_2') + x_1 n_1 (1 - c_2 n_2 - \gamma_2(x_2' - x_2)n_2')), \quad (6.5\text{b})$$
$$\dot{n}_2' = n_2'(-r_2(x_2') - d_2(n_2 + n_2') + x_1 n_1 (1 - \gamma_2(x_2 - x_2')n_2 - c_2 n_2')), \quad (6.5\text{c})$$

for the case of a mutant trait x_2' with population abundance n_2'. The resident-mutant
models (6.4) and (6.5) assume that intraspecific competition for commodities provided by the partner species is trait-dependent and described by the functions γ_1

and γ_2. In particular, $\gamma_1(0) = c_1$ and $\gamma_2(0) = c_2$, so that models (6.4) and (6.5) degenerate into the resident model (6.1) if the mutant is absent. Denoting by γ_{11} and γ_{21} the first derivatives of γ_1 and γ_2 at $x'_1 = x_1$ and $x'_2 = x_2$, respectively, parameters γ_{11} and γ_{21} measure the degrees of competitive asymmetry for commodities provided by the partner in species 1 and 2. Positive values of γ_{11} [γ_{21}] reflect a competitive advantage for slightly better mutualistic mutants in species 1 [2], i.e., a premium for providing more commodities; conversely, negative values of γ_{11} [γ_{21}] reflect a competitive advantage for slightly less mutualistic mutants (cheaters); if $\gamma_{11} = 0$ [$\gamma_{21} = 0$] competition is symmetric.

By assuming the timescale separation between demographic and evolutionary processes, and in the limit of infinitesimally small mutations, the AD approach provides a deterministic approximation of the underlying stochastic processes of mutation and selection. The final result is that the adaptive traits x_1 and x_2 vary on the evolutionary timescale in accordance with the following AD canonical equation:

$$\dot{x}_1 = k_1 \bar{n}_1(x_1, x_2) \left. \frac{\partial}{\partial x'_1} \lambda_1(x_1, x_2, x'_1) \right|_{x'_1 = x_1}, \tag{6.6a}$$

$$\dot{x}_2 = k_2 \bar{n}_2(x_1, x_2) \left. \frac{\partial}{\partial x'_2} \lambda_2(x_1, x_2, x'_2) \right|_{x'_2 = x_2}, \tag{6.6b}$$

where $\bar{n}_1(x_1, x_2)$ and $\bar{n}_2(x_1, x_2)$ are given by (6.3), parameters k_1 and k_2 are constant mutational rates (proportional to the frequency and variance of mutations in species 1 and 2), and λ_1 and λ_2 are the so-called mutant invasion fitnesses, defined as per-capita rates of increase from initial scarcity of the mutant populations n'_1 and n'_2 in a resident association (x_1, x_2) settled at the demographic equilibrium $(\bar{n}_1(x_1, x_2), \bar{n}_2(x_1, x_2))$ (see definition (3.19) in Chapter 3).

In formulas:

$$\lambda_1(x_1, x_2, x'_1) = \left. \frac{\dot{n}'_1}{n'_1} \right|_{\substack{n'_1 = 0 \\ n_1 = \bar{n}_1(x_1, x_2) \\ n_2 = \bar{n}_2(x_1, x_2)}}$$
$$= -r_1(x'_1) - d_1 \bar{n}_1(x_1, x_2) + x_2 \bar{n}_2(x_1, x_2) \left(1 - \gamma_1(x_1 - x'_1) \bar{n}_1(x_1, x_2) \right), \tag{6.7a}$$

$$\lambda_2(x_1, x_2, x'_2) = \left. \frac{\dot{n}'_2}{n'_2} \right|_{\substack{n'_2 = 0 \\ n_1 = \bar{n}_1(x_1, x_2) \\ n_2 = \bar{n}_2(x_1, x_2)}}$$
$$= -r_2(x'_2) - d_2 \bar{n}_2(x_1, x_2) + x_1 \bar{n}_1(x_1, x_2) \left(1 - \gamma_2(x_2 - x'_2) \bar{n}_2(x_1, x_2) \right), \tag{6.7b}$$

so that the evolutionary model (6.6) becomes

$$\dot{x}_1 = k_1 \bar{n}_1(x_1, x_2) \left(-\frac{d}{dx_1} r_1(x_1) + \gamma_{11} x_2 \bar{n}_1(x_1, x_2) \bar{n}_2(x_1, x_2) \right), \tag{6.8a}$$

$$\dot{x}_2 = k_2 \bar{n}_2(x_1, x_2) \left(-\frac{d}{dx_2} r_2(x_2) + \gamma_{21} x_1 \bar{n}_1(x_1, x_2) \bar{n}_2(x_1, x_2) \right), \tag{6.8b}$$

for $(x_1, x_2) \in \mathcal{X}$. Along an evolutionary trajectory $(x_1(t), x_2(t))$ of model (6.8) the population abundances n_1 and n_2 track the equilibrium abundances (6.3) corresponding to the current trait values, i.e., $(\bar{n}_1(x_1(t), x_2(t)), \bar{n}_2(x_1(t), x_2(t)))$. If the

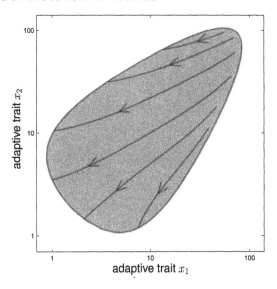

Figure 6.2 Evolutionary dynamics of the mutualistic traits x_1 and x_2 under symmetric competition for commodities provided by the partner species ($\gamma_{11} = \gamma_{21} = 0$, other parameter values as in Figure 6.1). Coevolution is characterized by mutualism disinvestment ($\dot{x}_1$ and $\dot{x}_2$ are negative for all (x_1, x_2) in the evolution set $\mathcal{X}$). The unviable set $\mathcal{U}$ (shaded area) covers the entire evolution set, so that the evolutionary suicide of the community is the final outcome of any ancestral condition.

evolutionary trajectory reaches the boundary of $\mathcal{X}$, both populations undergo evolutionary suicide. The set of all pairs $(x_1, x_2) \in \mathcal{X}$ giving rise, in the long run, to population extinction is called the unviable set of trait space and is denoted by $\mathcal{U}$ in the following. By contrast, the complementary set $\mathcal{X} - \mathcal{U}$ (viable set) identifies the evolutionary states that allow the long-term persistence of interspecific mutualisms (at one of the attractors of model (6.8)).

6.3 CATASTROPHIC DISAPPEARANCE OF EVOLUTIONARY ATTRACTORS

We now present the bifurcation analysis of the AD canonical equation (6.8) with respect to the degrees of competitive asymmetry for commodities γ_{11} and γ_{21} (the other parameters are kept constant at the values reported in the caption of Figure 6.1). In biological terms, the existence of an evolutionary attractor for positive values of γ_{11} and γ_{21} is consistent with the conjecture that a competitive premium for good mutualists is the key for the long-term persistence of interspecific mutualism. In fact, for nonpositive γ_{11} and γ_{21}, the rates of evolutionary change $\dot{x}_1$ and $\dot{x}_2$ given by model (6.8) are negative for all (x_1, x_2) in the evolution set $\mathcal{X}$, so that the evolutionary suicide of both partner populations is the inevitable long-term outcome ($\mathcal{U} = \mathcal{X}$, see Figure 6.2).

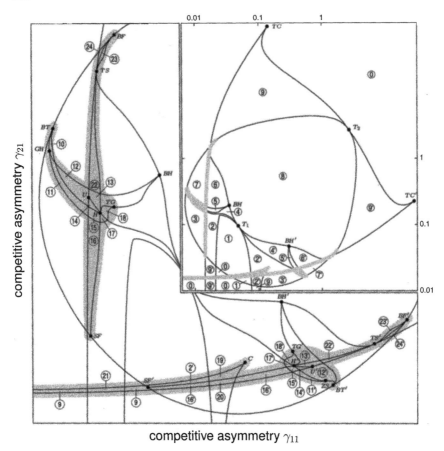

Figure 6.3 Bifurcation diagram of the AD canonical equation (6.8) in the $(\gamma_{11}, \gamma_{21})$ parameter plane (top-right panel) and stretched magnified view (left and bottom parts of the figure) illustrating the bifurcation structures in the two thin, shaded areas. Codimension-1 (curves) and -2 (filled circles) bifurcations are listed in Tables 6.1 and 6.2. Parameter values as in Figure 6.1.

Figure 6.3 shows the bifurcation diagram and unravels 25 different regions ⓪–㉔ in the parameter space $(\gamma_{11}, \gamma_{21})$. The corresponding 25 evolutionary state portraits are shown in Figure 6.2 for region ⓪ and in Figure 6.4 for regions ①–㉔. Figures 6.3 and 6.4 are self-explanatory. Curves and points (filled circles) in Figure 6.3 correspond to codimension-1 and -2 standard or extinction bifurcations, respectively. The type of bifurcation is indicated in Tables 6.1 and 6.2 (see Appendix A and Kuznetsov, 2004, for a detailed treatment). The two thin, shaded areas in the top-right panel hide complex bifurcation structures not visible at the scale of the panel and unraveled in the magnified view (left and bottom parts of the figure), where bifurcation curves are suitably stretched. In Figure 6.4, the evolution set $\mathcal{X}$ is the same in all panels because independent of γ_{11} and γ_{21} (which do not appear in the resident model (6.1)), shaded [white] areas indicate the unviable

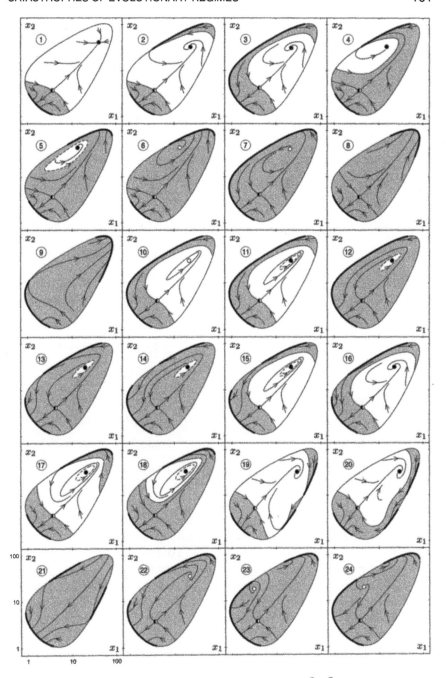

Figure 6.4 Evolutionary state portraits corresponding to regions ①–㉔ of the bifurcation diagram of Figure 6.3. Filled circles: stable evolutionary equilibria; half-filled circles: evolutionary saddles; empty circles: evolutionary repellors; solid [dashed] closed trajectories: stable [unstable] evolutionary cycles; shaded [white] areas: unviable [viable] set $\mathcal{U}$ [$\mathcal{X} - \mathcal{U}$]; thick segments: extinction segments; evolutionary trajectories are stretched for purpose of illustration.

Table 6.1 Codimension-1 bifurcation curves in Figure 6.3.

transition	bifurcation
(0)–(1), (0)–(3), (0)–(7), (0)–(9), (2)–(9'), (9)–(24)	saddle-node
(0)–(9), (1)–(2), (2')–(19), (2')–(20), (3)–(16), (7)–(22),	collision of tangent points
(7)–(24), (9)–(21), (9')–(21), (11)–(15), (12)–(13), (12)–(14),	
(16')–(20)	
(2)–(4), (13)–(14), (17)–(18)	saddle-tangent-point connection (unstable manifold)
(2)–(16), (5)–(13), (6)–(22), (15)–(17), (19)–(20), (23)–(24)	saddle-tangent-point connection (stable manifold)
(2)–(17), (3)–(11), (4)–(18), (15)–(16)	tangent of limit cycles
(3)–(10)	supercritical Hopf (stable cycle)
(3')–(12')	homoclinic (unstable cycle)
(4)–(5)	grazing (unstable cycle)
(4)–(8), (6)–(8)	boundary node/focus
(5)–(6), (7)–(12), (10)–(11), (13)–(22)	subcritical Hopf (unstable cycle)
(5)–(18)	grazing (stable cycle)
(7)–(10), (11)–(12), (14)–(15)	homoclinic (stable cycle)
(8)–(9)	boundary saddle

[viable] set $\mathcal{U}$ [$\mathcal{X} - \mathcal{U}$] corresponding to long-term extinction [persistence] of interspecific mutualism, while thick segments identify the parts of the boundary of the unviable set where extinction occurs (so-called extinction segments).

Notice that the evolutionary state portraits corresponding to regions (i') are not shown since they are almost symmetric copies, with respect to the diagonal $x_1 = x_2$, of those corresponding to regions (i) ($i = 2$–9, 11–18, 22–24; the symmetry, however, is broken because $c_1 \neq c_2$ and $d_1 \neq d_2$). Moreover, some of the regions are characterized by a state portrait, which is a rotated (thus qualitatively equivalent) version of that shown in Figure 6.4. For example, high degrees of competitive asymmetry (high γ_{11} and γ_{21}) yield community suicide due to the continuous increase of the mutualistic traits, i.e., a sort of 180-degree rotation of state portrait (0).

The bifurcation diagram of Figure 6.3 is rather complex and most of the details are confined to narrow regions of the plane $(\gamma_{11}, \gamma_{21})$. Although such regions might broaden when other parameters are varied, their biological relevance seems rather limited. However, the detection of all bifurcations is important because only when

Table 6.2 Codimension-2 bifurcation points in Figure 6.3.

point	bifurcation
BF, BF'	boundary saddle-node
BH, BH'	boundary Hopf
BT, BT'	Bogdanov-Takens (two vanishing eigenvalues)
C, TC, TC'	collision of tangent points cusp
GH	generalized Hopf (supercritical/subcritical transition)
H, H'	grazing homoclinic
SF, SF'	saddle-node-tangent-point connection (stable manifold)
T_1, T_2, TS, TS'	collision of tangent points at boundary equilibrium
TG, TG'	tangent of grazing cycles
U, U'	saddle-tangent-point-collision connection (unstable manifold)
ZS	zero saddle quantity (stable/unstable cycle transition)

the catalog of evolutionary scenarios is complete can one extract from it consistent biological conclusions. For example, the shaded regions of Figure 6.5 collect all pairs $(\gamma_{11}, \gamma_{21})$ for which the AD canonical equation (6.8) is characterized by at least one evolutionary attractor. This occurs in regions ①–⑤, ⑩–⑳, where the viable set $\mathcal{X} - \mathcal{U}$ (white area in the evolutionary state portraits of Figure 6.4) is not empty. The long-term evolutionary regime is stationary for pairs in the light gray area (union of regions ①–⑤, ⑫–⑭, ⑯, ⑲, and ⑳), cyclic for pairs in the black area (region ⑩), and stationary or cyclic for pairs in the dark gray areas (union of regions ⑪, ⑮, ⑰, and ⑱). Thus, Figure 6.5 concisely points out that asymmetric competition for commodities provided by the partner species, with a competitive premium for good mutualists, can indeed explain the evolutionary persistence of interspecific mutualism. However, the evolutionary suicide of the community is always possible for suitable ancestral conditions (shaded areas are present in all panels of Figure 6.4), and this may cause the empirical test of theoretical predictions to be problematic.

Four types of (codimension-1) extinction bifurcations introduced in Section A.7 are present in Figure 6.3, namely

- the collision of an evolutionary equilibrium with the boundary of the evolution set $\mathcal{X}$ (boundary equilibrium bifurcations); see transition from region ④ to region ⑧;

- the collision of an evolutionary cycle with the boundary of $\mathcal{X}$ (grazing bifurcation); see transition from ⑤ to ④;

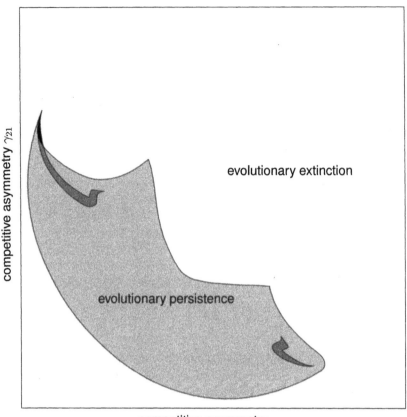

Figure 6.5 Evolutionary persistence of interspecific mutualism. Shaded areas: $(\gamma_{11}, \gamma_{21})$
pairs allowing persistence at an evolutionary attractor of the AD canonical equa-
tion (6.8). The attractor is an equilibrium in the light gray area (union of regions
①–⑤, ⑫–⑭, ⑯, ⑲, and ⑳ of Figure 6.3), a limit cycle in the black area (region ⑩), while both attractors are present in the dark gray areas (union of regions
⑪, ⑮, ⑰, and ⑱).

– the appearance/disappearance of an extinction segment through the collision
  of its delimiting points (so-called tangent points, where the evolutionary trajectory is tangent to the boundary of $\mathcal{X}$); see transition from ① to ②;

– the connection of the stable or unstable manifold of an evolutionary saddle
  with a tangent point; see transition from ② to ④.

Extinction bifurcations are important for understanding the dependence of evolutionary persistence upon parameter variations. For example, starting from a
$(\gamma_{11}, \gamma_{21})$ combination in region ① with mild competitive asymmetry in species
1, should the asymmetry in species 2 become harsher, the evolutionary state portrait will first see the appearance of a new extinction segment in the part of the $\mathcal{X}$-

Table 6.3 Bifurcations of Figure 6.3 involving the discontinuity of region $\mathcal{U}$.

transition	bifurcation
①–⑧	codimension-2 boundary equilibrium (stable node)
⑤–⑱	grazing (stable cycle)
①–⓪, ②–⑨', ③–⓪	saddle-node
⑩–⑦, ⑪–⑫, ⑮–⑭	homoclinic (stable cycle)

boundary characterized by high mutualistic trait values (see transition from panel ① to panel ② in Figure 6.4), with a consequent smooth expansion of the unviable set. Then, increasing further γ_{21}, the unstable manifold of the saddle comes in contact with one of the tangent points (see transition from panel ② to panel ④), giving rise to unviable initial conditions on both sides of the unstable manifold of the saddle. More asymmetry in species 2 would further restrict the viable set to be confined by the unstable evolutionary cycle which grazes the boundary of $\mathcal{X}$ passing from panel ⑤ to panel ④. The unstable cycle then shrinks around the stable evolutionary equilibrium and when the subcritical Hopf bifurcation is reached (transition from panel ⑤ to panel ⑥) the evolutionary persistence of interspecific mutualism is lost.

Figures 6.3 and 6.4 give us the opportunity to discuss a problem of general interest: the relationships between catastrophes of evolutionary attractors, namely bifurcations at which microscopic parameter perturbations beget transients toward macroscopically different evolutionary regimes (or extinction), and the discontinuities of the unviable set $\mathcal{U}$ with respect to parameters. The extent of the unviable set is obviously related to the risk of long-term extinction, so that abrupt expansions/contractions of the unviable set caused by microscopic parameter perturbations are of crucial importance in the planning of conservation strategies.

Our first remark is that there are transitions between various panels of Figure 6.4 that involve the discontinuity of the unviable set $\mathcal{U}$. For example, the transition from panel ① to panel ⓪ involves a huge variation of $\mathcal{U}$, but occurs when the saddle-node bifurcation curve separating the two regions (see Figure 6.3) is crossed by varying the parameters γ_{11} and γ_{21} even of an infinitesimally small amount. The complete list of transitions involving the discontinuity of $\mathcal{U}$ while crossing a bifurcation curve in parameter space is reported in Table 6.3. The list points out that only four bifurcations are responsible for the discontinuities of $\mathcal{U}$ and that they are all catastrophic. This is not a surprise, since a discontinuity of the unviable set requires a discontinuity of the evolutionary regime and, hence, by definition, the catastrophic bifurcation of an evolutionary attractor.

Looking at the problem from the opposite angle, we could first list all catastrophic bifurcations and then check if each one of them is associated with a discontinuity of $\mathcal{U}$. If we do so, we immediately discover that there are catastrophic bifurcations that do not involve a discontinuity of $\mathcal{U}$. For example, when the subcritical (i.e., catastrophic) Hopf bifurcation curve separating regions ⑤ and ⑥ is crossed, the transition is smooth since the viable set (white area in panel ⑤ of Figure 6.4) gradually shrinks to a point. In other words, the unviable set $\mathcal{U}$ becomes larger and larger until it finally covers the entire evolution set $\mathcal{X}$. However, a more systematic analysis would point out that all attractor catastrophes involving the boundary of $\mathcal{X}$ and a nonvanishing basin of attraction are associated with a discontinuity of $\mathcal{U}$ (see the first two rows of Table 6.3). This is exactly what one should a priori expect, since the basin of attraction of an evolutionary attractor, which is part of the viable set, suddenly becomes unviable if the evolutionary attractor disappears through a collision with the boundary of the evolution set $\mathcal{X}$. Indeed, the grazing of a stable evolutionary cycle is always accompanied by such a discontinuous expansion/contraction of the unviable set $\mathcal{U}$, while only codimension-2 boundary equilibrium bifurcations can have the same effect. In fact, when a stable evolutionary equilibrium collides with a tangent point (see again the transition from panel ④ to panel ⑧ in Figure 6.4), its basin of attraction vanishes while approaching the bifurcation, so that the transition is smooth. By contrast, when the boundary collision occurs concomitantly with the collision of two tangent points, as is the case of the codimension-2 bifurcation point T_1 (see the direct transition from panel ① to panel ⑧), the basin of attraction of the equilibrium does not vanish while approaching the bifurcation and suddenly becomes part of the unviable set at the bifurcation.

6.4 EVOLUTIONARY BRANCHING AND THE ORIGIN OF CHEATERS

The bifurcation analysis presented in the previous section has shown that intermediate degrees of asymmetric competition for commodities $(\gamma_{11}, \gamma_{21})$ allow the long-term persistence of interspecific mutualism at a stable evolutionary equilibrium $(\bar{x}_1, \bar{x}_2)$ (the filled circle in the evolutionary state portraits ①–⑤, ⑪–⑳ of Figure 6.4). At attracting evolutionary equilibria, selection may turn disruptive and open the evolutionary route to the coexistence of conspecific individuals characterized by different mutualistic attitudes, ranging from good mutualists that provide large amounts of commodities, to cheaters that are almost purely exploitative. This requires the evolutionary equilibrium to be a so-called branching point for the species under consideration, namely a point in trait space close to which resident and mutant conspecifics coexist while feeling opposite selection pressure.

The mathematical conditions to test if the evolutionary equilibrium $(\bar{x}_1, \bar{x}_2)$ is an evolutionary branching point (BP) or terminal point (TP) were derived in Section 3.7 and involve the second derivatives of the fitness functions (6.7). In particular, $(\bar{x}_1, \bar{x}_2)$ is a BP with respect to species i if

$$\frac{\partial^2}{\partial x_i'^2} \lambda_i(x_1, x_2, x_i') \Big|_{\substack{x_1=\bar{x}_1 \\ x_2=\bar{x}_2 \\ x_i'=\bar{x}_i}} > 0, \qquad (6.9a)$$

$$\frac{\partial^2}{\partial x_i \partial x_i'} \lambda_i(x_1, x_2, x_i') \Big|_{\substack{x_1=\bar{x}_1 \\ x_2=\bar{x}_2 \\ x_i'=\bar{x}_i}} < 0, \qquad (6.9b)$$

while it is a TP if either one or both conditions (6.9) hold with the opposite inequality sign. Straightforward algebra shows that both conditions (6.9) reduce to

$$
-\frac{\gamma_{i2}}{\gamma_{i1}} > \frac{\left.\dfrac{d^2}{dx_i^2}r_i(x_i)\right|_{x_i=\bar{x}_i}}{\left.\dfrac{d}{dx_i}r_i(x_i)\right|_{x_i=\bar{x}_i}},
\tag{6.10}
$$

where γ_{i2} is the second derivative at $x_i' = x_i$ of the function $\gamma_i(x_i' - x_i)$, which describes asymmetric intraspecific competition for commodities in species i, $i = 1, 2$. When condition (6.10) holds with the equality sign, the evolutionary equilibrium $(\bar{x}_1, \bar{x}_2)$ undergoes a so-called branching bifurcation. The corresponding bifurcation curve in the plane $(\gamma_{11}, \gamma_{21})$ is not known analytically, since $\bar{x}_1$ and $\bar{x}_2$ are not known in closed form, but can be easily produced numerically by continuing the parameter combinations that balance the two sides of (6.10).

Geometrically, γ_{i2} measures the curvature of $\gamma_i(x_i' - x_i)$ at $x_i' = x_i$. Positive curvature values mean that the main effect of the asymmetry is to punish cheater individuals that provide fewer commodities, since the deleterious effect $\gamma_i(x_i' - x_i)$ $[\gamma_i(x_i - x_i')]$ on the resident [mutant] population (see equations (6.4a) and (6.5b) [(6.4b) and (6.5c)]) increases more [decreases less] than linearly with x_i'. By contrast, negative curvature values characterize asymmetries which primarily reward good mutualists. Since γ_{i1} is positive (competition for commodities must favor good mutualists to allow the long-term persistence at $(\bar{x}_1, \bar{x}_2)$), as well as $dr_i(x_i)/dx_i = r_{i1}(1 + 2x_i)$ and $d^2r_i(x_i)/dx_i^2 = 2r_{i1}$, evolutionary branching can occur in both species provided the asymmetry of intraspecific competition for commodities is sufficiently rewarding.

When evolutionary branching is ascertained in species i, one should proceed as we did in Chapter 4, i.e., by deriving a new three-dimensional AD canonical equation describing the dimorphic mutualistic species i coevolving with its monomorphic partner. After the branching, the two conspecific mutualistic traits represent individuals of species i with similar but different mutualistic attitudes, which initially evolve in opposite direction, i.e., toward good mutualists and cheaters. Whether the spectrum of coexisting mutualistic attitudes will broaden or not in the course of evolution has to do with the branching properties of the stable evolutionary equilibria of the three- and higher-dimensional AD canonical equations. Instead of deriving and analyzing a sequence of AD canonical equations, here we follow the alternative provided by stochastic individual-based models (see Section 2.3). In contrast with the AD canonical equation, which is based on the separation between the demographic and evolutionary timescales (mutations are extremely small and rare events on demographic timescale), individual-based models more realistically consider the sole demographic timescale, over which each individual is described as an independent unit. In each small time interval $[t,\ t + dt]$ each individual can give rise to a birth, with possible mutation, or die, according to probabilities that scale by the time amount dt the per-capita birth and death rates specified by the resident model (6.1).

More precisely, an individual of species 1, characterized by trait value x_1, gives birth at rate

$$\sum_{k=1}^{I_2} x_{2k} \frac{1}{I} \left(1 - \sum_{j=1}^{I_1} \gamma_1(x_{1j} - x_1) \right),$$

and dies at rate $r_1(x_1) + cI_1/I$, where I_1 and I_2 are the number of individuals in species 1 and 2, x_{1j} and x_{2j} are the individual trait values, $j = 1, \ldots, I_1$, $k = 1, \ldots, I_2$, and I is a constant scaling factor between number of individuals and the population abundances appearing in the deterministic description (6.1). Similar expressions apply to the birth and death rates of species 2 individuals. In both species, offspring typically inherit the parental trait value (clonal reproduction), but mutations occur at small rates (the probability that a newborn is a mutant is set to 10^{-3}). The trait value of a mutant is normally distributed around the parental value (unit variance; negative trait values forced to zero). The functions γ_i, $i = 1, 2$, describing asymmetric competition are

$$\gamma_i(x_i' - x_i) = 2c_i \left(1 - \frac{1}{1 + \gamma_{i3} \exp\left(-\gamma_{i4}(x_i' - x_i)\right)} \right),$$

where parameters γ_{i3} and γ_{i4} can be adjusted to yield the prescribed values of γ_{i1} and γ_{i2}.

Figure 6.6 reports four realizations of the stochastic mutation-selection processes, corresponding to different combinations of parameters c_i, γ_{i1}, γ_{i2}, $i = 1, 2$, controlling intraspecific competition for commodities. At each time, the trait distribution in both species is displayed in scales of gray, where black represents a monomorphic species (the frequency of the corresponding trait value is equal to one) and white indicates the absence of the trait value in the population. Thus, a unimodal distribution well concentrated around its peak value represents a (quasi) monomorphic species and the evolution of the peak value is displayed as a sort of evolutionary time series (see Figure 6.6A, where the monomorphic community converges toward a TP). By contrast, in the polymorphic case, the trait distribution is concentrated around several peaks, one for each morph, and each evolutionary branching adds one morph to the branching species (see Figures 6.6B–D, where repeated branchings occur in species 1 (B), 2 (C), and both (D)).

As discussed in Chapter 2, if the rate and variance of mutations are sufficiently small, the trait distributions remain well concentrated around their peaks (as in Figure 6.6) and the dynamics of the peaks are well approximated by those of the corresponding AD canonical equation. Thus, the monomorphic evolutionary equilibria first reached by the simulations of Figure 6.6 and their branching properties are consistent with the analysis of the two-dimensional AD canonical equation. Indeed, such analysis (see, in particular, the branching condition (6.10)) allowed us to select the four parameter settings of Figure 6.6, yielding no (A), unilateral (B and C), and bilateral (D) branching. However, the derivation and the analysis of the three- and higher-dimensional AD canonical equations would have required a substantial amount of work, while stochastic individual-based simulations provide a quick and low-cost alternative. In particular, the simulations of Figure 6.6 show that when the monomorphic evolutionary equilibrium is a branching point,

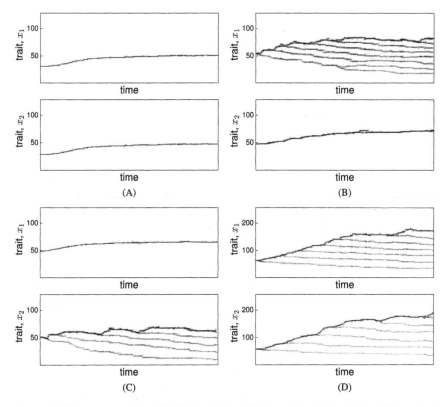

Figure 6.6 Evolutionary branching and the diversification of mutualism. Evolutionary time
series (x_1 and x_2 versus evolutionary time) show the trait distributions (gray-
scale: black, unit frequency of the corresponding trait value; white, absence)
across time. All simulations start with monomorphic initial conditions (x_1, x_2)
in the basin of attraction of the evolutionary equilibrium ($\bar{x}_1, \bar{x}_2$) of the two-
dimensional AD canonical equation (6.8). (A) No branching: $c_1 = 2.08$, $c_2 = 4$,
$\gamma_{11} = 0.04$, $\gamma_{21} = 0.02$, $\gamma_{12} = 0$, $\gamma_{22} = 0$. (B) Branching in species 1:
$c_1 = 1.1$, $c_2 = 4$, $\gamma_{11} = 0.4$, $\gamma_{21} = 0.02$, $\gamma_{12} = -7.2$, $\gamma_{22} = 0$. (C)
Branching in species 2: $c_1 = 2.08$, $c_2 = 2.05$, $\gamma_{11} = 0.04$, $\gamma_{21} = 0.4$, $\gamma_{12} = 0$,
$\gamma_{22} = -9.16$. (D) Branching in both species: $c_1 = 1.1$, $c_2 = 2.05$, $\gamma_{11} = 0.4$,
$\gamma_{21} = 0.4$, $\gamma_{12} = -7.2$, $\gamma_{22} = -9.16$ (reproduced from Ferrière et al., 2002).

the branching species undergoes a cascade of repeating branchings, suggesting that
condition (6.10), regulating the first branching, may also control the branchings to
higher polymorphism. The verification of this conjecture is left to the reader.

6.5 DISCUSSION AND CONCLUSIONS

The theoretical analysis presented in this chapter shows that asymmetrical intraspecific competition for the commodities offered by mutualistic partners provides a
simple and testable ecological mechanism that can account for the long-term per-

sistence of interspecific mutualisms. Cheating establishes a background against which better mutualists can display any competitive superiority. This can lead to the evolutionary coexistence of mutualist and cheater traits, even though natural selection can drive certain ancestral evolutionary states to the evolutionary suicide of the mutualistic association. These results are in agreement with empirical findings indicating that associations of mutualists and cheaters have existed over long spans of evolutionary time (Machado et al., 1996; Pellmyr et al., 1996; Pellmyr and Leebens-Mack, 1999; Després and Jaeger, 1999; Bronstein, 2001a), and that intraspecific competition for commodities is indeed asymmetrical and in favor of good mutualists (Addicott, 1985; Bull and Rice, 1991; Christensen et al., 1991; Mitchell, 1994; Iwasa et al., 1995; Anstett et al., 1998; Bultman et al., 2000).

More precisely, the biological interpretation of the analysis presented in this chapter is the following. First of all, short-term steady coexistence of the mutualistic pair is possible as long as the adaptive mutualistic traits are neither extremely low nor too high. At the boundary of the evolution set of trait space where short-term coexistence is possible (i.e., on the ovoid curve in Figure 6.1) the system undergoes a catastrophic bifurcation (saddle-node) and collapses abruptly. Short-term coexistence alone, however, by no means provides a sufficient condition for the long-term persistence of the mutualism: an evolutionary perspective is mandatory.

The bifurcation analysis of the AD canonical equation reported in Figures 6.3 and 6.4 shows that if resident and mutant individuals compete with equal success for the commodity provided by the other species, or if asymmetrical competition favoring good mutualists is too weak (see bottom-left part of region ⓞ in Figure 6.3), the mutualism erodes because cheating mutants are always able to invade, ultimately driving the partner species to extinction (see Figure 6.1). If the asymmetry is very strong at least in one species (see top and right parts of region ⓞ), the selective pressure favoring the provision of more commodities predominates, causing runaway selection until the costs incurred are so large that extinction is again the inexorable outcome (the corresponding evolutionary state portrait is a sort of 180-degree rotation of that of Figure 6.1). By contrast, at intermediate degrees of competitive asymmetry, the mutualistic association can evolve toward viable stationary or cyclic long-term evolutionary regimes (see regions ①–⑤, ⑩–⑳ in Figure 6.3 and corresponding evolutionary state portraits in Figure 6.4).

Whether stable evolutionary equilibria mark the terminal points of the evolutionary game or are branching points, where a strain of slightly better mutualists and a strain of slight cheaters coexist and start diverging in their mutualistic attitudes, depends on demographic and environmental parameter values (see the branching condition (6.10)). In the previous section we labeled the asymmetry of intraspecific competition for commodities in species i ($i = 1, 2$) as punishing ($\gamma_{i2} > 0$) [rewarding ($\gamma_{i2} < 0$)] if it confers a larger competitive disadvantage [advantage] to cheaters [good mutualists] than the advantage [disadvantage] induced in good mutualists [cheaters]. If we also say that the costs of increasing commodities production are accelerating [decelerating] if $d^2 r_i(x_i)/dx_i^2 > 0$ [< 0], we can interpret condition (6.10) as follows. Evolutionary branching occurs in species incurring a decelerating cost of mutualism if the asymmetry is rewarding or even slightly punishing. In this case, the competitive advantage to a slightly better mutualist is sufficient to overcome the increase in costs it experiences. At the same time, a slightly less mutualistic type can invade a population of better mutualists as long as the competitive disadvantage it suffers is not too large, because of the benefit from reduced costs (this sets a limit on how punishing the asymmetry can be). Likewise, condition (6.10) says that a species characterized by an accelerating cost of mutualism undergoes evolutionary branching only if competitive asymmetry is rewarding. In this case, a slightly less mutualistic type does not gain much through cost reduction and can invade a population of better mutualists only if its competitive disadvantage is small; a slightly better mutualist incurs a relatively large cost and needs a sufficient competitive advantage to invade successfully a population of cheaters. Both punishing and rewarding asymmetry, and accelerating and decelerating costs, appear to exist in mutualisms (Iwasa et al., 1995; Bultman et al., 2000).

The above discussion on the long-term persistence of interspecific mutualisms and on the evolutionary emergence of coexisting mutualistic and cheaters conspecific strategies shows, once more, that bifurcation analysis at various modeling stages (the resident model (6.1), the resident-mutant models (6.4) and (6.5), where the branching conditions (6.9) come from, and the AD canonical equation (6.8)) is a formidable tool from which interesting, sharp, and otherwise hardly justifiable conclusions can be drawn. In particular, the bifurcation analysis of the AD canonical equation allowed us to focus on the microscopic parameter perturbations which cause macroscopic rises or drops in the risk of evolutionary extinction, in terms of macroscopic expansions/contractions of the set of unviable evolutionary states leading, in the long run, to population extinction. Such perturbations must involve the catastrophic bifurcation of an evolutionary attractor whose basin of attraction suddenly becomes part of the unviable set at the bifurcation. When demographic and environmental parameters are close to such bifurcations, wide viable regions of ancestral evolutionary states cannot be associated with low risk of long-term extinction, since small parameter perturbations may radically change the long-term fate of the community.

Chapter Seven

Branching-Extinction Evolutionary Cycles

We show in this chapter that evolving communities can have a quite peculiar evolutionary attractor, called a branching-extinction evolutionary cycle. First, an adaptive trait characterizing a monomorphic species evolves toward a branching point, where the species turns dimorphic by splitting into two resident populations. Then, the two resident traits coevolve until one of the two populations goes extinct. Finally, the remaining population evolves back to the branching point, thus closing the evolutionary cycle. All this is shown by studying the evolution of cannibalistic traits in consumer populations and by focusing on the role of environmental richness. The evolution of "dwarf" (weakly cannibalistic) and "giant" (highly cannibalistic) coexisting morphs and the possibility of periodic evolutionary extinction of highly cannibalistic populations in rich environment is established among other evolutionary scenarios. Parts of this chapter are taken from Dercole and Rinaldi (2002) and Dercole (2003).

7.1 INTRODUCTION

Evolutionary cycles have captured the attention of theoretical ecologists and geneticists in the last decades (see, e.g., Abrams, 1992a; Marrow et al., 1992; Dieckmann et al., 1995; Iwasa and Pomiankowski, 1995, 1999; Marrow et al., 1996; Abrams and Matsuda, 1997; Gavrilets, 1997; see also Chapter 5). In all these cited works, the adaptive traits cyclically vary while the abundances of the resident populations track the demographic equilibrium corresponding to the current trait values. Other kinds of evolutionary cycles involve populations that switch between different demographic attractors, begetting evolutionary reversals (Doebeli and Ruxton, 1997; Khibnik and Kondrashov, 1997; see also Chapter 8), or characterized by wilder demographic dynamics (at least during part of the evolutionary cycle; see Khibnik and Kondrashov, 1997, and Chapter 9). Finally, there is also the possibility of evolutionary cycles due to alternating levels of polymorphism. Such cycles, called *branching-extinction evolutionary cycles* (see Section 1.8), are characterized by recurrent evolutionary branching and extinction, which periodically add and remove a population (or morph) to and from the community.

As we have seen in Chapter 3, at a branching point one of the resident populations, characterized by a particular trait value, coexists with a conspecific population of mutants characterized by a slightly different trait value. The two initially similar traits are under opposite selection pressures, so that the mutant population becomes a new resident population and the polymorphism of the species increases. Moreover, in sexual species, as remarked in Section 1.7, if the force of disruptive selection is strong enough and mechanisms of assortative mating induce individuals to mate with phenotypically similar partners, then interbreeding between the two resident populations ceases and two reproductively isolated subspecies actually evolve. At evolutionary extinction (see Section 1.8) an adaptive trait of a resident population reaches a critical value at which the corresponding population abundance vanishes or catastrophically collapses to zero, thus reducing the species polymorphism or even the number of coevolving species. Thus, in the simplest branching-extinction evolutionary cycle the evolutionary dynamics of a monomorphic species are characterized by a globally stable branching point, and the dimorphic evolutionary trajectory originating at the branching point ends with the evolutionary extinction of one of the two resident populations.

Branching-extinction evolutionary cycles have been observed by several authors (van der Laan and Hogeweg, 1995; Doebeli and Ruxton, 1997; Koella and Doebeli, 1999; Doebeli and Dieckmann, 2000) through stochastic simulations and individual-based models (see Chapter 2). However, it is hard to say if in these cases extinction is produced by demographic stochasticity, modeling accidental extinctions when the population abundances are relatively small, or by the deterministic mechanism of evolutionary extinction. Kisdi et al. (2001) presented a first example of branching-extinction evolutionary cycle where extinction occurs deterministically, but the trait dependence of the demographic parameters of their model is hardly defendable biologically and seems to be adopted simply in view of obtaining a branching-extinction cycle. Moreover, mathematically speaking, the long-term evolutionary behavior obtained by Kisdi et al. (2001) is not captured by a true cycle. In fact, the dimorphic evolutionary trajectory originating at the branching point converges to a critical point in trait space where the two resident populations coexist at one among an infinite number of neutrally stable demographic equilibria, ranging from the extinction of one population to that of the other, with high sensitivity with respect to the dimorphic initial condition. However, at this degenerate demographic equilibrium, small mutations cause the extinction of one of the two populations, while the remaining population evolves back to the branching point. Of course, which population goes extinct at the critical point is a matter of chance and, depending upon this random event, different monomorphic evolutionary transients lead back to the branching point. Thus, a stochastic simulation would show long-term evolutionary dynamics in which two different periods (from the branching point back to it) randomly alternate.

In this chapter we present an example of fully deterministic branching-extinction evolutionary cycle (first discussed in Dercole, 2003) concerning the evolution of cannibalistic traits in consumer populations. We show that a consumer monomorphic species evolves to an intermediate level of cannibalism at which it branches into two initially similar resident populations. Then, assuming that body size of adult individuals and cannibalism are positively correlated (as it is often the case; see Fox, 1975; Polis, 1981, 1988), we show that during the dimorphic evolutionary phase the two resident populations evolve into a weakly cannibalistic "dwarf" population and a highly cannibalistic "giant" population, until the giant population undergoes an evolutionary extinction. As we will see, the key point is that the giant population abundance does not gradually vanish on the evolutionary timescale while approaching extinction, but rather suddenly collapses (a case of so-called evolutionary suicide, as discussed in Sections 1.8 and 3.6). Such a discontinuous extinction event reverses the selection pressure on the dwarf population, which then begins to enhance its cannibalistic attitude.

The chapter is organized as follows. In the next two sections we describe the demographic dynamics of a polymorphic cannibalistic community and sketch the derivation of the corresponding monomorphic and dimorphic AD canonical equations. In Section 7.4 we derive the branching-extinction evolutionary cycle for a particular parameter setting. A discussion of the mechanisms leading to evolutionary attractors characterized by alternating levels of polymorphism and some comments on their robustness close the chapter.

7.2 A MODEL OF CANNIBALISTIC DEMOGRAPHIC INTERACTIONS

Cannibalism, defined as intraspecific predation, is a behavioral trait found in a wide variety of animals, ranging from protozoa and rotifers to birds and mammals (Fox, 1975). The most important studies based on field and laboratory data have been surveyed by Polis (1981, 1988), who has shown that pronounced cannibalism is a frequent feature in species that grow through a wide size range. Often cannibalism develops at a demographic timescale as a reaction of adult individuals to food scarcity (Fox, 1975). However, besides the evidence for dietary induction, several types of data and theoretical studies indicate that, for many species, there is a strong genetic component to cannibalism (see Polis, 1981, and references therein; Henson, 1997; Getto et al., 2005).

Cannibalistic consumer populations naturally call for relatively complex age- and/or size-structured demographic models (see, e.g., Diekmann et al., 1986; van den Bosch et al., 1988; Cushing, 1991; Claessen and de Roos, 2003; Claessen et al., 2004). However, to easily derive the AD canonical equation we use a strongly simplified model. In particular, we hide the size-structure of the population as well as all environmental heterogeneity and seasonalities, which are known to enhance cannibalism in many species (Fox, 1975). Thus, both resident and mutant populations are described with a first-order ODE with constant parameters. Although the model is only a caricature of the real world, it contains the basic ingredients for a sound discussion of the evolutionary emergence of cannibalism. In fact, the cannibalistic predation rate and the searching efficiency of common resources are described by trait-dependent demographic parameters. Moreover, the functional form of the model and the ranges of its admissible parameter values have been carefully selected to fit a paradigmatic case, namely that of the Eurasian perch (*Perca fluviatilis*), described in great detail in Claessen et al. (2000).

Assume that a cannibalistic consumer population is characterized by an adaptive trait x related to its cannibalistic attitude. Since we do not want to refer to a particular species, we cannot specify what x is. However, to facilitate the interpretation of the results, we take the liberty of assuming that the size of adult individuals is positively correlated with the cannibalistic trait. Thus, x can be simply identified with a suitable measure of body size, so that the coexistence of two populations, one with low and one with high cannibalism, should be revealed by the presence of dwarf and giant individuals in the same environment.

We now consider P consumer populations, characterized by abundances n_i and traits x_i, $i = 1, \ldots, P$, and assume that their demographic interactions are described by the following ODE system:

$$
\dot{n}_i = n_i \left(\frac{\sum\limits_{j=0}^{P} e_{ij} a_{ij} n_j}{1 + \sum\limits_{j=0}^{P} a_{ij} \tau_{ij} n_j} - \sum\limits_{j=1}^{P} \frac{a_{ji} n_j}{1 + \sum\limits_{l=0}^{P} a_{jl} \tau_{jl} n_l} - \sum\limits_{j=1}^{P} c_{ij} n_j \right), \quad (7.1)
$$

$i = 1, \ldots, P$, where the index 0 refers to a common resource available to all populations at constant abundance n_0 (from now on called *environmental richness*). The three terms on the right-hand side of equation (7.1) are natality due to food intake, mortality due to cannibalism, and mortality (or reduced natality) due to intraspecific competition. The first term is written in the form of a type II functional response (see Section 3.3) and takes into account that each individual has two alternative food sources: the common resource and its own conspecifics. In the case of the Eurasian perch, the common resource is zooplankton on which all perch feed, at least in the first stages of their life (Holcik, 1977). Thus, rich environments are those in which young perch have more access to food. The parameter e_{ij} is a conversion factor transforming food intake of type j into new biomass of type i, τ_{ij} is the handling time of the ith population associated with the food source of type j, and c_{ij} specifies the strength of intraspecific competition. Although all demographic parameters might depend upon various traits, to obtain a tractable problem we limit the analysis to the case in which the parameters e_{ij} and c_{ij} are constant ($e_{ij} = e$ and $c_{ij} = c$ for all $i, j = 1, \ldots, P$), while the attack rates a_{ij} and the handling times τ_{ij} are trait-dependent. As usual at this stage, other choices would be justifiable.

The attack rate a_{i0} specifies the consumption of the common resource and is assumed to be given by the same bell-shaped function of the trait x_i, for all $i = 1, \ldots, P$, because a consumer performs better when its body size is well tuned with the size of the local resource. In the analysis we use the following function:

$$
a_{i0}(x_i) = a_r(x_i) = \frac{2 a_{r0}}{\left(\dfrac{x_i}{a_{r1}} \right)^{a_{r2}} + \left(\dfrac{a_{r1}}{x_i} \right)^{a_{r2}}}, \quad (7.2)
$$

where a_{r0} is the maximum attack rate, achieved for $x_i = a_{r1}$, and a_{r2} controls the sharpness of the bell shape.

As for the cannibalistic attack rate a_{ij}, we assume it is shaped as in Figure 7.1. Along each ray at constant x_j / x_i ratio, the attack rate is bell-shaped and vanishes for x_i tending to both zero and infinity. Similarly, a_{ij} is a bell-shaped function of x_j / x_i, since the predation rate is higher when the body size of the victim is in a suitable ratio with that of the predator.

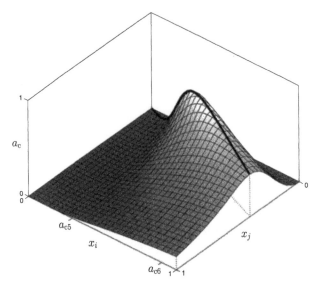

Figure 7.1 The cannibalistic attack rate a_{ij} as a function of the traits x_i and x_j (see equation (7.3)). The thick line indicates the restriction of a_{ij} on the ray $x_j = a_{c1}x_i$. Parameter values: $a_{c0} = 1$, $a_{c1} = 0.4$, $a_{c2} = 2$, $a_{c3} = 4$, $a_{c4} = 2$, $a_{c5} = 0.3$, $a_{c6} = 0.9$.

The function we use in our analysis is

$$a_{ij}(x_i, x_j) = a_c(x_i, x_j)$$

$$= \frac{2a_{c0}}{\left(\dfrac{a_{c1}x_i}{x_j}\right)^{a_{c2}} + \left(\dfrac{x_j}{a_{c1}x_i}\right)^{a_{c2}}} \frac{x_i^{a_{c3}}}{a_{c5}^{a_{c3}} + x_i^{a_{c3}}} \left(1 - \frac{x_i^{a_{c4}}}{a_{c6}^{a_{c4}} + x_i^{a_{c4}}}\right), \quad (7.3)$$

where a_{c0} is the maximum attack rate, $a_{c1} < 1$ is the optimum victim-predator body size ratio, $a_{c2} > 1$, $a_{c3} > 1$, $a_{c4} > 1$ control the sharpness of the bell shapes, while a_{c5} and a_{c6} are sort of thresholds indicating the body sizes at which cannibalism becomes physiologically significant and limited by habitat morphology, respectively (see Figure 7.1). To allow the survival of populations with negligible cannibalism ($x_i < a_{c5}$), we assume in the following $a_{r1} < a_{c5}$. Moreover, small values of a_{c2} imply high values of the cannibalistic attack rate a_{ii} (see equation (7.3) with $x_i = x_j$), i.e., great possibilities for individuals of trait x_i to predate individuals of the same trait. In the real world such a population would be characterized by a substantial change in size from juvenile to adult, so that adult individuals can easily predate young ones (Polis, 1981, 1988). For this reason the parameter $(1/a_{c2})$ is a sort of surrogate for the size range of the individuals in the population and will, indeed, be called *size range* in the following.

Finally, the handling times τ_{ij}, which can be estimated from feeding experiments performed under excessive food conditions (Byström and Garcia-Berthóu, 1999), are assumed to depend mainly upon the consumer trait x_i through the function

$$\tau_{ij}(x_i) = \tau(x_i) = \tau_1 x_i^{-\tau_2},$$

taken from Claessen et al. (2000).

7.3 COEVOLUTION OF DWARFS AND GIANTS

We now study the evolution of cannibalism starting from a single monomorphic consumer species, i.e., a single resident population characterized by abundance n and trait x. As we will see, the species can turn dimorphic through evolutionary branching, so that weakly cannibalistic dwarfs and highly cannibalistic giants coevolve during the dimorphic evolutionary phase.

Monomorphic Evolutionary Dynamics

Using model (7.1), with $P = 1$, one can easily show that its equilibrium $\bar{n}(x)$ is unique, stable, and given by the positive root of the second-order equation

$$
\begin{aligned}
0 = {} & ca_c(x, x)\tau(x)\bar{n}(x)^2 \\
& + \left(c(1 + a_r(x)\tau(x)n_0) + (1 - e)a_c(x, x) \right)\bar{n}(x) \\
& - ea_r(x)n_0.
\end{aligned}
$$

Then, consistently with the AD approach, we assume that the resident population is settled at its equilibrium $\bar{n}(x)$ when a mutant appears, that the trait x' of the mutant is only slightly different from x, and that the mutant population abundance n' is initially very small. Under these conditions, model (7.1), with $P = 2$, can be written in the standard form of a resident-mutant model:

$$
\begin{aligned}
\dot{n} &= nf(n, n', x, x'), \\
\dot{n}' &= n'f(n', n, x', x),
\end{aligned}
$$

from which, in accordance with definition (3.19) in Chapter 3, the invasion fitness of the mutation is obtained as

$$
\begin{aligned}
\lambda(x, x') &= f(0, \bar{n}(x), x', x) \\
&= \frac{e(a_r(x')n_0 + a_c(x', x)\bar{n}(x))}{1 + a_r(x')\tau(x')n_0 + a_c(x', x)\tau(x')\bar{n}(x)} \\
&\quad - \frac{a_c(x, x')\bar{n}(x)}{1 + a_r(x)\tau(x)n_0 + a_c(x, x)\tau(x)\bar{n}(x)} - c\bar{n}(x).
\end{aligned}
$$

The resulting one-dimensional AD canonical equation is therefore given by

$$
\dot{x} = k\bar{n}(x)\left.\frac{\partial}{\partial x'}\lambda(x, x')\right|_{x'=x}, \tag{7.4}
$$

where k is a constant parameter proportional to the frequency and variance of small mutations (the so-called mutational rate).

Equation (7.4) always admits the trivial solution $x = 0$, because both $\bar{n}(x)$ and $\partial\lambda/\partial x'|_{x'=x}$ are zero for $x = 0$ (as one can easily see by taking into account that $a_r(x)$ vanishes with x, see equation (7.2)). Moreover, the trivial solution $x = 0$ is always unstable (i.e., $\dot{x} > 0$ for small $x > 0$), since $\bar{n}(x)$ and $\partial\lambda/\partial x'|_{x'=x}$ are positive for small and positive values of x. In generic conditions, the nontrivial equilibria of (7.4) are either one or three, as shown in Figure 7.2 for three different combinations (see caption) of environmental richness (n_0) and size range ($1/a_{c2}$).

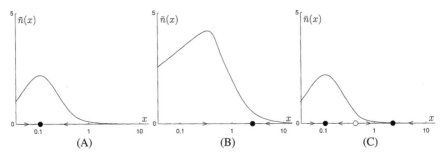

Figure 7.2 The equilibrium abundance $\bar{n}(x)$ of the resident population and the monomorphic evolutionary dynamics (on the horizontal axis in logarithmic scale), where
circles indicate evolutionary equilibria (filled [empty] circles: stable [unstable]
equilibria). (A) Evolution toward a weakly cannibalistic population of dwarfs
($n_0 = 10$, $a_{c2} = 2.5$). (B) Evolution toward a highly cannibalistic population of giants ($n_0 = 500$, $a_{c2} = 1.5$). (C) Evolutionary bistability ($n_0 = 10$,
$a_{c2} = 1.5$). Other parameter values: $c = 1$, $e = 0.6$, $a_{r0} = 1$, $a_{r1} = 0.1$,
$a_{r2} = 2$, $a_{c0} = 10$, $a_{c1} = 0.2$, $a_{c3} = 8$, $a_{c4} = 2$, $a_{c5} = 0.5$, $a_{c6} = 5$, $\tau_1 = 0.1$,
$\tau_2 = 0.25$.

In the case of Figure 7.2C, two stable evolutionary equilibria $\bar{x}^{(1)}$ and $\bar{x}^{(2)}$ (filled
circles on the x axis) are separated by an unstable equilibrium $\bar{x}^{(u)}$ (empty circle),
so that cannibalism can evolve toward either a low value $\bar{x}^{(1)}$, corresponding to a
dense population of dwarfs, or a high value $\bar{x}^{(2)}$, corresponding to a scarce population of giants. In the other two cases there is only one stable equilibrium: a low
trait value ($\bar{x}^{(1)}$) with high population abundance in case A, and a high trait value
($\bar{x}^{(2)}$) with low population abundance in case B. The transition from C to A [B] corresponds to a saddle-node bifurcation (see Appendix A) of the canonical equation
(7.4), characterized by the collision of $\bar{x}^{(u)}$ with $\bar{x}^{(2)}$ [$\bar{x}^{(1)}$].

Once monomorphic evolutionary dynamics have found a halt at a stable equilib-
rium ($\bar{x}^{(1)}$ or $\bar{x}^{(2)}$), one must establish if the equilibrium is a terminal point (TP) of
the mutation-selection process or a branching point (BP). For this, we use the test
presented in Chapter 3 (Section 3.7), i.e.,

$$\frac{\partial^2}{\partial x'^2} \lambda(\bar{x}^{(i)}, x') \bigg|_{x'=\bar{x}^{(i)}} \begin{cases} < 0, & \bar{x}^{(i)} \text{ is a TP} \\ > 0, & \bar{x}^{(i)} \text{ is a BP,} \end{cases} \quad i = 1, 2.$$

For example, the low equilibria $\bar{x}^{(1)}$ in Figure 7.2 are TPs, while the high equilibria
$\bar{x}^{(2)}$ are BPs, but other combinations are possible for other values of environmental
richness (n_0) and size range ($1/a_{c2}$).

The study of monomorphic evolutionary dynamics is completed by performing
the bifurcation analysis of the canonical equation (7.4) with respect to (n_0) and
($1/a_{c2}$), thus producing the diagram shown in Figure 7.3. The two saddle-node
bifurcation curves merging at the cusp point C are the combinations of parameter
values ($n_0, 1/a_{c2}$) for which the unstable equilibrium $\bar{x}^{(u)}$ collides with either $\bar{x}^{(1)}$
or $\bar{x}^{(2)}$. By contrast, the remaining curve represents the ($n_0, 1/a_{c2}$) combinations
separating $\bar{x}^{(2)}$-terminal points from branching points (branching bifurcation). The
parameter space ($n_0, 1/a_{c2}$) is subdivided into four regions, characterized by dif-

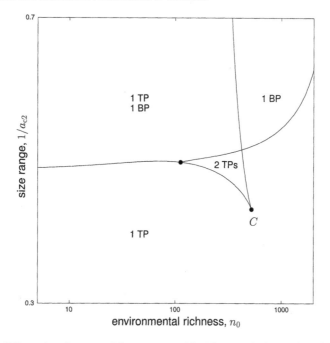

Figure 7.3 Bifurcation diagram of the monomorphic AD canonical equation (7.4) with respect to n_0 and $1/a_{c2}$. The curves identify four regions characterized by one or two stable evolutionary equilibria, which can be either evolutionarily terminal points (TP) or branching points (BP). Parameter values as in Figure 7.2.

ferent mixes of terminal and branching points. In particular, Figure 7.3 shows that in poor environments a TP always exists and that species dimorphism is a possible evolutionary option only in populations with wide size range. Moreover, dimorphism is the only option in populations with wide size range living in very rich environments.

Dimorphic Evolutionary Dynamics

We now focus on the coevolution of two coexisting populations, characterized by abundances n_1 and n_2 and cannibalistic traits x_1 and x_2. Recall that in sexual species, as is the Eurasian perch, which motivated the present study, dimorphic evolutionary dynamics describe the coevolution of two reproductively isolated resident populations, so that the analysis that follows assumes that individuals mate with partners of similar body size, so that population interbreeding practically stops as soon as the two traits x_1 and x_2 diversify.

Of course, the study of dimorphic evolutionary dynamics must be limited to the region of all pairs (x_1, x_2) for which model (7.1), with $P = 2$, has a stable and strictly positive equilibrium, i.e., to the so-called evolution set of the dimorphic community. Such a region can be computed by performing the bifurcation analysis of model (7.1) ($P = 2$) with respect to the traits x_1 and x_2 interpreted as constant

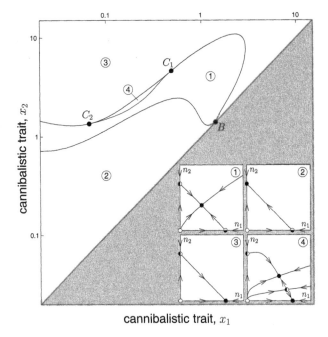

Figure 7.4 Bifurcation analysis of model (7.1) ($P = 2$) with respect to cannibalistic traits x_1 and x_2. Upper triangle: bifurcation diagram and regions ①–④ (filled circles indicate codimension-2 (cusp) bifurcation points; in particular, point B corresponds to the monomorphic branching point). Lower triangle: demographic state portraits of model (7.1) ($P = 2$) corresponding to regions ①–④ (filled circles: stable equilibria; half-filled circle: saddles; empty circles: repellors). The evolution set of the dimorphic community is the union of regions ① and ④ where demographic state portraits display a strictly positive equilibrium. Parameter values as in Figure 7.2 except for $n_0 = 500$ and $a_{c2} = 1.9$. Region ④ has been stretched for purpose of illustration.

parameters. Since the evolutionary trajectories in the space (x_1, x_2) are symmetric with respect to the diagonal $x_2 = x_1$, we limit the analysis to the region $x_1 < x_2$ and call populations 1 and 2 dwarf and giant populations, respectively. An example of a dimorphic evolution set obtained for wide size range and rich environments (i.e., for high values of environmental richness, n_0, and size range, $1/a_{c2}$, for which monomorphic evolutionary dynamics halt at a globally stable branching point, see Figure 7.3) is shown in Figure 7.4. Four qualitatively different demographic behaviors are identified (see the demographic state portraits in the lower part of the figure), corresponding to regions ①–④ of the bifurcation diagram (upper part of the figure). Since only in regions ① and ④ is there a stable and strictly positive demographic equilibrium, the dimorphic evolution set is the union of regions ① and ④.

The nature of a bifurcation curve separating two nearby regions in Figure 7.4 can be understood by comparing the two corresponding demographic state portraits.

For example, the bifurcation curve separating region ① from region ③ is characterized by the collision of a stable and strictly positive equilibrium with a saddle on the n_1-axis (transcritical bifurcation, see Appendix A). Thus, if a dimorphic evolutionary trajectory in region ① moves toward this boundary of the evolution set, the giant population vanishes and goes extinct when the trajectory hits the bifurcation curve (a case of so-called evolutionary murder, see Sections 1.8 and 3.6: what is ultimately responsible for the giants' evolutionary extinction is the evolution of the dwarf population). By contrast, the bifurcation curve separating region ③ from region ④ is characterized by the collision of a stable and strictly positive equilibrium with a strictly positive saddle (a saddle-node bifurcation that leaves the stable equilibrium on the n_1-axis as the only demographic attractor). Thus, if a dimorphic evolutionary trajectory in region ④ moves toward this boundary of the evolution set, the giant population abundance does not vanish, but catastrophically collapses (on the demographic timescale) at the bifurcation (evolutionary suicide: the giant population evolves to its self-destruction).

Let us now denote by $\bar{n}_1(x_1, x_2)$ and $\bar{n}_2(x_1, x_2)$ the resident population abundances at the stable and strictly positive equilibrium of model (7.1) ($P = 2$), for (x_1, x_2) in the dimorphic evolution set. The dimorphic AD canonical equation then reads

$$\dot{x}_1 = k\bar{n}_1(x_1, x_2) \left. \frac{\partial}{\partial x_1'} \lambda_1(x_1, x_2, x_1') \right|_{x_1' = x_1}, \qquad (7.5a)$$

$$\dot{x}_2 = k\bar{n}_2(x_1, x_2) \left. \frac{\partial}{\partial x_2'} \lambda_2(x_1, x_2, x_2') \right|_{x_2' = x_2}, \qquad (7.5b)$$

where $\lambda_i(x_1, x_2, x_i')$ is the fitness of the ith mutant population (abundance n_i', trait x_i'), $i = 1, 2$, which can be easily obtained from model (7.1) with $P = 3$. In fact, by writing model (7.1) ($P = 3$) in the form

$$\dot{n}_1 = n_1 f_1(n_1, n_2, n_i', x_1, x_2, x_i'),$$
$$\dot{n}_2 = n_2 f_2(n_1, n_2, n_i', x_1, x_2, x_i'),$$

plus

$$\dot{n}_1' = n_1' f_1(n_1', n_2, n_1, x_1', x_2, x_1)$$

for $i = 1$ and

$$\dot{n}_2' = n_2' f_2(n_1, n_2', n_2, x_1, x_2', x_2)$$

for $i = 2$, it follows that

$$\lambda_1(x_1, x_2, x_1') = f_1(0, \bar{n}_2(x_1, x_2), \bar{n}_1(x_1, x_2), x_1', x_2, x_1),$$
$$\lambda_2(x_1, x_2, x_2') = f_2(\bar{n}_1(x_1, x_2), 0, \bar{n}_2(x_1, x_2), x_1, x_2', x_2).$$

Notice that the resident demographic equilibrium $(\bar{n}_1(x_1, x_2), \bar{n}_2(x_1, x_2))$ is not known in closed form, but only implicitly through its definition

$$f_1(\bar{n}_1(x_1, x_2), \bar{n}_2(x_1, x_2), 0, x_1, x_2, \cdot) = 0,$$
$$f_2(\bar{n}_1(x_1, x_2), \bar{n}_2(x_1, x_2), 0, x_1, x_2, \cdot) = 0$$

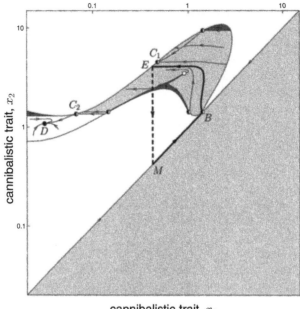

cannibalistic trait, x_1

Figure 7.5 Dimorphic evolutionary dynamics obtained through the numerical integration of the differential-algebraic system (7.6) (circles indicate dimorphic equilibria) and the branching-extinction evolutionary cycle (thick trajectory). Parameter values as in Figure 7.4.

(where the dot ($\cdot$) stands for any trait value), so that dimorphic evolutionary trajectories are actually defined by the following differential-algebraic system:

$$\dot{x}_1 = k n_1 \left. \frac{\partial}{\partial x_1'} f_1(0, n_2, n_1, x_1', x_2, x_1) \right|_{x_1' = x_1}, \tag{7.6a}$$

$$\dot{x}_2 = k n_2 \left. \frac{\partial}{\partial x_2'} f_2(n_1, 0, n_2, x_1, x_2', x_2) \right|_{x_2' = x_2}, \tag{7.6b}$$

$$0 = f_1(n_1, n_2, 0, x_1, x_2, \cdot), \tag{7.6c}$$

$$0 = f_2(n_1, n_2, 0, x_1, x_2, \cdot). \tag{7.6d}$$

7.4 THE BRANCHING-EXTINCTION EVOLUTIONARY CYCLE

The dimorphic evolutionary dynamics defined by model (7.5) within the evolution set shown in Figure 7.4 (the union of regions ① and ④) are sketched in Figure 7.5. The evolution set is partitioned into white, light gray, and dark gray regions. Trajectories starting in the white region tend toward a dimorphic evolutionary equilibrium D (which can be either a branching or terminal point, but this distinction is not investigated here). Trajectories starting in the light gray region hit the boundary of the evolution set between points C_1 and C_2, where the evolutionary suicide of the giant population takes place (see Figure 7.4). Notice that points C_1 and C_2 are evolutionary equilibria (saddles). In fact, $\bar{n}_2(x_1, x_2) = 0$ (i.e., $\dot{x}_2 = 0$) at such points (they lie on the intersection of regions ①, ③, and ④, see corresponding demographic state portraits in Figure 7.4) and $\dot{x}_1$ has opposite sign at opposite sides of C_1 and C_2 along the boundary of the evolution set. Finally, dark gray regions are those in which the dwarf [giant] population abundance smoothly vanishes when the evolutionary trajectory approaches the extinction boundary separating region ② [③] from region ① (see Figure 7.4).

Since the branching point B, where dimorphism originates, lies on the boundary of the light gray region, the evolutionary attractor of the community is the branching-extinction evolutionary cycle represented in Figure 7.5 by the thick trajectory. In words, when the dwarf and giant cannibalistic attitudes become sufficiently distinct, the giant population is not capable of sustaining itself by harvesting on the dwarf population. However, more cannibalistic mutants are at advantage and drive the population to self-extinction (see point $E \equiv (x_1^{(E)}, x_2^{(E)})$ in Figure 7.5). After the giant extinction (i.e., after the sudden (dashed) transition from E to M in Figure 7.5) the dwarf population evolves back to the branching point B, starting with a trait value $x_1 = x_1^{(E)}$, in accordance with the monomorphic canonical equation (7.4). Thus, starting from any ancestral monomorphic condition the final outcome of the mutation-selection process is the branching-extinction evolutionary cycle of Figure 7.5, characterized by two distinct evolutionary phases: a monomorphic evolution toward the branching point (from M to B) and a dimorphic evolution characterized by the temporary presence of a highly cannibalistic population of giants (from B to E).

7.5 DISCUSSION AND CONCLUSIONS

An evolutionary cycle characterized by alternating levels of polymorphism has been shown to be the evolutionary attractor of cannibalistic consumer populations with wide size range living in rich environments. The deterministic mechanisms that lead to such evolutionary cycles have been first addressed by Kisdi et al. (2001) and require the following minimal ingredients: (i) a monomorphic species that can turn dimorphic through an evolutionary branching; (ii) a dimorphic evolutionary phase originating at the branching point that leads to the evolutionary extinction of one of the two morphs, say morph 2; (iii) a postextinction monomorphic species (i.e., morph 1) in the basin of attraction of the branching point.

In the present study, condition (iii) forces the rate of evolutionary change of trait 1 ($\dot{x}_1$) to reverse in the transition from dimorphism to monomorphism. This is not possible if population 2 is murdered by population 1. In fact, in such a case, the abundance $\bar{n}_2(x_1, x_2)$ vanishes while approaching the boundary of the evolution set (transcritical bifurcation of the dimorphic demographic equilibrium), so that only population 1 is actually present at the boundary. This implies that just before and just after the extinction of population 2, population 1 feels the same biotic (and abiotic) environment, hence the same selection pressure, so that the extinction of population 2 cannot trigger a discontinuity of $\dot{x}_1$.

Thus, the key point of our branching-extinction evolutionary cycle is that population 2 undergoes an evolutionary suicide (saddle-node bifurcation of the dimorphic demographic equilibrium). Being present at nonvanishing abundances along the whole dimorphic evolutionary phase, population 2 has a relevant impact on the biotic environment felt by population 1, and its sudden disappearance abruptly changes the selection pressure on population 1. More precisely, in the case of Figure 7.5, just before the extinction of population 2, $\dot{x}_1$ is negative and given by equation (7.5a) evaluated at point E, where $\bar{n}_2(x_1, x_2)$ is strictly positive (and equal to the limit of $\bar{n}_2(x_1, x_2)$ along the dimorphic evolutionary trajectory approaching E). By contrast, after the sudden disappearance of the giant population, $\dot{x}_1$ is positive and given by the monomorphic canonical equation (7.4) evaluated at point M.

Kisdi et al. (2001) have considered a demographic model in which only transcritical bifurcations are possible (they used a bilinear competition model). Thus, to reverse the selection pressure on the postextinction population, they have been forced to consider a quite peculiar situation in which the dimorphic evolutionary trajectories converge to a codimension-2 bifurcation point, namely the point of intersection of two transcritical bifurcation curves. At this codimension-2 bifurcation point the dimorphic demographic equilibrium is neutrally stable and depends sensitively upon the dimorphic initial conditions. Moreover, further mutations determine the random extinction of one of the two populations.

In closing this chapter we would like to comment on the robustness of the result. In principle, a complete bifurcation analysis of the AD canonical equations (7.4) and (7.5) with respect to all couples of strategic parameters (like the environmental richness, n_0, and the size range, $1/a_{c2}$) would allow us to determine all possible qualitative evolutionary scenarios and their corresponding regions in parameter space. In particular, a measure of the region giving rise to branching-extinction evolutionary cycles would be indicative of the robustness of our conclusions. However, the bifurcation analysis of model (7.5) poses nontrivial technical problems, since the dimorphic demographic equilibrium $(\bar{n}_1(x_1, x_2), \bar{n}_2(x_1, x_2))$ is not known in closed form. The differential-algebraic system (7.6) should therefore be used in the analysis, though appropriate numerical techniques for the continuation of global bifurcations (see Appendix A) are not yet fully developed (Ascher and Spiteri, 1994).

A particular global bifurcation indeed characterizes the appearance/disappearance of the branching-extinction evolutionary cycle of Figure 7.5, namely the saddle-saddle connection B–C_2 (so-called heteroclinic bifurcation). In fact, the evolutionary dynamics of the community are trapped by the branching-extinction evolutionary cycle when the unstable manifold of the saddle point B reaches the boundary of the evolution set between points C_1 and C_2. However, if point E were below point C_2, the dimorphic evolutionary equilibrium D would be the evolutionary attractor reached by the community.

Despite technical difficulties, we checked, by means of extensive simulations of model (7.6) for various parameter settings, that the branching-extinction evolutionary cycle of Figure 7.5 is present in a relatively broad region of parameter space characterized by high environmental richness and size range. Moreover, a partial analysis carried out in Dercole and Rinaldi (2002) has shown that stable dimorphic evolutionary equilibria, possibly leading to higher polymorphism, and monomorphic low cannibalistic terminal points, perhaps reached after temporary dimorphic phases, represent alternative evolutionary scenarios when the conditions for branching-extinction evolutionary cycles are not matched.

Thus, our conclusion is that branching-extinction evolutionary cycles are robust evolutionary attractors and that their detection is of crucial importance for fully understanding evolutionary dynamics. Of course, in the case of cannibalistic communities, one should check if evolutionary cycles of the same kind are supported by more realistic age- and/or size-structured demographic models. However, the derivation of the AD canonical equation in such a case would be problematic. Well-organized simulations of a suitable individual-based model (see Section 2.3) may provide the best framework in which to answer this question.

Chapter Eight

Demographic Bistability and Evolutionary Reversals

In this chapter we show that when the evolutionary state of the community does not uniquely determine the resident demographic attractor, mutation-selection processes may force the resident populations to switch between alternative demographic attractors and cause abrupt changes in the selection pressure acting on the community. As a result, evolutionary cycles can develop even in the extreme case of a community composed of a single resident population characterized by a single adaptive trait. Indeed, periodic switches between alternative demographic equilibria may induce the periodic reversal of the rate of evolutionary change and force the trait to endlessly oscillate. We consider a renewable resource characterized by a single adaptive trait affecting both reproduction and competition for nutrients. The mechanism inducing evolutionary reversals is twofold. First, there exist trait values near which mutants can invade and yet fail to become fixed; although these mutants are eventually eliminated, their transitory growth causes the resident population to switch to an alternative demographic equilibrium. Second, asymmetric competition causes the direction of selection to revert between high and low abundances. When the conditions for periodic evolutionary reversals are not satisfied, the population evolves toward a steady state of either low or high abundance, depending on the degree of competitive asymmetry and on demographic and environmental parameters. A sharp transition between evolutionary stasis and cycling can occur in response to a smooth parameter change, and this may have implications for our understanding of size-abundance patterns. Most of this chapter is taken from Dercole et al. (2002a).

8.1 INTRODUCTION

The contrasting patterns of evolution in populations with chronically high or low abundances have captured the shared interests of ecologists and geneticists for a long time (Travis, 1990). There has been a considerable amount of verbal and mathematical theory on the subject (reviewed lucidly by Mueller, 1997). In this context, the distinction between the demographic and evolutionary dynamics and the discussion of their coupling (see Chapter 1, Section 1.5 in particular) emphasize that selective pressures exerted on individuals are shaped by population structure and abundance; in return, demographic dynamics are primarily determined by individual adaptive traits that are genetically based and molded by natural selection (Pimentel, 1968; Heino et al., 1998).

Over the last two decades, there has been intensive empirical investigation of the effect of population density on the nature, direction, and strength of selection on life-history traits. Exemplary studies include those by Wilbur (1984) and Travis et al. (1985) on larval anurans, who showed that the role of predatory salamanders as a selective force for rapid growth of tadpoles varies depending on tadpole abundance: at very low densities this force is weak because growth rates are high and encounter rates with predators are low; a greater numerical impact of predators occurs only at higher tadpole densities.

Although much empirical effort has been devoted to understand how crowding influences patterns of selection on life-history traits (Wilbur, 1984; Travis et al., 1985), the backward effect of selection on population abundances, i.e., the other arch of the demographic-evolutionary feedback, remains poorly understood, even theoretically (Metz et al., 1992; Rand et al., 1994; Heino et al., 1998). One important effect is the propensity to generate complex demographic dynamics, including periodic and chaotic oscillations. Much attention has been paid to the question of whether mutation-selection processes could drive adaptive traits from values leading to stable demographic equilibria, to values corresponding to cycles or chaos (see Gatto, 1993; Ferrière and Gatto, 1993, 1995; Doebeli and Koella, 1995; Ferrière and Fox, 1995; Ebenman et al., 1996; see also Chapter 9); contrasting theoretical predictions are awaiting to be tested empirically. A crucial assumption in all models developed around this issue is that any combination of the traits uniquely determines the demographic attractor of the resident populations (De Feo and Ferrière, 2000). However, demographic models can have alternative attractors associated with different patterns of population abundances that can be reached for the very same trait values (demographic bi- or multistability; see Scheffer et al., 2001, for a survey). How the demographic-evolutionary feedback works in these cases is the question we consider in this chapter.

We focus on a very simple single-population, single-trait model characterized by two alternative stable demographic equilibria, one at high and one at low population abundance. At the former, there is selection for trait values that reduce population abundance, while at the latter there is selection for trait values that increase population abundance. Under suitable conditions that we identify by means of the AD approach, these opposing patterns of selection force the population to switch from the high [low] to the low [high] equilibrium abundance, thus causing a so-called *evolutionary reversal*, namely the sudden change of sign, on the evolutionary timescale, of the trait evolutionary rate of change. A salient result (first noticed by Doebeli and Ruxton, 1997, by relying on alternative stable demographic cycles in spatially structured models) is that there are broad ecological conditions under which periodic evolutionary reversals drive endless trait oscillations (i.e., cyclic Red Queen dynamics, see Section 1.6) of a single adaptive trait.

The chapter is organized as follows. In the next section we give some empirical support to demographic bistability and the biological motivations of our theoretical investigation. Then, in Section 8.3, we describe the model in detail and show that periodic evolutionary reversals indeed occur for suitable parameter settings. This result is reviewed in Section 8.4 through the AD canonical equation, while its discussion is given in the concluding section.

8.2 BIOLOGICAL BACKGROUND

We consider a community composed of a single monomorphic species, namely a renewable resource feeding on constantly available nutrients. The demographic dynamics are determined by a set of fixed ecological (demographic and environmental) parameters, and by life-history parameters that are functions of an adaptive trait. This trait is assumed to be genetically determined and heritable, but it may vary due to mutations of small effects.

We assume that for suitable ecological parameters there can be values of the trait for which the resident resource population is characterized by two alternative stable demographic equilibria. Demographic bistability was first put forward by Holling (1973), Noy-Meir (1975), and May (1977). Subsequently, alternative low- and high-density states have been found in a host of ecological models. Although the first experimental examples that were proposed were criticized strongly (Connell and Sousa, 1983), more recent studies (reviewed in Scheffer et al., 2001) support the view that multiple stable states can characterize natural systems.

Here we consider a simple phenomenological model to generate alternative stable demographic equilibria, akin to the minimal models proposed by Scheffer et al. (2001). One possible interpretation of this model takes into account that mating encounters are more likely in crowded populations. Alternatively, the resource might be harvested by a consumer that exploits a spectrum of other resource species (so-called generalist consumer). The option among various food sources ensures that the consumer demographic and evolutionary dynamics are weakly affected by those of one among many available resources, so that resource-consumer coevolution can be neglected and the abundance of the consumer population considered as constant.

We also assume that the adaptive trait is related to competitive performance, and we call it, for this reason, *competitive ability*. Thus, variation in individual trait results in asymmetric resident-mutant competition for nutrients. Moreover, individuals that are at competitive advantage are nonetheless expected to suffer some physiological cost: this is why competitive ability is assumed to be costly to some components of reproductive success. A common case involving asymmetric intraspecific competition and physiological costs of higher energy demands occurs when competing individuals differ in size. Often, larger individuals enjoy a competitive advantage because of their superior ability at obtaining limited nutrients (Brooks and Dodson, 1965; Wilson, 1975; Persson, 1985; Calder, 1996), but the opposite can also be true, like in the case of small zooplanktivorous fish outcompeting large ones for food (Persson et al., 1998). However, this dichotomy does not influence our findings, since the only assumption that counts is that a positive variation of the trait is accompanied by an advantage in competition (or growth) and by a counterbalancing disadvantage in growth (or competition).

The body size interpretation of competitive ability may be useful because the relationship between body size and population abundance arguably has attracted great attention. Yet the search for general patterns has not been conclusive (Lawton, 1989; Blackburn and Lawton, 1994; Rosenzweig, 1995). Data from extant populations show that low abundance, under some circumstances, is associated with large body size (Brown, 1995; Gaston and Kunin, 1997), but the pattern is far from general. For example, Navarette and Menge (1997) found no tendency for small species to occur at higher densities than larger species in the tropical intertidal communities of Panama, in sharp contrast with the pattern in intertidal communities in temperate Chile (Marquet et al., 1990). Thus, though direct measures of competitive performance are difficult to obtain from real communities, there is hope that our theoretical predictions could be tested against the results of experimental studies (like those reported by Lenski and Bennett, 1993, for bacteria) on the evolution of body size.

8.3 ASYMMETRIC COMPETITION AND THE OCCURRENCE OF EVOLUTIONARY REVERSALS

The Resident Community

We start by considering a logistically growing resource population, described by its abundance n and by the competitive ability x of its individuals, where

$$\dot{n} = r(x)n - cn^2. \qquad (8.1)$$

The net growth rate (per capita) $r(x)$ is assumed to decrease exponentially with competitive ability x (i.e., $r(x) = r_0 \exp(-r_1 x)$, with r_0 and r_1 constant parameters) to express the cost of enhanced competitive performance, while cn measures the extra mortality (or reduced natality) caused by intraspecific competition. Competitive ability x scales between 0 and 1. This scaling can easily be achieved by means of the transformation $x = \log(z/z_{\min}) / \log(z_{\max}/z_{\min})$, where z is the "real" trait value (e.g., body size), and $z_{\min}$ and $z_{\max}$ are minimum and maximum trait values, respectively (Schwinning and Fox, 1995).

Demographic bistability is triggered by discounting the population growth rate in (8.1) by an additional trait-independent demographic factor $H(n)$. As discussed in the previous section, two important examples of such a factor are the reduction of reproduction due to a shortage of mating encounters in sparse population, and the mortality due to the harvesting of a generalist consumer. Both examples can be accounted for by the same discounting function $H(n) = \tau^{-1}n^2/(h^2 + n^2)$ (Dennis, 1989, and references therein; Stephens and Sutherland, 1999), which is called the Holling type III functional response. As the type II functional response $\tau^{-1}n/(h + n)$ introduced in Section 3.3, the discounting function $H(n)$ saturates at τ^{-1} for high abundances and reaches half of its maximum when n matches the so-called half-saturation constant h. However, at low abundances $H(n)$ has a vanishing slope, while the type II functional response has a nonzero initial slope τ^{-1}/h. Thus, the type III response provides a simple way of accounting for spatial heterogeneity in the chance of mating (the probability of mating encounters goes with n^2) and for the fact that a generalist consumer tends to harvest on its most available food sources.

With this choice of $H(n)$, the resident model of the community, therefore becomes

$$\dot{n} = n\left(r(x) - cn - \frac{\tau^{-1}n}{h^2 + n^2}\right). \qquad (8.2)$$

Notice that if body size is a measure of competitive ability and $H(n)$ is consumption rate, it would be more realistic to assume a trait-dependent discounting function, since prey body size typically influences the functional response of the consumer. However, this would give rise to an analytically intractable model even if the final result would be qualitatively the same. In any case, model (8.2) applies when competitive ability does not influence predation or when body size has only a weak influence on the predator functional response.

Further notice that there are two causes of density dependence in (8.2), namely two terms in the population per-capita growth rate (the term in parenthesis) affected by n: a negative density dependence due to intraspecific competition ($-cn$ decreases with n), and a positive density dependence due to the additional discounting factor ($-\tau^{-1}n/(h^2 + n^2)$ increases with n).

Finally, it is also worth noticing that there is no apparent advantage for the population to evolve toward higher competitive abilities, since this would result only in reduced reproduction rates. However, as we will soon see, a more competitive mutant population can be at advantage and replace a less competitive resident population, thus resulting in an overall disadvantage for the population. This clearly points out once more that mutation-selection processes do not imply any sort of benefit or optimization at community level.

Demographic Bistability

As shown in Figure 8.1, there is a lower threshold on competitive ability (x_l) below which the resident model (8.2) has a single stable equilibrium $\bar{n}_h(x)$ at high abundance, and an upper threshold (x_h) above which model (8.2) has a single stable equilibrium $\bar{n}_l(x)$ at low abundance. For x values between x_l and x_h, there are two alternative stable equilibria ($\bar{n}_h(x)$ and $\bar{n}_l(x)$) corresponding to high and low abundance, respectively, separated by an unstable equilibrium at intermediate abundance. Thus, the high-density equilibrium $\bar{n}_h(x)$ is defined only for $x < x_h$, while the low-density equilibrium $\bar{n}_l(x)$ only for $x > x_l$. These two subsets of the one-dimensional trait space therefore identify the so-called evolution sets associated with the possible demographic attractors of the resident community.

The transition from three to one demographic equilibria, which occurs when x is at the threshold value x_l [x_h], corresponds to a saddle-node bifurcation (see Appendix A), through which the low- [high-] density stable equilibrium and the unstable equilibrium collide and disappear. In any trait-parameter plane (see, e.g., the (x, c) plane in Figure 8.2), the two saddle-node bifurcation curves determine the thresholds x_l and x_h for each value of c within an appropriate range. These two curves meet at a cusp point (see Appendix A), which identifies the c value beyond which no demographic bistability can occur. By writing the standard conditions for saddle-node bifurcations, we find

$$x_l = -\frac{1}{r_1} \ln\left(\frac{\bar{n}_l(x_l)}{r_0}\left(c + \frac{\tau^{-1}}{h^2 + \bar{n}_l^2(x_l)}\right)\right), \tag{8.3a}$$

$$x_h = -\frac{1}{r_1} \ln\left(\frac{\bar{n}_h(x_h)}{r_0}\left(c + \frac{\tau^{-1}}{h^2 + \bar{n}_h^2(x_h)}\right)\right), \tag{8.3b}$$

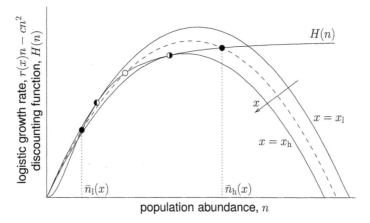

Figure 8.1 Demographic bistability. The equilibria of the resident model (8.2) are given by the intersection of the trait-dependent logistic growth rate $r(x)n - cn^2$ (the family of parabola) and the discounting function $H(n)$. For x between the lower threshold x_l and the upper threshold x_h (see dashed parabola), the resident population has three demographic equilibria: two stable equilibria (filled circles), $\bar{n}_l(x)$ at low abundance and $\bar{n}_h(x)$ at high abundance; and an unstable equilibrium (empty circle). For x at the two critical values x_l and x_h, one of the stable equilibria collides with the unstable equilibrium (see half-filled circles). Parameter values: $r_0 = 1, r_1 = 1, c = 0.1, \tau = 1, h = 0.75$.

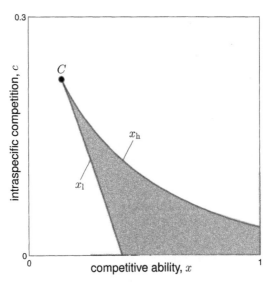

Figure 8.2 Bifurcation diagram of the resident model (8.2). Two saddle-node bifurcation curves meet at the cusp point C. Three equilibria of model (8.2) (two stable, one unstable) are present in the cusp (shaded) region, while a single stable equilibrium is present outside the cusp region. Parameter values as in Figure 8.1.

where

$$\bar{n}_{\mathrm{l}}(x_{\mathrm{l}}) = \left(\frac{\tau^{-1} - 2ch^2 - \sqrt{\tau^{-1}(\tau^{-1} - 8ch^2)}}{2c} \right)^{1/2}, \tag{8.4a}$$

$$\bar{n}_{\mathrm{h}}(x_{\mathrm{h}}) = \left(\frac{\tau^{-1} - 2ch^2 + \sqrt{\tau^{-1}(\tau^{-1} - 8ch^2)}}{2c} \right)^{1/2}. \tag{8.4b}$$

The cusp point is given by

$$x = \frac{1}{r_1} \ln \left(\frac{8r_0 \tau h}{3\sqrt{3}} \right), \tag{8.5a}$$

$$c = \frac{1}{8\tau h^2}. \tag{8.5b}$$

Resident-Mutant Interactions

We now specify the demographic dynamics of interacting resident and mutant populations, characterized by traits x and x' and abundances n and n', respectively, by the following resident-mutant model:

$$\dot{n} = n \left(r(x) - cn - \gamma(x, x')n' - \frac{\tau^{-1}(n + n')}{h^2 + (n + n')^2} \right), \tag{8.6a}$$

$$\dot{n}' = n' \left(r(x') - \gamma(x', x)n - cn' - \frac{\tau^{-1}(n + n')}{h^2 + (n + n')^2} \right). \tag{8.6b}$$

We assume that there is no discrimination between resident and mutant populations with regard to discounting (e.g., individual traits do not affect mating encounters or consumption), so that the discounting function in (8.6a) is given by

$$\frac{n}{n + n'} \frac{\tau^{-1}(n + n')^2}{h^2 + (n + n')^2} = \frac{\tau^{-1}n(n + n')}{h^2 + (n + n')^2}$$

(similarly in (8.6b), where n and n' are exchanged).

By contrast, the logistic growth rate (per capita) must incorporate the effect of intraspecific competition between two similar traits. This is achieved by the trait-dependent competition coefficient $\gamma(x, x')$, which is taken to depend only on the difference $x' - x$ and is equal to c at $x' = x$. It should increase as the difference $x' - x$ increases, expressing a stronger deleterious effect on the x-population growth of better x'-competitors. The positive slope

$$c_1 = \frac{\partial}{\partial(x' - x)} \gamma(x, x') \Big|_{x' = x}$$

provides a measure of the strength of the asymmetry, and is called *competitive asymmetry* in the following.

In the numerical analysis, we use

$$\gamma(x, x') = 2c \left(1 - \frac{1}{1 + \exp(\gamma_1(x' - x))} \right),$$

which has all the properties mentioned above ($c_1 = c\gamma_1/2$).

Periodic Evolutionary Reversals

As we saw in Chapter 3, all possible outcomes of the resident-mutant competition can be classified by means of a bifurcation analysis of the resident-mutant model (8.6) in the trait space (x, x'). Under the assumption that mutations have small effects, it is legitimate to restrict the analysis to the vicinity of the diagonal $x' = x$, that is, to the region of the trait space where mutant individuals differ only slightly from residents.

The result is shown in Figure 8.3, which unravels seven regions above the diagonal that have at least one point in common with it. For $x' < x$ the bifurcation diagram is not shown, because it is given by a symmetric copy with respect to the diagonal. The seven corresponding demographic state portraits are also shown in the figure, and must be read in the space (n, n'), if corresponding to a point (x, x') above the diagonal, and in the space (n', n), if $x' < x$ and the symmetric point (x', x) lies in one of the regions ①–⑦.

In region ①, the resident population has only one stable equilibrium, namely the high-density equilibrium $\bar{n}_h(x)$. Small mutations characterized by slightly larger trait values invade and replace the resident type (side panel ①). The evolutionary random walk takes on a small step toward higher competitive abilities, and this results in a slightly decreased equilibrium abundance (see Figure 8.1). The same outcome is observed in regions ②–⑤ (see corresponding side panels). In all these regions, a mutant with a slightly larger trait value can invade and eliminate the resident type. Thus, starting from any ancestral state of high abundance for which x is less than x_h, evolution proceeds through the gradual increase of individual competitive ability, causing the concurrent, slow, decrease of population abundance.

A new phenomenon is observed in region ⑥, where the resident trait is still slightly less than x_h while the mutant trait is now slightly greater than x_h. In this case, the mutant first invades, but the resident "strikes back" (side panel ⑥): after its initial increase, the mutant reaches a peak abundance, then it starts declining and eventually goes extinct. This unexpected temporary invasion arises from the existence, for the resident population, of two alternative demographic equilibria that differ in their vulnerability to invasion by the mutant. Initially the resident population stands on its high-density equilibrium, which appears to be invasible. As the mutant population grows, the resident population cannot sustain its high density any longer and starts to approach its low-density state. While doing so, it creates a competitive environment that becomes noninvasible by the mutant. Now the latter is doomed, and as it goes extinct the population is sent back to its original evolutionary state; but in the meantime the demographic state has changed: the resident population experienced a sudden transition from the state of high abundance to the state of low abundance (switch of the resident demographic attractor; see also Mylius and Diekmann, 2001).

The same picture holds for the dual case, starting from an ancestral state of low abundance and intermediate or high competitive ability (x larger than x_l). Here mutation-selection processes drive the slow decrease of the trait and the concomitant slight increase of population abundance. When competitive ability comes close to the low threshold x_l, any invasion attempt by mutants smaller than x_l triggers

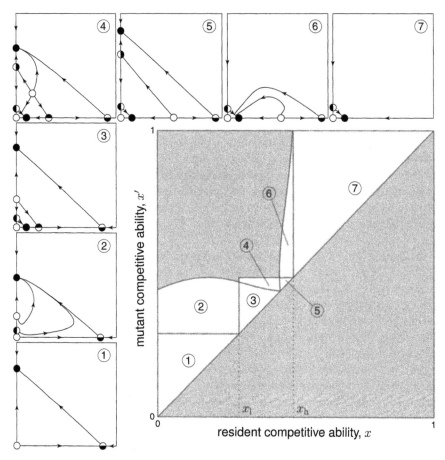

Figure 8.3 Bifurcation diagram of the resident-mutant model (8.6) in the resident-mutant trait space (x, x'), restricted to a neighborhood of the diagonal $x' = x$. The diagram (main panel) is symmetric with respect to the diagonal; only the upper part is shown. Seven regions ①–⑦ touching the diagonal correspond to qualitatively different resident-mutant demographic behaviors described in the attached side panels (in state space (n, n'); the behavior corresponding to a point (x, x') below the diagonal with (x', x) in region ⓘ is given by panel ⓘ read in space (n', n)). Filled circles indicate stable equilibria; half-filled circles: saddles; empty circles: repellors. Trait values in side panels: $(0.1, 0.2)$ in panel ①, $(0.2, 0.4)$ in ②, $(0.35, 0.4)$ in ③, $(0.4, 0.48)$ in ④, $(0.46, 0.48)$ in ⑤, $(0.46, 0.6)$ in ⑥, $(0.6, 0.7)$ in ⑦. Other parameters as in Figure 8.1 and $\gamma_1 = 6.6$.

the switch of the resident demographic equilibrium, from $\bar{n}_{\mathrm{l}}(x)$ to $\bar{n}_{\mathrm{h}}(x)$.

Altogether this analysis shows that the evolutionary dynamics never come to a halt. The competitive ability characterizing the population oscillates between a minimum (x_{l}) and a maximum (x_{h}) value (cyclic Red Queen dynamics). The abundance of the population switches between the low- and the high-density equilibria every time the adaptive process reaches either of these threshold trait values.

Invasion Does Not Imply Fixation

The analysis performed above points out a case in which the invasion of the mutant population does not imply its "fixation" in the community, namely its permanence with resident status. By contrast, in Chapter 3 we have seen that an invading mutant population generically substitutes the similar resident population, or coexists with it in the vicinity of evolutionary equilibria. No other possibilities are available, under the assumption that both x and x' correspond to evolutionary states in the evolution set associated with the current resident demographic attractor. However, this is not in conflict with the resident-strikes-back scenario because in that case the resident population is settled at its high- [low-] density demographic equilibrium and x is slightly smaller [greater] than x_h [x_l], while x' is slightly greater [smaller] than x_h [x_l]. In other words, when the resident strikes back, only the resident trait corresponds to an evolutionary state in the evolution set associated with the current resident demographic attractor. This implies that the resident substitution cannot occur, since the equilibrium of the resident-mutant model (8.6) at which the substitution transient should end does not exist (see Figure 8.3, panel ⑥ in space (n, n') [② in space (n', n)]).

Finally, notice that when mutational effects are limited, the resident-strikes-back scenario is not the generic cause of the resident attractor switch. In fact, the boundary (x_h or x_l) of the evolution set associated with the resident demographic equilibrium ($\bar{n}_h(x)$ or $\bar{n}_l(x)$) corresponds to a saddle-node bifurcation of the resident model (8.2), so that when x is close to such a boundary, the resident demographic equilibrium is close to the unstable equilibrium of model (8.2) and, hence, to the boundary of its basin of attraction. Small demographic fluctuations of the resident population abundance, not necessarily involving mutations, can therefore result in a demographic state (n) in the basin of attraction of the other demographic equilibrium, which would then be reached on the demographic timescale.

8.4 SLOW-FAST APPROXIMATION OF THE
AD CANONICAL EQUATION

One can achieve the same conclusions obtained in the previous section by using a slow-fast approximation of the AD canonical equation, which reads

$$\dot{x} = k\, n\, s(n, x), \tag{8.7a}$$

$$\epsilon \dot{n} = n\left(r(x) - cn - \frac{\tau^{-1}n}{h^2 + n^2}\right), \tag{8.7b}$$

where ϵ is a small timescaling factor separating the demographic and evolutionary timescales, k is a constant mutational rate (proportional to the frequency and variance of mutations), and $s(n, x)$ is the so-called selection derivative, i.e., the slope of the mutant fitness landscape (see (8.6b)) in the vicinity of the resident trait

$$s(n, x) = \frac{\partial}{\partial x'}\left(r(x') - \gamma(x', x)n - cn' - \frac{\tau^{-1}(n + n')}{h^2 + (n + n')^2}\right)\Bigg|_{\substack{n'=0 \\ x'=x}}$$

$$= -r_0 r_1 \exp(-r_1 x) + \frac{c\gamma_1}{2}n.$$

The dynamics of model (8.7) involve a fast component, that of the resident population abundance n governed by the first equation, and a slow component, that of the competitive ability x described by the second equation. The analysis of such a slow-fast system can be performed through a timescale separation method popularized in ecology by May (1977) (see also Rinaldi and Scheffer, 2000, for a recent survey, and Matsuda and Abrams, 1994b; Khibnik and Kondrashov, 1997, for applications in the context of evolutionary dynamics). The analysis requires two steps. First, the two stable equilibria $\bar{n}_h(x)$ and $\bar{n}_l(x)$ of the fast component are determined for each frozen value of the slow variable x. This yields the thin curves shown in the first column of Figure 8.4. Then the sign of $\dot{x}$ at each point $(x, \bar{n}_h(x))$ and $(x, \bar{n}_l(x))$ is determined so as to predict the direction of evolutionary change. This amounts merely to finding the isocline $s(n, x) = 0$ in the space (n, x), which is defined by

$$n = \frac{2r_0 r_1}{c\gamma_1} \exp(-r_1 x).$$

Figure 8.4, where the boundary of each gray region is the isocline $s(n, x) = 0$, displays the three possible evolutionary outcomes. In Figure 8.4A, starting from point 0, we observe a first, fast, demographic transient $(0 \rightarrow 1)$, followed by a slow evolutionary transient $(1 \rightarrow 2)$, entraining a slow demographic change yet leaving the population with low abundance. At point 2, a fast demographic transient abruptly brings the population to high abundance $(2 \rightarrow 3)$, and from point 3 a final slow evolutionary transient takes place, driving the population toward an evolutionary equilibrium at point 4, where low competitive ability is associated with high population abundance. At this point, selection may turn disruptive, causing evolutionary branching, but this phenomenon is not investigated here. Figure 8.4B shows a similar pattern whereby the adaptive process drives the population toward a state of permanent low abundance and promotes high competitive ability.

In contrast, the adaptive process never comes to a halt in the case shown in Figure 8.4C (which corresponds to the parameter setting of Figure 8.3). After a first, fast demographic transient $(0 \rightarrow 1)$, and a slow evolutionary transient $(1 \rightarrow 5)$, the system is trapped forever on a slow-fast limit cycle $(5 \rightarrow 2 \rightarrow 3 \rightarrow 4 \rightarrow 5)$.

One can see by comparing Figures 8.4A, B, and C that the necessary and sufficient condition for the existence of such an evolutionary cycle is that the selection isocline $s(n, x) = 0$ separates the two stable branches of resident demographic equilibria. This is a general condition for the existence of slow-fast limit cycles, known as separation principle (Muratori and Rinaldi, 1991). In the present case, elementary algebra yields the following conditions:

$$\bar{n}_l(x_1) < \frac{2r_0 r_1}{c\gamma_1} \exp(-r_1 x_1) < \frac{r_0 h^2}{c \bar{n}_l^2(x_1)} \exp(-r_1 x_1), \quad (8.8a)$$

$$\frac{r_0 h^2}{c \bar{n}_h^2(x_h)} \exp(-r_1 x_h) < \frac{2r_0 r_1}{c\gamma_1} \exp(-r_1 x_h) < \bar{n}_h(x_h) \quad (8.8b)$$

(where x_1, x_h, $\bar{n}_l(x_1)$, and $\bar{n}_h(x_h)$ are given in (8.3) and (8.4)), which respectively correspond to the fact that in Figure 8.4C the isocline $s(n, x) = 0$ passes between points 2 and 3 at $x = x_1$ and between points 4 and 5 at $x = x_h$.

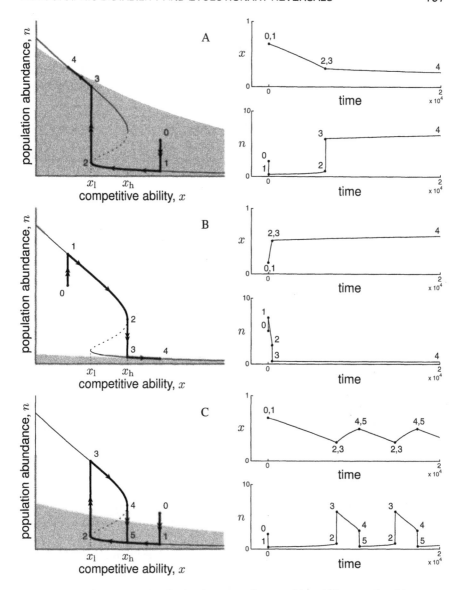

Figure 8.4 Evolutionary dynamics in the space of competitive ability x and resident population abundance n (first column) and corresponding time evolution of x and n (second column). In the first column, thin curves indicate the set of demographic equilibrium abundances as a function of x (solid [dotted] portions: stable [unstable] equilibria); shaded region: $s(n,x) < 0$; thick curve: exemplary trajectory (single arrows: slow (evolutionary) dynamics, double arrows: fast (demographic) dynamics). (A) Evolution toward low competitive ability and permanent high abundance ($\gamma_1 = 2.4$). (B) Evolution toward high competitive ability and permanent low abundance ($\gamma_1 = 32$). (C) Periodic evolutionary reversals ($\gamma_1 = 6.6$). Other parameter values as in Figure 8.1 and $k = 10^{-4}$.

The occurrence of periodic evolutionary reversals depends on six ecological parameters: the maximum growth rate (per capita) r_0, the physiological cost of enhanced competitive performance r_1, the intraspecific competition coefficient c, the competitive asymmetry c_1, and the discounting parameters τ and h. Evolutionary reversals can develop only when the resident population has two alternative demographic equilibria, which imposes the condition

$$c < \frac{1}{8\tau h^2} \tag{8.9}$$

(see (8.5b) and Figure 8.2). Thus, demographic bistability requires that the competition coefficient is lower than a threshold set by the discounting parameters; this constraint becomes weaker as the discounting pressure increases (smaller τ and/or smaller h).

Conditions (8.8) and (8.9) are analytically known bifurcation conditions of model (8.7), and are graphically represented in Figure 8.5 with respect to parameters c, c_1, τ, and h. In bistable resident populations, the occurrence of periodic evolutionary reversals is determined by the strength of competitive asymmetry (c_1), which has to lie within a range of intermediate values such that the selective pressure on the adaptive trait reverts at low and high abundances. Figures 8.5A and B show that this range shrinks under harsher conditions expressed by a larger competition coefficient c, whereas for fixed c, it expands if discounting is enhanced (smaller τ and/or smaller h). Altogether, a consistent pattern is that evolutionary reversals are more likely to occur in species characterized by significant competitive asymmetry, and facing strong discounting pressures.

When conditions for periodic evolutionary reversals are not satisfied, permanent low and high abundance represent alternative by-products of mutation-selection processes operating on competitive ability. Although there is a narrow parameter range for which permanent high abundance evolves concomitantly with intermediate competitive ability ($x_l < x < x_h$, see Figure 8.5C), there is a clear tendency for high abundance to evolve most often along with small trait values, whereas permanent low abundance can be associated with the evolution of large as well as intermediate trait values. This may help to understand why the association of scarcity and large body size in groups of closely related species remains difficult to establish empirically.

The ecological parameters may vary in response to environmental change. It turns out that even small and smooth changes in parameters may have a dramatic impact on the evolutionary dynamics of the population. Here we focus our discussion on the transition between periodic evolutionary reversals and permanent low abundance. A transition of this kind caused by a slight environmental deterioration should imply a dramatic rise of the extinction risk. Evolutionary cycling may suddenly disappear and be replaced with a state of permanent low abundance when selection at low density ceases to be directional (toward the critical trait value x_l at which evolutionary reversal occurs) and becomes stabilizing instead (transition from Figure 8.4C to B). By inspection of Figure 8.5, this can happen as a result of (i) slightly more competition (larger competition coefficient c), (ii) a small increase in competitive asymmetry (c_1), (iii) slightly less discounting (small increase of τ or h).

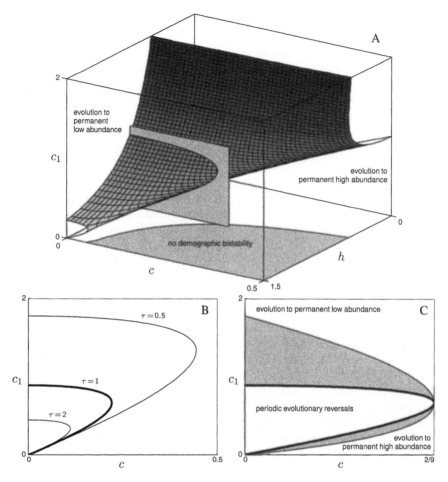

Figure 8.5 Bifurcation diagram of model (8.7) with respect to parameters c (intraspecific competition coefficient), c_1 (competitive asymmetry), and τ and h (discounting parameters). (A) Effect of c, c_1, and h. The volume bounded by the displayed surface contains parameter combinations that give rise to periodic evolutionary reversals. Permanent low abundance evolves above the surface. Permanent high abundance evolves below the surface. Demographic bistability occurs only for values of c and h lying outside the shaded region in the horizontal plane. (B) Effect of τ on the region (inside curves) of the (c, c_1) plane corresponding to periodic evolutionary reversals. (C) Effect of c and c_1 on the transition between periodic evolutionary reversals and either permanent low abundance with high competitive ability or permanent high abundance with low competitive ability. The shaded region indicates parameter combinations for which permanent low and high abundance evolve along with intermediate trait values ($x_1 < x < x_h$). In (B) and (C), the bold curve bounds the (c, c_1) region where periodic evolutionary reversals develop, and corresponds to the cross section indicated in (A). Fixed parameters as in Figure 8.1.

8.5 DISCUSSION AND CONCLUSIONS

How density can influence and be influenced by evolutionary processes has concerned biologists ever since Darwin suggested that abundant species were more
likely than rare species to be the sources of evolutionary novelties. Nonetheless, despite considerable attention to the matter, evolutionary biologists still hold widely
varying views on the consequences of population density on evolution (Orians,
1997) and on the role of evolution in determining population abundance and dynamics (Holt, 1997). The theoretical study reported in this chapter has investigated
the interaction between population density and life-history evolution by focusing
on an individual trait related to competitive performance that is potentially under
different selection pressures depending on the population abundance. The model
allows for multiple demographic equilibria, which entails that the population may
rest on a state of low abundance or on a state of high abundance under the very
same conditions.

Evolutionary Reversals and Population Dynamics

In ecological systems characterized by alternative equilibria, the demographic equilibrium on which the population settles depends on ecological history, and recurrent jumps between equilibria have been traditionally explained by means of environmental or ecological factors (see, e.g., Southwood and Comins, 1976; Hanski,
1985; Crawley, 1992). In contrast, our analysis shows that the periodic alterna-
tion between low-density and high-density demographic equilibria can develop at
an evolutionary timescale as a result of a purely endogenous mechanism: periodic
reversals of the selection pressure on an adaptive trait.

The simple, deterministic model considered here produces perfectly regular evolutionary cycles in trait value and population abundance. Random variability, however, is inherent to all populations. In a stochastic environment, the regularity of
such slow-fast cycles is likely to break down (Rinaldi and Scheffer, 2000), in lieu
of which more irregular alternations of phases of low and high abundance should
be expected. Documenting such fluctuations in population abundance requires ex-
ceptional data sets collected at a timescale seldom accessible to ecologists. Perhaps
the best example of such data is provided by fossil deposits of pelagic fish, from
which several millennia of demographic dynamics have been reconstructed with
time resolution as fine as a decade (see, e.g., Figure 1.8 in Section 1.9).

In the Pacific sardine *Sardinops caerulea* (Soutar and Isaacs, 1974; Baumgartner
et al., 1992), population data show that the species has persisted at very low density
over more than 55% of the time, through phases of irregular durations, from 10
to 200 years. Alternations between low and high abundance occurred rapidly and
more or less in unison across the whole species range. Interestingly, a competitor,
and possibly predator species, the Northern anchovy *Engraulis mordax*, has been
fluctuating more smoothly without reaching extreme population densities, suggesting that direct environmental factors may not be solely responsible for the dynamics
of the sardine. Similar patterns have been documented in *Sardinops sagax* over the
whole Holocene period, with bouts of very low abundance spanning up to 500 years

(De Vries and Pearcy, 1982); and in *Sardinops melanistica* at the timescale of the last four centuries (Tsuboi, 1984; Kondô, 1986). Tens and even hundreds of consecutive generations at either high or very low abundance may have given ample time for contrasting selective pressures to operate on traits related to competitive performance, and cause the alternations of high and low abundance observed in the data.

The combination of multiple demographic equilibria and evolutionary reversals between them offers an alternative explanation, previously suggested in a verbal model by Cury (1988), to purely ecological models of such long-term fluctuations between high and low density in fish communities (Ferrière and Cazelles, 1999). It would therefore be interesting to reanalyze the fossil fish scale data to investigate the occurrence of concomitant changes in fish body size (or other adaptive traits related to competitive abilities), which could be indicative of density fluctuations driven by evolutionary reversals.

Mechanism Causing Evolutionary Reversals

The biological mechanism causing evolutionary reversals involves two ingredients. First, there must exist trait values near which mutant traits can invade and yet fail to go to fixation, while their transitory presence causes the resident population to switch between two demographic attractors. This phenomenon brings new evidence that mutant invasion does not imply mutant fixation, as often assumed in evolutionary models (see the discussion on the fitness landscapes approach in Section 2.7). Except for Dercole et al. (2002a), on which this chapter is based, the few examples known to contradict the "invasion implies fixation" principle had been constructed by resorting to mutations of unlikely large effect (Doebeli, 1998; Diekmann et al., 1999; De Feo and Ferrière, 2000; Mylius and Diekmann, 2001; see also Case, 1995, and Abrams and Shen, 1989, in the context of species invasion).

The second key ingredient for evolutionary reversals is that the selective pressure operates in opposite directions on the same trait values, depending on whether the population density is high or low. Why this contrasting effect of selection arises in our model is easily understood: at high population density, a mutant is engaged in many competitive contests; thus, it has much to gain by investing more into competitive efficiency (through an increased trait value) at the expense of reducing its reproductive rate (provided that the trade-off between competition and reproduction is not too steep). In contrast, at low density encounters are few, and there is little to be gained from improved competitive ability; thus, one can expect advantageous mutants to be characterized by a larger intrinsic reproductive rate, achieved with reduced trait values.

Doebeli and Ruxton (1997) showed that cyclic evolution involving sharp transitions between multiple resident demographic attractors can occur in spatially structured communities, and Khibnik and Kondrashov (1997) found similar Red Queen dynamics of multispecies coevolution. Our study demonstrates that spatial heterogeneity and coevolutionary scenarios are not necessary to explain the reversal of selective pressures between different demographic attractors and the endless variations in adaptive traits; purely local intraspecific interactions are sufficient.

Matsuda and Abrams (1994a) also considered the evolution of body size under asymmetric competition, but the uniqueness of the demographic resident equilibrium at any trait value prevented the mechanism causing evolutionary reversals from operating in their model. As a consequence, they could only observe the evolutionary runaway to large body size and low population density, a process that could lead to population extinction (see Section 1.8). In contrast, if the conditions for evolutionary reversals are met in our model, the evolution toward high competitive abilities terminates at a trait value where the population abundance quickly and dramatically drops, whereas the selective pressure reverts and "rescues" the population from extinction by promoting the reduction of competitive ability and the increase of population abundance as a by-product.

Concluding Remarks

Any ecological scenario that makes the number of demographic attractors dependent upon an adaptive trait sets the stage for attractor switches driven by the evolutionary dynamics of the trait. This has been clearly recognized and exploited by Matsuda and Abrams (1994b), who showed that the evolution of antipredator ability may result in self-extinction of a prey population. Here we have used the same principle to demonstrate that the occurrence of evolutionary reversals and consequent evolutionary cycling is a likely property of ecological systems that possess alternative demographic equilibria (see also Doebeli and Ruxton, 1997; Khibnik and Kondrashov, 1997). We stress that this property is not bound to the assumption that multiple demographic equilibria are caused by a special ecological mechanism. A discounting function of the form of Holling type III response was chosen here to make the analysis mathematically tractable. Qualitatively similar results would be obtained with a Holling type II-like discounting function; the low-abundance stable equilibrium would simply be replaced with the zero equilibrium, thus making the evolutionary extinction (actually a case of evolutionary suicide, see Sections 1.8 and 3.6) of the population almost unavoidable (Gyllenberg and Parvinen, 2001).

Interestingly, some compelling evidence for evolutionary reversals of social traits has been gathered recently (Velicer et al., 1998; Hibbett et al., 2000). Sociality can induce an Allee effect (i.e., a positive density dependence in the population per-capita growth rate; Allee, 1931) that may be responsible for a discounting factor of the type considered here (Courchamp et al., 1999). Thus, adaptive switches between multiple demographic equilibria may point to a purely endogenous mechanism responsible for the repeated evolutionary rise and fall of social behavior.

Our study delineates two ways in which selection may depend on resident population density. In the first and classical case, each value of a trait uniquely determines the population density. If the density varies monotonically (e.g., decreases) with the trait value, the direction of adaptation may change with density as a mere consequence of directional selection toward an intermediate trait value: at a lower trait value, hence higher density, selection would favor an increase of the trait; at a higher trait value, hence lower density, selection would promote the decrease of the trait. In contrast, when multiple demographic equilibria are feasible, the direction and strength of selection may differ at the same trait value, depending upon the demographic state (e.g., high versus low density) on which the population is actually resting. Real systems possessing multiple demographic equilibria should therefore be useful to set up laboratory experiments in which the effect of density on selection is strictly isolated.

Chapter Nine

Slow-Fast Populations Dynamics and Evolutionary Ridges

We show in this chapter that contrasting timescales in the demographic dynamics of the resident populations may generate sharp transitions from stationary to cyclic demographic dynamics, resulting in abrupt variations of the selection pressure acting on the community. Such variations raise so-called evolutionary ridges in trait space, where resident populations are poised between stationary and cyclic coexistence. The main result is that evolutionary trajectories may slide along evolutionary ridges and even be trapped at special points called evolutionary pseudo-equilibria. The novel phenomena of evolutionary sliding and confinement of traits at evolutionary pseudo-equilibria should be generic to all communities in which the nature of demographic interactions causes the selection pressure to vary in trait space almost discontinuously. The model studied specifically describes a resource-consumer interaction, with one adaptive trait for each population. The AD canonical equation is derived for both cases of stationary and cyclic coexistence and its analysis points out a number of interesting features: from evolutionary extinction, to various forms of cyclic Red Queen dynamics. One of these forms is characterized by intermittent bouts of demographic oscillations of the two populations. An evolutionary justification of the well-known "paradox of enrichment" is also provided. Most of this chapter is taken from Dercole et al. (2006).

9.1 INTRODUCTION

The derivation of the AD canonical equation is straightforward when the resident populations settle on stable demographic equilibria for all trait values in the relevant trait space. This is particularly true when the equilibrium is unique and known in closed form, as we have seen in Chapters 4, 5, and 6. This is why most AD applications are restricted to evolving communities where this is the case (Abrams, 1992a; Marrow et al., 1992; Dieckmann et al., 1995; Iwasa and Pomiankowski, 1995; Marrow et al., 1996; Abrams and Matsuda, 1997; Gavrilets, 1997; Dieckmann and Doebeli, 1999; Iwasa and Pomiankowski, 1999; Geritz et al., 1999; Kisdi, 1999; Doebeli and Dieckmann, 2000; Ferrière et al., 2002; Dercole et al., 2003; Dercole, 2005). By contrast, when resident demographic dynamics do not settle at equilibria, in particular, when cyclic short-term coexistence of the populations is possible, the construction of the canonical equation becomes problematic, because, in general, demographic cycles are not known in closed form. A significant exception to this predicament is the case of so-called *slow-fast systems*, namely systems composed of populations whose demographic fluctuations develop on contrasting timescales. Indeed, in such cases one can approximate the population cycle with the so-called *singular cycle* (see Rinaldi and Scheffer, 2000, and references therein), which is nothing but the cycle corresponding to completely separated timescales. Usually, the singular cycle can be easily constructed using very simple geometric rules, which, in the case of two populations, boil down to a straightforward manipulation of the system isoclines (see Appendix A and Muratori and Rinaldi, 1991).

In this chapter we show that once the singular cycle has been detected, it is possible to derive the AD canonical equation, or a satisfactory approximation of it, in closed form. When the community is settled at a demographic equilibrium, the AD canonical equation is a system of standard (i.e., smooth) ODEs, because the equilibrium depends smoothly upon the traits. By contrast, the transition from a demographic equilibrium to a singular cycle along an evolutionary trajectory is often discontinuous (Rinaldi and Scheffer, 2000). Thus, the evolutionary trajectory obtained just after the transition may be radically different from the trajectory followed up to the transition. For example, for two coevolving populations each characterized by a single trait, the two-dimensional evolution set (the trait pairs allowing stationary or cyclic short-term coexistence) is partitioned into two subregions, say, S corresponding to stationary coexistence and C to cyclic coexistence. The two ODEs composing the AD canonical equation are continuous inside each region but are, in general, discontinuous at the boundary separating S from C. In fact, two different evolutionary gradients are associated with each point of the boundary: one is the vector tangent to the evolutionary trajectory obeying the canonical equation valid in region S and the other is the vector tangent to the evolutionary trajectory obeying the canonical equation valid in region C (see Figure 9.1). If the transversal components of these two vectors with respect to the discontinuity boundary have the same sign, as in the dotted part of the discontinuity boundary in Figure 9.1A, the trajectory crosses the boundary and the populations switch from stationary to cyclic demographic regime (or vice versa). On the contrary, if the transversal components of the two vectors are of opposite sign, i.e., if the two evolutionary gradients are "pushing" in opposite directions (solid part of the discontinuity boundary in Figure 9.1), the traits are forced to remain on the boundary and "slide" on it. Evolutionary sliding can be temporary, as in Figure 9.1A where the sliding motion terminates at point T, or permanent, when the sliding motion halts at a so-called evolutionary *pseudo-equilibrium*, namely at a point P on the boundary (see Figure 9.1B) where the two evolutionary gradients align. A pseudo-equilibrium has all the properties of an equilibrium (in particular, it can be an attractor, a saddle, or a repellor) even if the selection pressures do not vanish at that point. Of course, the above results are formally correct only in the limit case of completely separated demographic timescales, while in systems with contrasting timescales the traits will evolve along trajectories that are very close to the boundary. In other words, the boundary separating the two possible demographic regimes raises an attractive *ridge* in the evolution set, along which evolutionary trajectories from various ancestral conditions are canalized.

Systems of discontinuous ODEs have been studied by mathematicians since the

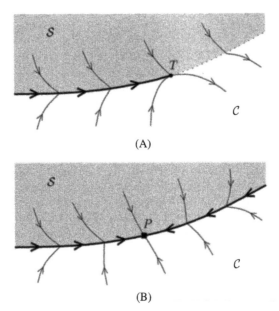

(A)

(B)

Figure 9.1 Evolutionary dynamics in the neighborhood of the boundary separating stationary coexistence (region S) from cyclic coexistence (region C). (A) Evolutionary
sliding toward T (solid boundary) and crossing (dotted boundary). (B) Evolutionary sliding toward the pseudo-equilibrium P.

early 1960s, but have seldom arisen in biological modeling so far (Křivan, 1998;
Křivan and Sikder, 1999; Genkai-Kato and Yamamura, 1999; Dercole et al., 2002b,
2007a). Details on the definition and computation of *sliding motions* can be found
in Filippov (1988), who has played a major role in the development of this mathematical field (see also Dercole et al., 2007a, for a review of discontinuous ecological models and their analysis).

We can summarize this brief overview by saying that the study of the interplay
of demographic and evolutionary processes simplifies greatly if the demographic
dynamics of the resident populations have different characteristic timescales, and
that a likely outcome is that evolution forces the populations to remain poised for
very long periods of time, if not forever, between stationary and cyclic coexistence
along evolutionary ridges. Whether evolutionary ridges are found in trait space, of
course, depends upon the specific demographic model and upon the values assigned
to its parameters.

In this chapter we consider the coevolution of slow-fast resource-consumer communities. Our first goal is to show how the various steps of the method of analysis
outlined above can be performed in practice. The second target is to achieve a unified view of the richness of resource-consumer coevolutionary dynamics, extending
the results drawn in Chapter 5 by allowing cyclic coexistence of the resident populations. Focusing on slow-fast populations may sound rather restrictive. However,
contrasting demographic timescales in resource-consumer communities have been

documented in many instances. Resources are often faster than consumers in growing and reproducing. Well-known examples are found among aquatic ecosystems: in the plankton food chain, the turnover of algae is faster than that of most zooplankton species, which, in turn, grow faster than fish (Scheffer, 1998). Among terrestrial ecosystems the Boreal forest is also rich in examples: plants (forbs and grasses) have fast dynamics in comparison with most herbivores (hares, squirrels, and small rodents), which reproduce faster than their consumers (lynx, coyote, and red fox) (Stenseth et al., 1997). But the opposite case, namely that of slow resource and fast consumer, is also frequently observed in nature. Spruce budworm (Ludwig et al., 1978) and larch budmoth (Baltensweiler, 1971) provide typical examples among plant-insect interactions. Here we present the analysis of the first case only, namely that of fast resource and slow consumer. Notice that the conclusions obtained in this context most likely do not apply to the case of slow resource and fast consumer. However, our approach is general and should, in principle, be applicable to that case as well.

9.2 BIOLOGICAL BACKGROUND

Understanding the determinants of population dynamics is a common denominator to many areas of ecology and evolutionary biology. Individual adaptive traits influence life-history parameters and should thereby impact on population size and structure; in return, the characteristics of a population influence the selective pressures that operate on the genetic variation of the traits. Thus, as discussed in Chapter 1 (see Section 1.5 in particular), demographic and evolutionary changes are entangled in a feedback loop. There are serious motivations to investigate the consequences of this feedback loop on population dynamics. In the context of conservation biology, there is growing concern about the evolutionary consequences of anthropic threats on ecological interactions (Ferrière et al., 2004a). An important challenge of virulence management is to find how we could operate on the ecological structure of host-pathogen interactions to achieve control of the evolution of pathogen virulence and host resistance (Dieckmann et al., 2002). On a more fundamental side, the issue of how evolutionary processes shape the demographic patterns we observe today is attracting much attention from population and community biologists (Thompson, 1998).

In general, the questions one can ask about the dynamical interplay of demographic and evolutionary processes are of two kinds (May and Anderson, 1983; Ferrière and Gatto, 1993, 1995; Abrams, 2000): (1) How does evolution or coevolution of adaptive traits affect the demographic stability of a population or community? Are there conditions under which evolution should favor demographic instability? (2) Under which conditions can we expect evolutionary dynamics to drive temporal fluctuation in the genetic composition of a population (the so-called Red Queen dynamics, see Section 1.6)? How do evolutionary dynamics respond to changes in ecological factors? Theoretical studies of these issues for resource-consumer interactions are the most prominent, and there seem to be two main reasons for that. First, prey-predator and host-parasite interactions have long been known for their potential to generate a whole spectrum of demographic dynamics in response to variation in individual trait values, from stable equilibria to cycles and chaos (Murdoch et al., 2003). Second, models describing the coevolution-ary dynamics of prey-predator and host-parasite communities often predict that the genotypic state of both species should cycle forever (Dieckmann et al., 1995).

Although a significant number of studies have dealt with some aspects of the main questions raised above, Dercole et al. (2006) recently addressed them simul-taneously in a unified framework based on the following ingredients: (i) The de-mographic interaction under consideration is known to be influenced by individual traits under genetic control. (ii) A realistic model for the effect of individual traits on the demographic interaction can be postulated. The timescales of demographic and evolutionary processes can be separated, so that (iii) an appropriate measure of fitness can be defined and derived explicitly from the demographic model, and (iv) the effect of demographic and evolutionary factors on the population dynamics can be clearly teased apart.

The natural modeling framework in Dercole et al. (2006) was therefore the AD canonical equation, which, however, needed not to be restricted to the case of short-term stationary coexistence considered in Chapter 3. The demographic model con-sidered in Dercole et al. (2006) is the Rosenzweig-MacArthur resource-consumer model that we have already described in Section 3.3 and Chapter 5. The resource is described by a logistic growth, while the consumer is characterized by a saturat-ing (Holling-type II) harvesting functional response. The presence of a saturating functional response is in contrast with most previous studies of resource-consumer coevolution and implies that stationary and cyclic short-term coexistence are pos-sible for different parameter settings. Species-specific demographic parameters are assumed to depend upon two adaptive traits, one for each species. In particular, the consumption rate is supposed to be influenced by both traits in such a way that the resource can reduce consumption risk by increasing or decreasing its trait value relative to a most vulnerable state, which depends on the consumer trait. In accor-dance with Abrams (2000), this property is called the "bidirectional axis of resource vulnerability." Although, mathematically speaking, this property is neither neces-sary nor sufficient to give rise to cyclic evolutionary dynamics, it is recognized to do so (Abrams, 2000).

The chapter is organized as follows. In the next section we derive the AD canon-ical equation describing the evolutionary dynamics of the resource-consumer com-munity. We first present the demographic model and obtain the singular limit cycle in the case of slow consumer and fast resource. Then, we specify the trait de-pendence of the relevant demographic parameters and extend the AD canonical equation to the case of a general demographic attractor. Its computation, in the case of the singular limit cycle, requires a substantial amount of algebra, which is presented in several subsections. We then present the method of analysis, based on simulation and bifurcation analysis of the canonical equation, and in the following section we interpret the results in the light of previous studies. A final discussion of the limitations and merits of the obtained results and a list of possible extensions close the chapter.

9.3 THE AD CANONICAL EQUATION FOR GENERAL DEMOGRAPHIC ATTRACTORS

In this section we derive the AD canonical equation corresponding to general resident demographic attractors and, in particular, to the stationary and periodic attractors of the slow consumer and fast resource resident populations. The section is long and dense with analytical details. The reader interested only in the results might go directly to Section 9.4.

The Resource-Consumer Demographic Model

The community we consider is a resource-consumer interaction described by the standard Rosenzweig-MacArthur model (Rosenzweig and MacArthur, 1963). For technical reasons, the model is written in a slightly different form than in Chapter 5 (see Section 3.3 for comparison), namely

$$\dot{n}_1 = rn_1 \left(1 - \frac{n_1}{K} \right) - \frac{\tau^{-1}n_1}{h + n_1} n_2, \tag{9.1a}$$

$$\dot{n}_2 = bn_2 + e\frac{\tau^{-1}n_1}{h + n_1} n_2 - dn_2, \tag{9.1b}$$

where n_1 and n_2 are resource and consumer population abundances, respectively. In the absence of consumer the resource population grows logistically, with net growth rate r and carrying capacity K, while in the absence of resource the consumer population decays exponentially with positive net death rate $d-b$ (we assume that the intrinsic birth rate b is smaller than the death rate d but that the maximum birth rate $(b + e\tau^{-1})$ is greater than d). Moreover, consumers are characterized by a Holling type II functional response, with maximum consumption rate τ^{-1} (τ is the consumer handling time) and half-saturation constant h, and the extra natality resulting from consumption is simply proportional to consumption through an efficiency coefficient e.

In this paragraph, we repeat the results derived in Appendix A (see "Example: Rosenzweig-MacArthur Resource-Consumer Model" in Section A.4), recasting them to the model formulation (9.1). For any parameter setting, model (9.1) has a globally stable attractor in the positive quadrant of the (n_1, n_2) plane. If the resource carrying capacity is low, i.e.,

$$K < \frac{h(d - b)}{b + e\tau^{-1} - d}, \tag{9.2}$$

the attractor is the trivial equilibrium $(K, 0)$, i.e., the consumer population goes extinct. By contrast, for intermediate values of the carrying capacity

$$\frac{h(d - b)}{b + e\tau^{-1} - d} < K < \frac{h(-b + e\tau^{-1} + d)}{b + e\tau^{-1} - d}, \tag{9.3}$$

stationary coexistence occurs at the strictly positive equilibrium

$$(\bar{n}_1, \bar{n}_2) = \left(\frac{h(d - b)}{b + e\tau^{-1} - d}, \frac{erh\left(K\left(b + e\tau^{-1} - d\right) - h(d - b)\right)}{K\left(b + e\tau^{-1} - d\right)^2} \right) \tag{9.4}$$

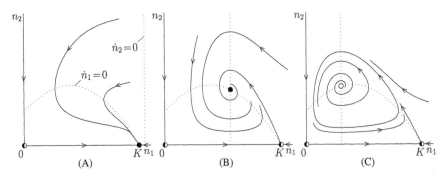

Figure 9.2 The three state portraits of model (9.1): (A) consumer extinction (see (9.2));
(B) stationary coexistence (see (9.3)); (C) cyclic coexistence (see (9.5)). Filled
circles: stable equilibria; half-filled circles: saddles; open circle: repellor; dotted
lines: nontrivial isoclines.

(which collides and exchanges stability with the trivial equilibrium $(K, 0)$ through
a transcritical bifurcation at the left-hand K value of (9.3)), while for high values
of K, i.e.,

$$K > \frac{h(-b + e\tau^{-1} + d)}{b + e\tau^{-1} - d}, \tag{9.5}$$

populations coexist on a limit cycle, as shown in Figure 9.2. When condition (9.5)
is satisfied with the equality sign, model (9.1) undergoes a supercritical Hopf bifurcation (geometrically characterized by the vertical consumer isocline passing
through the vertex of the resource nontrivial isocline, the parabola; see dotted lines
in Figure 9.2), so that the equilibrium $(\bar{n}_1, \bar{n}_2)$ given by (9.4) is critically stable and
the populations are balanced between stationary and cyclic coexistence.

The limit cycle is not known analytically, and this implies that the evolutionary dynamics cannot be described explicitly when the resource carrying capacity
is high. However, as already said in the Introduction, this may be remedied if the
resource grows at a much faster rate than consumers. Indeed, in that case the limit
cycle can be fairly well approximated by the singular limit cycle, which can be easily derived from the resource isocline, as shown in Figure 9.3. The singular cycle
is composed of two fast and two slow phases. The first fast motion is the collapse
of the resource population: it occurs when the consumer abundance is at its highest value $n_{2,\max}$. Then, in the absence of resource, consumer abundance slowly
decays until a lower threshold $n_{2,\min}$ is reached. At this point, the resource population rises quickly to $n_{1,\max}$ while consumer abundance remains at its lowest value.
Finally, the second slow motion takes place: the consumer slowly regenerates and
the resource slowly decays along the resource isocline.

The maximum consumer abundance $n_{2,\max}$ is given by

$$n_{2,\max} = \frac{r\tau(K + h)^2}{4K}, \tag{9.6}$$

while the minimum consumer abundance $n_{2,\min}$ is the solution of the following

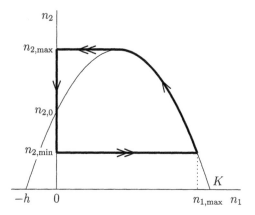

Figure 9.3 Resource isocline (parabola) and singular limit cycle (thick line); single [double] arrows indicate slow [fast] motion.

transcendental equation (Rinaldi and Muratori, 1992b):

$$n_{2,\max} - n_{2,\min} = n_{2,0} \log \left(\frac{n_{2,\max}}{n_{2,\min}} \right), \tag{9.7}$$

where $n_{2,0} = hr\tau$ is the intersection of the resource isocline with the vertical axis (see Figure 9.3).

The Community Evolution Set

We now assume that the resource-consumer interactions involve two species-specific adaptive traits denoted by x_1 for the resource and x_2 for the consumer. We consider positive trait values and, as done in Chapter 5, we imagine x_1 and x_2 as scaled body sizes of adult resource and consumer individuals. The resource trait x_1 controls the effectiveness of the resource in exploiting the available nutrients. In the absence of consumers, the logistic term

$$rn_1 \left(1 - \frac{n_1}{K} \right)$$

in (9.1a) arises from linearly density-dependent resource birth and death rates $b_1(n_1) = b_{10} - b_{11}n_1$ and $d_1(n_1) = d_{10} + d_{11}n_1$, so that $r = b_{10} - d_{10}$ and $K = (b_{10} - d_{10})/(b_{11} + d_{11})$, where b_{10}, b_{11}, d_{10}, and d_{11} may depend upon x_1. We imagine that the birth rate is density- and trait-independent (i.e., constant b_{10} and $b_{11} = 0$), while the death rate has a density-dependent component controlled by x_1. Thus, in (9.1a) r is constant while K depends on x_1. We further assume that K peaks at a particular trait value (K_1 in the following), which represents the body size at which the resource is most effective.

Similarly, the consumer intrinsic birth rate b is constant, while its death rate d depends upon x_2 and is minimum when consumers are best adapted to their environment ($x_2 = d_2$ in the following). The consumption rate is a function of both traits, and the consumer [resource] benefits [loses] most from the interaction when traits are balanced, i.e., when resource and consumer body sizes are in a suitable relationship, which defines a bidirectional axis of resource vulnerability (Abrams, 2000). This mechanism is present if, for example, the searching effectiveness of the consumer depends upon both traits but with a certain degree of plasticity, so that the same effectiveness can be achieved for a continuum of pairs (x_1, x_2). Since the half-saturation constant h is inversely related to searching effectiveness, $h(x_1, x_2)$ must be minimum when x_1 and x_2 are balanced, i.e., $x_1 = x_2$ provided both traits are measured on an appropriate scale.

These are standard assumptions for resource-consumer community modeling (Abrams, 2000), which have the advantage of involving the minimum possible number of demographic parameters. In the analysis we use

$$K(x_1) = K_0 \frac{2}{\left(\dfrac{x_1}{K_1}\right)^2 + \left(\dfrac{K_1}{x_1}\right)^2}, \tag{9.8a}$$

$$d(x_2) = d_0 \frac{\left(\dfrac{x_2}{d_2}\right)^2 + \left(\dfrac{d_2}{x_2}\right)^2}{2}, \tag{9.8b}$$

$$h(x_1, x_2) = h_0 + h_2(x_1 - x_2)^2, \tag{9.8c}$$

where, as said above, K_1 and d_2 are the values of x_1 and x_2 at which resource and consumer are most effective, respectively. Although the functional forms (5.4) do not have a specific empirical underpinning, they satisfy all the requirements discussed above, they are smooth and can be easily shaped by varying their parameters (two for each function). Together with model (5.1) they completely specify the resident model of the resource-consumer community.

Given the possible asymptotic regimes of the resident model, we can immediately infer that the trait space (x_1, x_2) is partitioned into three regions $\mathcal{E}$, $\mathcal{S}$, and $\mathcal{C}$ characterized by inequalities (9.2), (9.3), and (9.5), respectively. For traits in region $\mathcal{E}$ the consumer population cannot persist and goes extinct on the demographic timescale, while for traits in regions $\mathcal{S}$ and $\mathcal{C}$ the populations persist in stationary or cyclic demographic regimes, respectively. Thus, the union of regions $\mathcal{S}$ and $\mathcal{C}$ gives the evolution set of the community. Figure 9.4 shows an example of the three regions for the particular parameter setting indicated in the caption. For other parameter settings some of these three regions can disappear or become unbounded. Notice that nonlinear transformations of the traits appear on the axes of Figure 9.4, under which the point (K_1, d_2) becomes the origin.

At the boundary separating region $\mathcal{S}$ from region $\mathcal{C}$ all quantities associated with the asymptotic regime of the slow-fast system are discontinuous. In fact, close to the boundary in region $\mathcal{S}$ the vertical consumer isocline is on the right of the vertex of the resource isocline but close to it. Thus, the asymptotic regime is stationary and the equilibrium $(\bar{n}_1, \bar{n}_2)$ given by (9.4) is characterized by a value of $\bar{n}_2$ very close to $n_{2,\max}$ (the n_2 coordinate of the vertex of the resource isocline). By contrast, for pairs (x_1, x_2) close to the boundary in region $\mathcal{C}$ the attractor of the slow-fast system is well approximated by the singular limit cycle shown in Figure 9.3, which is characterized by a mean value of the consumer population much lower than $n_{2,\max}$

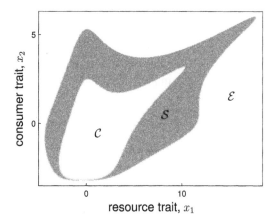

Figure 9.4 Regions $\mathcal{E}$, $\mathcal{S}$ (gray region), and $\mathcal{C}$, in the trait space (x_1, x_2). For $(x_1, x_2) \in \mathcal{E}$ the consumer population goes extinct (see (9.2) and Figure 9.2A); for $(x_1, x_2) \in \mathcal{S}$ the attractor of model (9.1) is a strictly positive equilibrium (stationary coexistence, see (9.3) and Figure 9.2B); for $(x_1, x_2) \in \mathcal{C}$ the attractor is a limit cycle (cyclic coexistence, see (9.5) and Figure 9.2C). For purpose of illustration, the transformations $\log(x_1/K_1) + x_1/K_1 - 1$ and $\log(x_2/d_2) + x_2/d_2 - 1$ are applied. Parameter values: $b = 0.001$, $e\tau^{-1} = 0.5$, $K_0 = 1$, $K_1 = 1$, $d_0 = 0.01$, $d_2 = 3$, $h_0 = 0.02$, $h_2 = 0.02$.

(De Feo and Rinaldi, 1997). This means that an abrupt loss of the mean consumer population will accompany any evolutionary transition from region $\mathcal{S}$ to region $\mathcal{C}$, while a sharp increase will occur when evolution proceeds in the opposite direction. Notice that these results are correct for the limit case of complete separation of resource and consumer demographic timescales. However, *singular perturbation theory* (Tikhonov, 1952; Hoppensteadt, 1966) guarantees that for the general and more realistic case of contrasting (but not completely separated) timescales, the evolutionary transition from $\mathcal{S}$ to $\mathcal{C}$, though formally continuous, will be associated with a sharp and marked loss of mean consumer population.

Resident-Mutant Interactions

After a mutation has occurred in the resource, the demographic dynamics of the community are described by the following resident-mutant model:

$$\dot{n}_1 = r n_1 \left(1 - \frac{n_1}{K(x_1)} - \frac{n_1'}{K(x_1)} \alpha(x_1, x_1') \right) - \frac{a(x_1, x_2) n_1}{1 + a(x_1, x_2)\tau n_1 + a(x_1', x_2)\tau n_1'} n_2,$$
$$(9.9a)$$

$$\dot{n}_1' = r n_1' \left(1 - \frac{n_1}{K(x_1')} \alpha(x_1', x_1) - \frac{n_1'}{K(x_1')} \right) - \frac{a(x_1', x_2) n_1'}{1 + a(x_1, x_2)\tau n_1 + a(x_1', x_2)\tau n_1'} n_2,$$
$$(9.9b)$$

$$\dot{n}_2 = n_2 \left(b + e \frac{a(x_1, x_2) n_1 + a(x_1', x_2) n_1'}{1 + a(x_1, x_2)\tau n_1 + a(x_1', x_2)\tau n_1'} - d(x_2) \right),$$
$$(9.9c)$$

where $\alpha(x_1, x_1')$ describes resource intraspecific competition ($\alpha(x_1, x_1) = 1$) and $a(x_1, x_2) = \tau^{-1}/h(x_1, x_2)$ is the consumer attack rate (see Section 3.3), which is useful for writing the functional response when more than one type of resource is available to the consumer. We assume symmetric resource intraspecific competition (i.e., $\alpha(x_1, x_1') = \alpha(x_1', x_1)$) and, in particular, that $\alpha(x_1, x_1')$ is an even function of the difference $x_1 - x_1'$.

Analogously, after the mutation has occurred in the consumer population, the resident-mutant model is

$$\dot{n}_1 = r n_1 \left(1 - \frac{n_1}{K(x_1)}\right) - \frac{\tau^{-1} n_1}{h(x_1, x_2) + n_1} n_2 - \frac{\tau^{-1} n_1}{h(x_1, x_2') + n_1} n_2', \quad (9.10a)$$

$$\dot{n}_2 = n_2 \left(b + e \frac{\tau^{-1} n_1}{h(x_1, x_2) + n_1} - d(x_2)\right), \quad (9.10b)$$

$$\dot{n}_2' = n_2' \left(b + e \frac{\tau^{-1} n_1}{h(x_1, x_2') + n_1} - d(x_2')\right). \quad (9.10c)$$

The resource and consumer selection derivatives are therefore given by

$$\frac{\partial}{\partial x_1'} \left(\frac{\dot{n}_1'}{n_1}\right)\Bigg|_{\substack{n_1'=0 \\ x_1'=x_1}} = \frac{r}{K(x_1)^2} \frac{d}{dx_1} K(x_1) n_1$$

$$+ \frac{\tau^{-1}}{h(x_1, x_2)} \frac{\partial}{\partial x_1} h(x_1, x_2) \frac{n_2}{h(x_1, x_2) + n_1}, \quad (9.11a)$$

$$\frac{\partial}{\partial x_2'} \left(\frac{\dot{n}_2'}{n_2}\right)\Bigg|_{\substack{n_2'=0 \\ x_2'=x_2}} = -e\tau^{-1} \frac{\partial}{\partial x_2} h(x_1, x_2) \frac{n_1}{(h(x_1, x_2) + n_1)^2}$$

$$- \frac{d}{dx_2} d(x_2). \quad (9.11b)$$

Notice that due to the above assumptions, the competition function α does not appear in the resource selection derivative.

If the mutant population does not invade, the trait x_1' [x_2'] is ruled out from the game, while in the opposite case the mutant population grows, at least temporarily. The possibility of a temporary growth of the mutant population followed by its extinction can be excluded in our case, because this would require the existence of multiple attractors of the resident model (as we have seen in Chapter 8), while model (9.1) has a single attractor. Thus, only the fate of the resident population after invasion of the mutant remains to be established. In general, this is not always easy to settle. It might be that the resident goes extinct, so that the mutant replaces the former resident and the trait x_1 [x_2] is replaced, in the end, by x_1' [x_2']. But it might also be that the resident and the mutant populations achieve stationary or cyclic coexistence. For large classes of models (in particular, Lotka-Volterra models), both stationary and cyclic coexistence can be excluded, but in other models coexistence is possible. In our case, if the mutant is the consumer, we have two consumer populations (the resident and the mutant) competing exploitatively for the same (logistic) resource. Even if two consumers and one resource can coexist only on a limit cycle (Koch, 1974; Hsu et al., 1978), the conditions for cyclic coexistence in the case of slow consumer and fast resource are not satisfied if the mutations are small (Muratori and Rinaldi, 1989). Therefore, we can conclude that an invading consumer replaces the former resident.

As far as resource mutations are concerned, the invasion implies substitution theorem (see Appendix B) generically guarantees that, if the resident populations are at demographic equilibrium, an invading resource mutant substitutes the resident population. Exceptions can occur through resource evolutionary branching only when traits reach an evolutionary halt, namely an equilibrium of the AD canonical equation. Resource evolutionary branching has been studied in Chapter 5 and will not be investigated here. Finally, extensive simulations of model (9.9) have shown that resource mutations invading oscillating resident populations also lead to the substitution of the resource resident population. Thus, if mutations are small and occur rarely enough, evolutionary dynamics within each population remain monomorphic.

The AD Canonical Equation

We are now in the position of deriving the AD canonical equation describing the dynamics of the two traits. Evolutionary processes generally assume two main ingredients: genetically based variations of individual traits generated through reproduction, and selection on this variation resulting from demographic interactions. This is a complex process because individual traits under consideration may affect both the birth process and the demographic interactions. As we have seen in Chapter 3, the assumption of rare mutations of small effects allows one to approximate the dynamics of population abundances and trait distributions with deterministic models.

In the limit of rare mutations of small effects, the rate of change of an adaptive trait x over evolutionary time is given by the AD canonical equation (3.24), which we here rewrite as

$$\epsilon^2 \dot{x} = \left(\frac{1}{2} \epsilon^2 \sigma^2(x) \right) \left(\epsilon\mu(x) \frac{1}{\epsilon} b(\bar{n}(x,X), 0, \bar{N}(x,X), x, \cdot, X) \bar{n}(x,X) \right)$$

$$\times \left(\frac{\left. \frac{\partial}{\partial x'} f(0, \bar{n}(x,X), \bar{N}(x,X), x', x, X) \right|_{x'=x}}{b(\bar{n}(x,X), 0, \bar{N}(x,X), x, \cdot, X)} \right), \qquad (9.12)$$

where ϵ is a timescaling factor separating the demographic and evolutionary timescales,

$$f(n, n', N, x, x', X) = b(n, n', N, x, x', X) - d(n, n', N, x, x', X)$$

is the per-capita growth rate $\dot{n}/n$ of the resident population characterized by trait x (the balance between birth rate b and death rate d, per capita, see (3.1)), $\bar{n}(x,X)$ is its equilibrium abundance, and $\bar{N}(x,X)$ and X pack all other population abundances and traits characterizing the community. The probability of a mutation in trait x is $\epsilon\mu(x)$, while the standard deviation of the mutational step distribution is $\epsilon\sigma(x)$, so that (9.12) is correct in the limit $\epsilon \to 0$, where mutations become extremely rare (on the demographic timescale) and have an infinitesimal effect on the trait. Three terms compose (9.12) (see Section 3.5 for details): half of the variance of the mutational step distribution (first parenthesis), the mutational birth output per unit of evolutionary time (second parenthesis), and the selective advantage of the mutation (third parenthesis), namely the first-order derivative with respect to the mutational step $(x' - x)$ of the probability of resident substitution (see (3.35)). Notice that the two birth rates in (9.12) cancel out, so that the AD canonical equation for stationary demographic attractors does not depend on the mutant population birth rate, but only on its growth rate.

Here we need an extension of (9.12) to the case of more general resident demographic attractors. The rigorous derivation of such an extension is a hard mathematical exercise that lies beyond the scope of this book (see Dieckmann and Law, 1996, for a heuristic discussion of the problem). However, as explained below, averaging the mutant birth rate

$$\mu(x)b(n, 0, N, x, \cdot, X)n \tag{9.13}$$

and the selection term

$$\frac{1}{b(0, n, N, x, x, X)} \left. \frac{\partial}{\partial x'} f(0, n, N, x', x, X) \right|_{x'=x} \tag{9.14}$$

over the resident demographic attractor is appropriate in the case considered in this chapter. In formulas, this results in

$$\dot{x} = \frac{1}{2}\mu(x)\sigma^2(x)\langle b(n, 0, N, x, \cdot, X)n\rangle \left\langle \frac{\left. \frac{\partial}{\partial x'} f(0, n, N, x', x, X) \right|_{x'=x}}{b(n, 0, N, x, \cdot, X)} \right\rangle, \tag{9.15}$$

where $\langle \rangle$ indicates temporal averaging over the resident demographic attractor corresponding to trait values (x, X).

At this point, the analysis of evolutionary dynamics through equation (9.15) would remain problematic because, in general, the resident demographic attractor is not known analytically in closed form. Slow-fast systems represent a significant exception to this predicament, since any slow-fast demographic attractor can be approximated with the so-called *singular attractor* corresponding to completely separated demographic timescales, and this permits explicit calculation of the averages in (9.15). The case of slow-consumer-fast-resource limit cycles is particularly favorable because the singular cycle has been easily identified (see Figure 9.3). Moreover, the singular cycle is composed of two long phases of slow motion of both populations alternated with two fast phases of significant resource variation. Thus, slow-fast cycles are very long and mutant populations experience little variations in the resident state during their initial phase of growth or decline (with the only exception of particular mutations occurring during the short episodes of fast resource variation). This supports the use of equation (9.15), which, indeed, takes the expectation of the mutant birth rate (9.13) and the selection term (9.14) over all possible resident states at the time of mutant arising.

Writing equation (9.15) for the resource and consumer traits x_1 and x_2 and taking into account that resource birth rate (per capita) is density-independent, while consumer birth rate (per capita) is given by $(b + e\tau^{-1}n_1/(h + n_1))$, we obtain

$$\dot{x}_1 = k_1 \langle n_1 \rangle \left\langle \left. \frac{\partial}{\partial x_1'} \left(\frac{\dot{n}_1'}{n_1} \right) \right|_{\substack{n_1'=0 \\ x_1'=x_1}} \right\rangle, \tag{9.16a}$$

$$\dot{x}_2 = k_2 \left\langle \left(b + \frac{e\tau^{-1}n_1}{h(x_1,x_2)+n_1} \right) n_2 \right\rangle$$

$$\times \left\langle \left(b + \frac{e\tau^{-1}n_1}{h(x_1,x_2)+n_1} \right)^{-1} \frac{\partial}{\partial x_2'} \left(\frac{\dot{n}_2'}{n_2} \right) \Big|_{\substack{n_2'=0 \\ x_2'=x_2}} \right\rangle, \qquad (9.16b)$$

where $k_i = 1/2\mu_i\sigma_i^2$, $i = 1,2$, are constant mutational rates. Substituting the selection derivatives (9.11) into (9.16), we finally obtain

$$\dot{x}_1 = k_1 \langle f_1 \rangle$$

$$\times \left(\frac{r}{K(x_1)^2} \frac{d}{dx_1} K(x_1) \langle f_1 \rangle + \frac{\tau^{-1}}{h(x_1,x_2)} \frac{\partial}{\partial x_1} h(x_1,x_2) \langle f_2(x_1,x_2) \rangle \right),$$
$$(9.17a)$$

$$\dot{x}_2 = k_2 d(x_2) \langle f_3 \rangle$$

$$\times \left(-\frac{d}{dx_2} d(x_2) \langle f_4(x_1,x_2) \rangle - e\tau^{-1} \frac{\partial}{\partial x_2} h(x_1,x_2) \langle f_5(x_1,x_2) \rangle \right),$$
$$(9.17b)$$

where $\langle f_i \rangle$, $i = 1,\dots,5$, are the average values on the resident demographic attractor corresponding to traits (x_1,x_2) of the functions

$$f_1 = n_1, \qquad (9.18a)$$

$$f_2(x_1,x_2) = \frac{n_2}{h(x_1,x_2)+n_1}, \qquad (9.18b)$$

$$f_3 = n_2, \qquad (9.18c)$$

$$f_4(x_1,x_2) = \frac{h(x_1,x_2)+n_1}{bh(x_1,x_2)+(b+e\tau^{-1})n_1}, \qquad (9.18d)$$

$$f_5(x_1,x_2) = \frac{n_1}{(b+e\tau^{-1})n_1^2 + h(x_1,x_2)(2b+e\tau^{-1})n_1 + bh(x_1,x_2)^2}. \qquad (9.18e)$$

Each term $\langle f_i \rangle$ can be computed in the region of stationary coexistence, i.e., for $(x_1,x_2) \in \mathcal{S}$, by replacing n_1 and n_2 with their equilibrium values given by (9.4). After some algebra one obtains

$$\langle f_1 \rangle = \frac{h(x_1,x_2)(d-b)}{b+e\tau^{-1}-d(x_2)}, \qquad (9.19a)$$

$$\langle f_2(x_1,x_2) \rangle = r\tau \left(1 - \frac{h(x_1,x_2)}{K(x_1)} \frac{d(x_2)-b}{b+e\tau^{-1}-d(x_2)} \right), \qquad (9.19b)$$

$$\langle f_3 \rangle = h(x_1,x_2)r\tau \left(1 - \frac{h(x_1,x_2)}{K(x_1)} \frac{d(x_2)-b}{b+e\tau^{-1}-d(x_2)} \right)$$

$$\times \left(1 + \frac{d(x_2)-b}{b+e\tau^{-1}-d(x_2)} \right), \qquad (9.19c)$$

$$\langle f_4(x_1,x_2) \rangle = \frac{1}{d(x_2)}, \qquad (9.19d)$$

$$\langle f_5(x_1,x_2) \rangle = \frac{(d(x_2)-b)(b+e\tau^{-1}-d(x_2))}{e^2\tau^{-2}h(x_1,x_2)d(x_2)}. \qquad (9.19e)$$

By contrast, the terms $\langle f_i \rangle$ in the region of cyclic coexistence, i.e., for $(x_1, x_2) \in \mathcal{C}$, can be computed through a series of approximations. The computation is performed under the assumption that the resource population has fast dynamics in comparison with the consumer population, and that the half-saturation constant is small with respect to the carrying capacity of the resource, i.e., $h/K \ll 1$. Thus, $\langle f_i \rangle$ can be identified with the average value of f_i on the singular limit cycle (see Figure 9.3). Since this cycle is composed of two slow and two fast segments, the computation can be limited to the slow segments. More precisely, if the duration of the first slow phase (along the n_2 axis) is T' and the duration of the second slow phase (along the parabola) is T'', we can write

$$\langle f_i \rangle = \frac{1}{T' + T''} \left(\int_0^{T'} f_i dt + \int_0^{T''} f_i dt \right),$$

where the two integrals are computed along the slow segments of the singular cycle. After a substantial amount of algebra, reported in the following subsections, the result is

$$\langle f_1 \rangle = \frac{K(x_1) \tau (d(x_2) - b)}{2e} \Big(p_1(x_1, x_2) + h(x_1, x_2) r \tau$$
$$\times \frac{p_2(x_1, x_2) \left(p_4(-p_2, x_1, x_2) - p_4(p_2, x_1, x_2) \right) - 2 p_3(x_1, x_2)}{n_{2,\max}(x_1, x_2) - n_{2,\min}(x_1, x_2)} \Big),$$

$$\text{(9.20a)}$$

$$\langle f_2(x_1, x_2) \rangle = \frac{1}{e} \Big(r \tau^2 (b + e\tau^{-1} - d(x_2)) + \frac{h(x_1, x_2) r^2 \tau^3 (d(x_2) - b)}{n_{2,\max}(x_1, x_2) - n_{2,\min}(x_1, x_2)} \Big)$$
$$\times (p_3(x_1, x_2) + p_1(x_1, x_2) p_4(p_1, x_1, x_2)), \qquad \text{(9.20b)}$$

$$\langle f_3 \rangle = h(x_1, x_2) r \tau, \qquad \text{(9.20c)}$$

$$\langle f_4(x_1, x_2) \rangle = \frac{2b + e\tau^{-1} - d(x_2)}{b(b + e\tau^{-1})}, \qquad \text{(9.20d)}$$

$$\langle f_5(x_1, x_2) \rangle = \frac{r \tau^2 (d(x_2) - b)}{e(n_{2,\max}(x_1, x_2) - n_{2,\min}(x_1, x_2))(b + e\tau^{-1})} \Big(p_1(x_1, x_2)$$
$$\times p_4(p_1, x_1, x_2) + \frac{h(x_1, x_2)}{K(x_1)} p_4(-p_2, x_1, x_2) - p_4(p_2, x_1, x_2) \Big),$$

$$\text{(9.20e)}$$

where

$$p_1(x_1, x_2) = 1 - \frac{h(x_1, x_2)}{K(x_1)}, \qquad \text{(9.21a)}$$

$$p_2(x_1, x_2) = 1 + \frac{h(x_1, x_2)}{K(x_1)}, \qquad \text{(9.21b)}$$

$$p_3(x_1, x_2) = \sqrt{\frac{4}{r \tau K(x_1)} (n_{2,\max}(x_1, x_2) - n_{2,\min}(x_1, x_2))}, \qquad \text{(9.21c)}$$

$$p_4(p, x_1, x_2) = \log \left(\frac{p}{p_3(x_1, x_2) + p} \right), \qquad \text{(9.21d)}$$

and

$$n_{2,\max}(x_1, x_2) - n_{2,\min}(x_1, x_2) = \frac{r\tau(h(x_1, x_2) + K(x_1))^2}{4K(x_1)} - h(x_1, x_2)r\tau$$
$$\times \left(p_5(x_1, x_2) - \sqrt{p_5(x_1, x_2)^2 - 2}\right),$$

(9.22a)

$$p_5(x_1, x_2) = \frac{\exp\left(\dfrac{(h(x_1, x_2) + K(x_1))^2}{4h(x_1, x_2)K(x_1)}\right)}{\dfrac{(h(x_1, x_2) + K(x_1))^2}{4h(x_1, x_2)K(x_1)}} - 1.$$

(9.22b)

Computation of T' and T''

The first slow phase of the singular cycle is characterized by $n_2(0) = n_{2,\max}$ and $n_1(t) = 0$, so that $\dot{n}_2 = (b - d)n_2$, i.e., $n_2(t) = n_{2,\max} \exp((b-d)t)$. Since $n_2(T') = n_{2,\min}$, we obtain

$$T' = \frac{1}{d - b} \log\left(\frac{n_{2,\max}}{n_{2,\min}}\right),$$

(9.23)

where $n_{2,\max}$ is given by (9.6) and $n_{2,\min}$ can be derived from (9.7).

The second slow phase of the singular cycle is characterized by $n_1 \gg h$, since $h \ll K$. Hence, (9.1b) can be approximated by $\dot{n}_2 = (b + e\tau^{-1} - d)n_2$, which for $n_2(0) = n_{2,\min}$ has solution $n_2(t) = n_{2,\min} \exp((b + e\tau^{-1} - d)t)$. Since $n_2(T'') = n_{2,\max}$, we obtain

$$T'' = \frac{1}{b + e\tau^{-1} - d} \log\left(\frac{n_{2,\max}}{n_{2,\min}}\right).$$

(9.24)

Computation of $\langle f_1 \rangle$

During the first slow phase of the singular limit cycle the function f_1 is zero (see (9.18a)). This implies that

$$\langle f_1 \rangle = \frac{1}{T' + T''} \int_0^{T''} n_1(t)dt,$$

where the integration must be performed along the parabola, where

$$n_1(t) = \frac{K - h}{2} + \frac{K}{2}\sqrt{\left(\tau^{-1} + \frac{h}{K}\right)^2 - \frac{4\tau^{-1}}{rK}n_2(t)}, \qquad (9.25a)$$

$$n_2(t) = n_{2,\min} \exp((b + e\tau^{-1} - d)t). \qquad (9.25b)$$

This integration can be performed explicitly, without introducing any further approximation. Taking (9.23) and (9.24) into account, the result is (9.20a), where $p_1, \ldots, p_4$ are given by (9.21).

Computation of $\langle f_2 \rangle$

From (9.18b), we can write

$$\langle f_2 \rangle = \frac{1}{T' + T''} \left(\frac{1}{h} \int_0^{T'} n_2(t) dt + \int_0^{T''} \frac{n_2(t)}{h + n_1(t)} dt \right).$$

In the first integral $n_2(t) = n_{2,\max} \exp((b - d)t)$, while in the second integral $n_1(t)$ and $n_2(t)$ are as in (9.25). If h is neglected with respect to n_1 in the second term, the two integrals can be performed analytically and the result, using (9.23) and (9.24), is (9.20b), with p_1, p_3, and p_4 as in (9.21).

Computation of $\langle f_3 \rangle$

From (9.18c) it follows that

$$\langle f_3 \rangle = \frac{1}{T' + T''} \left(\int_0^{T'} n_2(t) dt + \int_0^{T''} n_2(t) dt \right),$$

where in the first integral $n_2(t) = n_{2,\max} \exp((b - d)t)$, while in the second integral $n_2(t)$ can be approximated by the exponential function (9.25b). A straightforward integration gives (9.20c).

Computation of $\langle f_4 \rangle$

During the first slow phase of the singular limit cycle, f_4 is constant and equal to $1/b$, while during the second slow phase it is approximately constant and equal to $1/(b + e\tau^{-1})$ (see (9.18d)). Thus,

$$\langle f_4 \rangle = \frac{T'}{T' + T''} \frac{1}{b} + \frac{T''}{T' + T''} \frac{1}{b + e\tau^{-1}},$$

from which, taking (9.23) and (9.24) into account, (9.20d) follows.

Computation of $\langle f_5 \rangle$

During the first slow phase of the singular limit cycle, f_5 is zero, while it can be approximated with $1/((b + e\tau^{-1})n_1)$ during the second slow phase (see (9.18e)). Hence,

$$\langle f_5 \rangle = \frac{1}{T' + T''} \frac{1}{b + e\tau^{-1}} \int_0^{T''} \frac{1}{n_1(t)} dt,$$

where $n_1(t)$ is given by (9.25a) with $n_2(t)$ as in (9.25b). This integral can also be computed, using classical decomposition techniques, and the result is (9.20e), with p_1, p_2, and p_4 as in (9.21).

Computation of $n_{2,\min}$

The extreme values $n_{2,\max}$ and $n_{2,\min}$ of consumer abundance along the singular limit cycle are given by (9.6) and (9.7). While (9.6) defines $n_{2,\max}$ explicitly,

(9.7) is a transcendental equation that can be solved with respect to $n_{2,\min}$ only numerically. To avoid such a computation each time, the right-hand side of the AD canonical equation must be evaluated for different trait and parameter values, we use an approximated formula for $n_{2,\min}$, obtained by neglecting the terms of order higher than two in the Taylor expansion

$$\exp\left(\frac{n_{2,\min}}{n_{2,0}}\right) = 1 + \frac{n_{2,\min}}{n_{2,0}} + \frac{1}{2}\left(\frac{n_{2,\min}}{n_{2,0}}\right)^2 + O\left(\left(\frac{n_{2,\min}}{n_{2,0}}\right)^3\right).$$

The corresponding approximated, but explicit, formula for $(n_{2,\max} - n_{2,\min})$ is given by (9.22). Of course, the approximation is good if $n_{2,\min} \ll n_{2,0}$, which is always the case for $h \ll K$.

Validity of the Approximations

All the approximations we have introduced are a priori justified if $h \ll K$. To evaluate to which extent our approximations are valid, we have systematically compared the values of $\langle f_i \rangle$, $i = 1, \dots, 5$, given by (9.20)–(9.22), with the true values computed through numerical integration of the function f_i along the singular cycle. The result of this analysis, carried out for many parameter settings, is that the approximation is definitely satisfactory for $\langle f_2 \rangle, \dots, \langle f_5 \rangle$ (errors of the order of 1% if $h/K \leq 0.2$). By contrast, the approximation of $\langle f_1 \rangle$ is more crude but still acceptable ($\leq 1\%$ for $h/K \leq 0.1$). Of course, to be sure that the impact of our approximations on the final results is not too heavy we should look at the values of h/K in region C. For this, we can first notice that for a particularly meaningful pair of traits, namely the optimal pair (K_1, d_2) (i.e., point $(0, 0)$ in all state portraits), h/K is simply given by (see (9.8))

$$\frac{h}{K} = \frac{h_0 + h_2(K_1 - d_2)^2}{K_0}$$

and that this value is definitely low (between 0.05 and 0.1 and exceptionally 0.2) in our examples. A more meaningful indicator is perhaps the portion of region C in which $h/K \leq 0.2$. Computed on our state portraits this indicator is quite satisfactory: at 95% of the points $(x_1, x_2) \in C$ the ratio h/K is lower than 0.2.

9.4 EVOLUTIONARY SLIDING AND PSEUDO-EQUILIBRIA

The discontinuous AD canonical equation derived in the previous section is now used for studying the evolutionary dynamics of the resource-consumer community. Numerical simulations provide a straightforward approach. A typical evolutionary state portrait is reported in Figure 9.5 (parameter values are specified in the caption). As anticipated, the evolution set is composed of regions S and C corresponding to stationary and cyclic demographic regimes, while region $\mathcal{E}$ corresponds to consumer extinction.

Figure 9.5 displays a small region where the consumer population evolves to extinction (dark region in the figure). Since the boundary separating region S from

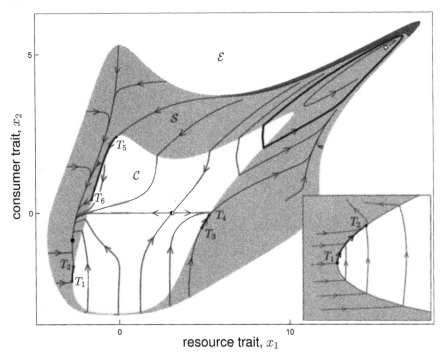

Figure 9.5 An evolutionary state portrait of the AD canonical equation (9.17); parameter
values as in Figure 9.4 and $r = 1$, $e = 0.1$, $k_1 = 0.1$, $k_2 = 1$. There are three
equilibria, a stable node (filled circle) and an unstable focus (empty circle) in
region $\mathcal{S}$ and a saddle (half-filled circle) in region $\mathcal{C}$, and one limit cycle (thick
trajectory) partly in $\mathcal{S}$ and partly in $\mathcal{C}$; there are two attractors, the node and the
cycle, and their basins of attraction are separated by the stable manifold of the
saddle; there are three sliding (thick) segments, one attracting ($T_1 T_2$, stretched
and magnified in the lower right panel) and two repelling ($T_3 T_4$ and $T_5 T_6$). Consumer evolutionary extinction occurs for ancestral conditions in the dark region.

region $\mathcal{E}$ is a transcritical bifurcation of the resident model (9.1), this is a case of
evolutionary murder (see Sections 1.8 and 3.6), where resource evolution ultimately
drives the consumer population toward vanishing abundances. Two evolutionary attractors are present: an equilibrium with low resource and consumer traits in region
$\mathcal{S}$ and a limit cycle with high resource and consumer traits partly in $\mathcal{S}$ and partly in
$\mathcal{C}$. The two basins of attraction are separated by the stable manifold of the saddle
lying in region $\mathcal{C}$. Thus, if the ancestral conditions are on the left of this manifold, the traits converge to the equilibrium and, after an evolutionary transient, the
populations stationarily coexist. However, for some ancestral conditions, one piece
of the evolutionary trajectory lies in region $\mathcal{C}$: this means that during the corresponding period of time the populations oscillate along a demographic cycle that
is slowly drifting on the evolutionary timescale. By contrast, population dynamics
associated with evolutionary trajectories in the other basin of attraction are radically different. Indeed, long periods of time characterized by slowly varying pop-

ulations recurrently alternate with long periods of time during which populations fluctuate on a demographic cycle. This phenomenon of recurrent bouts of demographic oscillations driven by nonstationary evolutionary dynamics is perhaps the most complex mode of behavior of coevolving populations, as explained in Khibnik and Kondrashov (1997), who have named it "ecogenetically driven Red Queen dynamics."

Figure 9.5 shows the possibility of evolving along evolutionary ridges, i.e., segments of the boundary separating cyclic from stationary demographic regimes (thick segments in the figure). In the case of segment T_1T_2, the sliding is stable (because the two evolutionary gradients are pointing toward the boundary) and is visited by many evolutionary trajectories, so that segment T_1T_2 indeed defines an evolutionary ridge in the evolution set; while in the other two cases (segments T_3T_4 and T_5T_6) the sliding is unstable (because the two evolutionary gradients are pointing away from the boundary). Figure 9.5 also points out that the initial and final points of a sliding segment are so-called *tangent points*, at which one of the two evolutionary gradients is tangent to the boundary (see, e.g., points T_1 and T_2 in the lower right panel of Figure 9.5). The numerical detection of sliding segments is actually rooted in this simple geometric property (Kuznetsov et al., 2003; Dercole and Kuznetsov, 2005). Sliding is a novel type of evolutionary dynamics with far-reaching ecological implications: when traits are sliding along an evolutionary ridge, resource and consumer are poised between stationary and cyclic coexistence, i.e., coevolution drives the populations toward and maintains them at the onset of their most complex dynamical behavior, which is also the path along which the mean consumer population changes abruptly.

As discussed in detail in Section 3.8, how sensitive the above phenomena are to relevant demographic and environmental parameters can be investigated by means of a thorough bifurcation analysis of the AD canonical equation. This means that all the invariant sets of the evolutionary state portrait (e.g., the three equilibria and the limit cycle of Figure 9.5) must be "continued" with respect to a parameter (see Appendix A), in order to detect the critical parameter values at which invariant sets undergo a structural change. To make this notion more intuitive, let us describe one such bifurcation in some detail. In response to a continuous increase of some parameter the limit cycle of Figure 9.5 will grow up to the point where it touches the saddle and simultaneously disappears. This is an example of a so-called homoclinic bifurcation. Detecting this bifurcation would have important biological implications. First, we would know that the duration of the recurrent bouts of fast demographic oscillations would increase with the parameter (toward an infinite time at the bifurcation). This is because, close to the saddle, the rates of evolutionary change almost vanish, i.e., coevolution almost stops and the time spent in region C by the evolutionary cycle becomes very large. But we would also know that for a further small increase of the parameter, i.e., after the disappearance of the evolutionary cycle, there would be only two long-term possibilities for the community, namely extinction of the consumer and stationary coexistence at the evolutionary equilibrium in region S.

The bifurcation analysis of the discontinuous AD canonical equation is much more complex than that of a standard canonical equation, because discontinuous systems, besides all standard bifurcations, also have special bifurcations involving some sliding on the discontinuity boundary (such bifurcations are for this reason called *sliding bifurcations*). Of special biological interest is the issue of the appearance or disappearance of an evolutionary ridge (sliding segment). This bifurcation is characterized by the collision of two tangent points, and can therefore be continued with respect to two parameters. In other words, the points in a two-parameter space associated with the appearance of a sliding segment lie on a curve that can be produced through numerical continuation. Another biologically relevant sliding bifurcation is the collision of an equilibrium with the discontinuity boundary. Under suitable conditions, this bifurcation can give rise to an evolutionary pseudo-equilibrium.

Many sliding bifurcations are possible and their comprehensive classification has been obtained only in the special case of second-order systems (Kuznetsov et al., 2003). The methods for numerical bifurcation analysis of standard systems can be extended to discontinuous systems without major difficulties (Dercole and Kuznetsov, 2005). The use of numerical bifurcation analysis is essential in this study, because the bifurcations and attractors involved in the AD canonical equation (9.17) are far too many for one to reach a useful synthesis only through the simulation of evolutionary trajectories. However, a comprehensive presentation of the full bifurcation analysis is beyond the scope of this chapter. Instead, we will concentrate on the most biologically insightful bifurcations and present them through a series of suitably selected evolutionary state portraits.

9.5 RESULTS AND DISCUSSION

In this section, we discuss the series of evolutionary state portraits reported in Figure 9.6. Portraits A–D reported in the first column have been obtained for increasing values of the parameter h_2 (see (9.8c)), i.e., for increasing sensitivity of the consumption rate to the resource and consumer traits. Thus, the advantage for the resource to unbalance its trait with respect to the consumer trait increases from top to bottom. For the lowest value of h_2 (portrait A) the system has two alternative evolutionary attractors, an equilibrium in region S and a pseudo-equilibrium (squared point), separated by the stable manifold of the saddle in region C (half-filled point). Generic trajectories approaching the pseudo-equilibrium first reach the boundary separating S from C and then slide on it toward the pseudo-equilibrium. Thus, the fate of the two populations is to remain trapped forever in a sort of uncertain mode of coexistence, between stationary and cyclic demographic regimes. An increase of h_2 gives rise to a supercritical Hopf bifurcation of the evolutionary equilibrium, which becomes unstable and surrounded by a stable cycle; just after the Hopf bifurcation, this evolutionary cycle is small and entirely contained in region S (see portrait B). This is the most frequently discussed example of cyclic Red Queen dynamics, namely long evolutionary cycles entraining the resident population abundances, which, however, are always at equilibrium on the demographic timescale. For higher values of h_2 (portrait C) the pseudo-equilibrium is replaced with an equilibrium in region S (very close to the boundary between S and C), while the

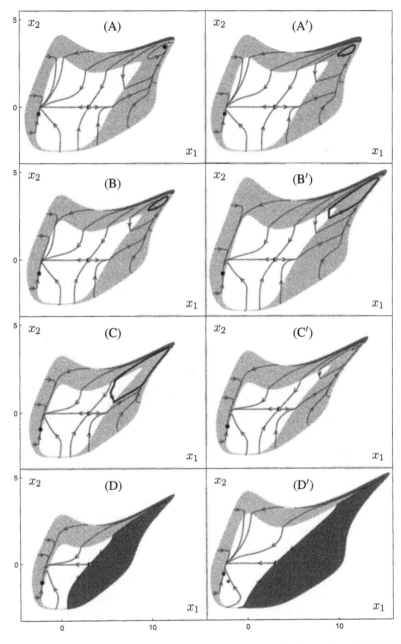

Figure 9.6 Eight evolutionary state portraits of the AD canonical equation (9.17). In the first column parameters are $k_1 = 0.1$, $k_2 = 1$, $r = 1$, $b = 0.001$, $e = 0.1$, $\tau^{-1} = 1.5$, $K_0 = 1$, $K_1 = 1$, $d_0 = 0.01$, $d_2 = 3$, $h_0 = 0.02$, and $h_2 = 0.01$ in A, $h_2 = 0.017$ in B, $h_2 = 0.03$ in C, $h_2 = 0.05$ in D; portrait A′ is obtained from A by increasing k_1 from 0.1 to 0.145; B′ from B by increasing τ^{-1} from 1.5 to 3; C′ from C by increasing d_2 from 3 to 3.3; D′ from D by increasing K_0 from 1 to 5. All dark regions correspond to consumer evolutionary extinction.

evolutionary cycle becomes larger and penetrates region $\mathcal{C}$. As anticipated in the previous section, the resulting dynamics are characterized by slow variations of population abundances entrained by the evolutionary cycle (in region $\mathcal{S}$) and by recurrent bouts of fast (demographic) oscillations (in region $\mathcal{C}$). A further increase of h_2 forces the evolutionary cycle to hit the saddle in region $\mathcal{C}$ (homoclinic bifurcation), thus destroying the Red Queen dynamics and leaving the community with a single evolutionary attractor in region $\mathcal{S}$ (portrait D). Increasing h_2, therefore, prolongs coexistence in the demographic oscillatory regime without ever locking the populations permanently in this regime.

We can now look at the second column of Figure 9.6 where each evolutionary state portrait has been obtained from the corresponding one in the first column by varying one parameter. Portrait A$'$, compared with A, shows the effect of an increase of the resource mutational rate (k_1). Accelerating the pace of resource evolution relative to the consumer has the power of generating Red Queen dynamics: the evolutionary stable equilibrium of portrait A undergoes a supercritical Hopf bifurcation and becomes a small evolutionary cycle entirely contained in region $\mathcal{S}$.

Portrait B$'$ is obtained from B by increasing the maximum consumption rate (τ^{-1}); this may be caused by an environmental change (e.g., a change in habitat structure or temperature) modifying the consumer searching behavior or metabolism. The result is the enhancement of Red Queen dynamics, as the evolutionary cycle now enters region $\mathcal{C}$. Insisting in that direction, the cycle disappears through a homoclinic bifurcation (i.e., the coexistence of resource and consumer with high trait values becomes impossible for large consumption rates).

Portrait C$'$ is obtained from C by making the consumption-independent optimal trait values K_1 and d_2 more different. As predicted by Abrams (2000), this reduces the number of attractors and, indeed, after the perturbation the system is left with a single evolutionary attractor.

Finally, portrait D$'$ is obtained from D by enriching the resource habitat, i.e., by increasing the maximum carrying capacity (K_0). If before enriching (portrait D) the system is at its evolutionary equilibrium in region $\mathcal{S}$, the populations coexist in the stationary demographic regime. Thus, in agreement with the paradox of enrichment (Rosenzweig, 1971), if evolutionary processes were absent (i.e., the traits x_1 and x_2 were kept frozen), significant enrichment would destabilize the population. This is clearly recognizable from portrait D$'$, where the point $*$ in region $\mathcal{C}$ is the copy of the evolutionary equilibrium of portrait D. Interestingly, after enrichment the evolutionary processes act in the opposite direction and the final result (portrait D$'$) is that the traits tend to an evolutionary pseudo-equilibrium. In other words, the full destabilization of the population triggered by enrichment is opposed by the counteracting forces of evolution. This is in line with a conjecture first formulated by Rosenzweig and Schaffer (1978b), and has been verified to hold even for large variations of K_0.

Summarizing the results yields a series of eight statements: the first three are general, while the others are specific of resource-consumer communities. Besides being supported by the above interpretation of specific evolutionary state portraits, these statements provide the synthesis of a thorough bifurcation analysis of the AD canonical equation (9.17).

1. *Evolutionary sliding and pseudo-equilibria.* Evolutionary sliding along the boundary separating stationary from cyclic coexistence occurs for many parameter settings. The evolutionary sliding can be temporary along evolutionary ridges (sliding segment) or halt at an evolutionary pseudo-equilibrium. When the adaptive traits are sliding, or resting at a pseudo-equilibrium, the populations are in critically stable demographic states and their mean characteristics (abundance, density-dependent parameters) can vary abruptly for small changes of individual traits. Evolutionary sliding and pseudo-equilibria on evolutionary ridges are novel theoretical phenomena of general significance for biological systems subject to discontinuous selection pressures.

2. *Evolutionary extinction.* There is always a subregion (dark area in all evolutionary state portraits) where evolutionary trajectories tend toward the boundary of region $\mathcal{E}$. This causes the consumer population to go extinct in the long run, a phenomenon that is not predictable on the basis of purely ecological arguments. Evolution to extinction has been also noted in Chapters 5, 6, and 7. These examples highlight a common mechanism. Evolutionary change is driven by the "marginal" benefit of performing better on interactions (consumption, competition, cooperation) than other conspecifics. Yet the "direct" physiological cost to the individual can become so great that eventually the population growth rate becomes negative, causing extinction.

3. *Multiple evolutionary attractors.* Although in some cases (not shown in the portraits of Figure 9.6) there is only one evolutionary equilibrium, most often there are several (attractors, repellors, and saddles), in addition to evolutionary cycles and consumer extinction. Two general implications can be drawn. First, even if different populations share the very same history of selective pressures, differences in current evolutionary state and differences in contemporary demographic dynamics can trace back to ancient differences in their genotypic state. Experimental evolution in *Escherichia coli* provides strong empirical support to this prediction (Travisano et al., 1995). Second, the co-occurrence of consumer evolutionary extinction and other viable evolutionary attractors provides a firm mathematical basis for the notion of evolutionary trapping suggested from empirical observations (Colas et al., 1997; Schlaepfler et al., 2002): under given environmental conditions, the consumer population can be trapped on an evolutionary trajectory heading to extinction whereas alternative, ecologically safe, evolutionary attractors could have been reached. Schlaepfler et al. (2002) and Ferrière et al. (2004b) have discussed the implications of evolutionary trapping in a conservation perspective.

4. *Two forms of cyclic Red Queen dynamics.* The first one (evolutionary cycle in region $\mathcal{S}$, see portraits B and A$'$) corresponds to slow periodic variations of the traits entraining slow population cycles. This is the form that has been discussed predominantly in the literature (see Abrams, 2000, for a review, and Chapter 5), often on the basis of Lotka-Volterra models that did not allow for other forms of Red Queen dynamics. The second form (evolutionary

cycle partly in S and partly in C, see portraits C and B$'$) corresponds to slow periodic variations of the traits accompanied by recurrent and long bouts of demographic oscillations. This form, predicted by Khibnik and Kondrashov (1997), is the most complex form of cyclic Red Queen dynamics. Although empirical data demonstrating this type of dynamics in resource-consumer interaction remain scant (Lively, 1993; Abrams, 2000), this could change rapidly in the wake of recent experimental work on rotifers and their resource (algae) (Yoshida et al., 2003).

5. *Factors enhancing Red Queen dynamics.* Our study confirms that a bidirectional axis of resource vulnerability is a potent mechanism for generating coevolutionary cycles (Abrams, 2000). The typical sequence of evolutionary attractors detected by increasing the impact of the traits on vulnerability is the following (see portraits A, B, C): first an evolutionary equilibrium associated with a stationary demographic regime, then an evolutionary cycle with entrained slow population oscillations, and finally an evolutionary cycle associated with recurrent bouts of fast demographic oscillations. For a further increase of the vulnerability mechanism (see portraits D), Red Queen dynamics suddenly disappear, a phenomenon that has gone unnoticed in previous studies. Other factors, including the frequency of mutations, the variance of mutational effects, and the maximum consumption rate can also beget and enhance Red Queen dynamics (see portraits A, A$'$ and B, B$'$).

6. *The consumer chases the resource.* All evolutionary cycles we have detected are counterclockwise. This means that the consumer trait increases when the resource trait is large and decreases in the opposite case. This property is a consequence of the presence of a bidirectional axis of resource vulnerability, and is, indeed, present in all studies where the resource has a most vulnerable form depending upon the consumer trait (see Marrow et al., 1992; Marrow and Cannings, 1993; Dieckmann et al., 1995; Dieckmann and Law, 1996; Gavrilets, 1997; Abrams and Matsuda, 1997; Khibnik and Kondrashov, 1997).

7. *Evolution toward demographic stability: The paradox of enrichment.* Ecological theory predicts that stable resource-consumer interactions should yield to large amplitude demographic cycles in richer environments (Rosenzweig, 1971). The "paradox of enrichment" emphasizes that this does not occur in nature (see, e.g., Murdoch et al., 1998). Abrams and Walters (1996) found an ecological solution to the paradox for certain types of resource-consumer communities, later confirmed by experimental findings (McCauley et al., 1999). Rosenzweig and Schaffer (1978b) took a general evolutionary approach to the problem, arguing that coevolution should tend to restore the demographic stability lost through enrichment. Evolution may actually play such a significant role in light of, e.g., the findings of Yoshida et al. (2003) on rapid evolutionary change in resource-consumer systems. Our work substantiates, refines, and broadens Rosenzweig and Schaffer's view in the case of slow consumer and fast resource (see portraits D, D$'$).

8. *Evolution opposes permanent demographic oscillations.* There seems to be no realistic environmental condition under which an evolutionary attractor is entirely in region C, though evolutionary trajectories are often trapped on the boundary between S and C (see statement 1). Only if the consumption rate is almost independent of resource and consumer traits (i.e., if h_2 is of the order of 10^{-3}), is there an evolutionary equilibrium in C (notice that for $h_2 = 0$, point (K_1, d_2) is a stable evolutionary equilibrium in C). Thus, evolution seems to oppose permanent demographic oscillations.

9.6 CONCLUDING REMARKS

In this chapter we have studied the interplay of demographic and evolutionary dynamics in an analytical model for slow-fast populations. The study has been conducted through simulation and bifurcation analysis of the AD canonical equation, which is discontinuous in trait space on the boundary that separates region S, corresponding to stationary demographic coexistence of the resident populations, from region C, where the coexistence is cyclic. Focusing on slow-fast populations may be not as restrictive as one could a priori imagine, because interacting populations often involve contrasting timescales. Furthermore, the novel phenomena of evolutionary sliding along evolutionary ridges, and the confinements of traits at evolutionary pseudo-equilibria, should be generic to all systems in which smooth changes in adaptive traits can cause abrupt changes in demographic dynamics, thereby causing sharp variations in the selection pressure.

The example of slow-fast populations we have considered is that of a resource-consumer interaction, with one adaptive trait for each population. A triplet of demographic parameters is influenced by the coevolving traits: two of them (resource carrying capacity and consumer death rate) depend upon only one trait, while the third one (half-saturation constant of consumer functional response) depends upon both traits in such a way that resources can reduce their risk by increasing or decreasing their trait value relative to a most vulnerable form, which depends on the consumer trait. This mechanism, called the bidirectional axis of resource vulnerability, is quite powerful in generating Red Queen dynamics, as stressed by Abrams (2000).

Our analysis points out a number of other properties: enhancing of Red Queen dynamics through the increase of the relative mutational rate of the resource, generic occurrence of evolutionary extinction in the consumer, and coevolution acting against demographic destabilization resulting from enrichment; these conclusions are not new, but they had been obtained previously from independent studies involving different modeling approaches and assumptions that make them sometimes difficult to compare (see Abrams, 2000). Our analysis also unravels novel evolutionary phenomena whose scope extends beyond resource-consumer coevolution. This includes the possibility that coevolution canalizes evolutionary trajectories along evolutionary ridges formed by segments of the boundary between regions S and C (evolutionary sliding), or comes to a halt at special points of that boundary (evolutionary pseudo-equilibria). Although the absence of evolutionary attractors in region C lends weight to the view that the coevolution of resource-consumer populations acts against sustained demographic oscillations, convergence to an evolutionary pseudo-equilibrium implies that the populations are poised between stationary and cyclic coexistence. Also, recurrent long bouts of demographic oscillations are typical of the most complex Red Queen regimes.

How general our conclusions for resource-consumer coevolution are is likely to be influenced by the specific model we have formulated. Our approach, however, is broad in scope and should, in principle, be repeatable for any demographic model involving slow-fast dynamics. In particular, it would be interesting to conduct a similar study for the dual case, namely that of slow resource (e.g., plants) and fast consumer (e.g., insects). Considering, for example, how common recurrent insect-pest outbreaks are in natural or exploited forests, we would not be surprised if in that case coevolution would have just the opposite effect on demographic dynamics, namely that of favoring cyclic regimes. Yet the analysis is likely to be complicated by the fact that a plant-insect model capable of explaining periodic pest outbreaks should contain at least two extra sources of consumer mortality: intraspecific competition and consumption by insectivores (Ludwig et al., 1978; Rinaldi and Muratori, 1992a). The more general picture would include nutrients for the resource species and higher-level consumers. Models of resource and consumer behavioral optimization can predict quite different outcomes when the two-species interaction is embedded in its four-level trophic chain (Abrams, 1992b). Abrams' (2000) observation that the analysis of coevolutionary dynamics in systems with three or more species represents empirical and theoretical "terra incognita" remains valid and this study gives one more reason to open this research avenue (see Chapter 10).

Finally, we cannot neglect mentioning that the conjecture formulated by Ellner and Turchin (1995) on the basis of their analyses of population time series, namely that "ecosystems might evolve toward the edge of chaos," may here find some support. In fact, formulated in different words, the conjecture becomes "ecosystems might evolve toward the edge of their most complex dynamical behavior," and, indeed, our findings are much in line with this statement, since the most complex dynamical behavior of the Rosenzweig-MacArthur resource-consumer model is cyclic coexistence. But the support could become even stronger if we could extend the present analysis to tritrophic food chains with dynamically diversified trophic levels. This is virtually possible since singular cycles and singular homoclinic bifurcations (responsible of chaotic dynamics) have already been found in this system (De Feo and Rinaldi, 1998; Deng, 2001). What remains to be done, after identifying potential adaptive traits for each trophic level, is to derive and then study the corresponding AD canonical equation. If the results were to be the natural extension of what we have found for ditrophic food chains, then we might discover that coevolution tends to oppose chaotic population dynamics but favors, at least under certain conditions, evolutionary sliding along the boundary of the chaotic region. Bifurcations leading to chaotic demographic oscillations are indeed likely to be a common cause of discontinuity in the selection pressure acting on the community. Complex Red Queen dynamics giving rise to intermittent bouts of chaotic demographic oscillations of the resident populations would then be the expected outcome, as predicted by Ellner and Turchin's conjecture.

Chapter Ten

The First Example of Evolutionary Chaos

We present in this chapter the first example of chaotic Red Queen dynamics. We consider a Lotka-Volterra tritrophic food chain composed of a resource, its consumer, and a predator species, each characterized by a single adaptive trait, and we show that for suitable modeling and parameter choices the evolutionary trajectories of the corresponding AD canonical equation approach an evolutionary strange attractor in the three-dimensional trait space. Most of this chapter is taken from Dercole and Rinaldi (2008).

10.1 INTRODUCTION

Until now we have examined numerous examples of evolutionary dynamics, which, however, were all concerned with the evolution of at most two traits. This limitation was motivated by a technical point: the evolution of two traits is described by a second-order AD canonical equation, and second-order dynamical systems enjoy a number of remarkable properties that do not hold for higher-dimensional systems. The bifurcations of second-order systems are also a particular subclass of the bifurcations of general n-dimensional systems and this has allowed us to use throughout the book only the most simple notions and methods of bifurcation analysis.

Physically speaking, the most important difference between two- and higher-dimensional systems is that in the former class the most complex regimes are cyclic, while in the latter (random-like) complex aperiodic regimes are also possible (see, e.g., Strogatz, 1994). This behavior is called *deterministic chaos* because it is generated by purely deterministic differential equations. When the system is chaotic, its trajectories tend toward and then remain on a so-called *strange* or *chaotic attractor* (Ruelle and Takens, 1971), which resembles a tangle and has fractal geometry (Mandelbrot, 1977; Peitgen et al., 2004). Trajectories starting from two very close points within the strange attractor diverge (*stretching*) at an average speed measured by the so-called *Lyapunov exponent* (see, e.g., Alligood et al., 1996), but are forced to remain within the strange attractor by a *folding* mechanism. The result of these two conflicting forces is a high sensitivity to initial conditions, often referred to as *unpredictability*.

In some of our applications (see Chapters 5, 6, and 9) we have detected cyclic evolutionary regimes, which were therefore the most complex possible regimes due to the limited number of adaptive traits. By assuming rapid evolution (behavioral adaptation or very high mutational rates), Abrams and Matsuda (1997) found chaotic dynamics in a three-dimensional space composed of a single adaptive trait, characterizing a resource harvested by a nonevolving consumer, and the two abundances of the resource and consumer populations. However, by letting adaptive traits and population abundances evolve on the same timescale, one cannot say if complex evolutionary dynamics are the genuine consequence of mutation-selection processes or if they are induced by particular demographic interactions. Thus, the following basic question spontaneously arises: Have evolutionary mechanisms involving a sufficiently large number of adaptive traits the power of generating deterministic chaos?

In principle, an answer to this question could be given if rich sets of field or laboratory data would be available. Unfortunately, time series of evolutionary traits are too short (with a few exceptions, among which are the paleontological time series, see Section 1.9) to justify the use of the statistical techniques proposed in the last decades for revealing the existence of deterministic chaos (Ott et al., 1994; Abarbanel, 1996). On the other hand, paleontological time series often refer to the evolution of systems driven by highly chaotic climatic variations. Therefore, the statistical tests applied to these series simply say that the biological response to climate variability is chaotic, but do not reveal, however, if evolutionary mechanisms have the power of generating their own chaos. By contrast, evolutionary models with constant parameters virtually mimic the ideal conditions of an absolutely not varying physical environment, and are therefore perfectly suited for answering our question. Moreover, given as granted that more details (for example, on sex or age, stages, and space structures) will increase the chances of generating chaos in a model, we can reasonably pretend to positively answer the question if we can show that a simple three-dimensional evolutionary model can be chaotic. This is, actually, what has been done in various fields of science, starting with meteorology (Lorenz, 1963), where the first strange attractor has been found, and proceeding with mechanics (Hayashi et al., 1970), chemistry (Rössler, 1976), electronics (Madan, 1993), epidemiology (Schwartz and Smith, 1983), and ecology (Hastings and Powell, 1991). Of course, we should a priori expect a positive answer to our question if we believe that evolution is at least as complex as the just mentioned fields.

This chapter is devoted to the presentation of the first chaotic evolutionary attractor. In the next section, a very simple tritrophic food chain is considered and the corresponding three-dimensional AD canonical equation is derived. Then, in the following section it is shown that for a suitable parameter setting the evolutionary dynamics are chaotic, and the main characteristics of the evolutionary strange attractor are pointed out. In Section 10.4 it is also shown how the evolutionary regimes change from stationary to cyclic and then from cyclic to chaotic when the mutational rate of the population at the lowest trophic level is increased. This numerically points out the most frequent "route to chaos," namely the famous Feigenbaum cascade of period-doubling bifurcations (Feigenbaum, 1980). A few comments on the value and limitations of our findings close the chapter.

10.2 A TRITROPHIC FOOD CHAIN MODEL AND ITS

AD CANONICAL EQUATION

The demographic model we use for pointing out evolutionary chaos is a classical Lotka-Volterra tritrophic food chain composed of resource, consumer, and predator populations, each characterized by a single adaptive trait (e.g., body size, as in Chapter 5). There are three reasons for this choice. First, the model is very simple and therefore appropriate for pointing out the first evolutionary strange attractor. Second, for given values of the traits the three populations can coexist only at a unique and stable demographic equilibrium. In other words, we will show that evolution can be chaotic without requiring wild demographic dynamics of the populations. Third, the model is the natural extension of the ditrophic food chain model analyzed in Chapter 5, which was shown to have the most complex evolutionary regimes (namely cyclic regimes because of the limitation to two adaptive traits). This suggests that a two-species model with cyclic evolutionary dynamics could easily become chaotic by adding a third coevolving species.

The Lotka-Volterra tritrophic food chain model (resident model) has the form

$$\dot{n}_1 = rn_1 - cn_1^2 - a_2 n_1 n_2, \tag{10.1a}$$

$$\dot{n}_2 = e_2 a_2 n_1 n_2 - d_2 n_2 - a_3 n_2 n_3, \tag{10.1b}$$

$$\dot{n}_3 = e_3 a_3 n_2 n_3 - d_3 n_3, \tag{10.1c}$$

where n_1, n_2, and n_3 are resource, consumer, and predator population abundances, r and c are resource net growth rate and intraspecific competition, and a_i, e_i, and d_i are attack rate, efficiency, and net death rate (not due to consumption) of consumer $(i = 2)$ and predator $(i = 3)$.

Model (10.1) has a unique nontrivial equilibrium

$$\bar{n}_1 = \frac{1}{c}\left(r - \frac{a_2 d_3}{e_3 a_3}\right), \tag{10.2a}$$

$$\bar{n}_2 = \frac{d_3}{e_3 a_3}, \tag{10.2b}$$

$$\bar{n}_3 = \frac{e_2 a_2}{a_3 c}\left(r - \frac{a_2 d_3}{e_3 a_3}\right) - \frac{d_2}{a_3}, \tag{10.2c}$$

which is positive if and only if $\bar{n}_3 > 0$, i.e.,

$$\frac{r}{c} - \frac{a_2 d_3}{ce_3 a_3} - \frac{d_2}{e_2 a_2} > 0. \tag{10.3}$$

Moreover, the equilibrium (10.2) is always globally stable (in the positive orthant $n_i \geq 0$, $i = 1, 2, 3$), which means that under condition (10.3) the resident model has only one asymptotic mode of behavior, namely stationary coexistence. Condition (10.3) marks the extinction of the predator population and technically corresponds to a transcritical bifurcation (see Appendix A) of model (10.1), at which the equilibrium (10.2) collides and exchanges stability with the trivial equilibrium lying on the face $n_3 = 0$ of the demographic state space.

If we now imagine that a mutant population is also present, we must enlarge model (10.1) by adding a fourth ODE for the mutant population and by specifying how the demographic parameters depend upon the traits x_1, x_2, x_3, x_1', x_2', x_3'. The number of possibilities is practically unlimited because even for well-identified species there are many meaningful options. To be consistent with the analysis of the ditrophic food chain carried out in Chapter 5, we assume that the parameters r, e_i, and d_i, $i = 1, 2$, are trait-independent. Thus, in the case of a mutation in the resource population, the resident-mutant model is

$$\dot{n}_1 = n_1(r - c(x_1)n_1 - \gamma(x_1, x_1')n_1' - a_2(x_1, x_2)n_2),$$
$$\dot{n}_1' = n_1'(r - \gamma(x_1', x_1)n_1 - c(x_1')n_1' - a_2(x_1', x_2)n_2),$$
$$\dot{n}_2 = n_2(e_2 a_2(x_1, x_2)n_1 + e_2 a_2(x_1', x_2)n_1' - d_2 - a_3(x_2, x_3)n_3),$$
$$\dot{n}_3 = n_3(e_3 a_3(x_2, x_3)n_2 - d_3),$$

where $\gamma(x_1, x_1')$ is the competition coefficient characterizing reduced birth rate and/or increased death rate in the resource resident population due to the competition with the resource mutant population (necessarily $\gamma(x_1, x_1) = c(x_1)$). As in Chapter 5, we assume a constant (therefore symmetric) competition function $\alpha(x_1, x_1') = \gamma(x_1, x_1')/c(x_1)$ (see Section 3.3 for a discussion on competition symmetry), so that $\gamma(x_1, x_1') = c(x_1)$.

Similarly, the two other resident-mutant models, describing the demographic interactions in the cases of mutations in the consumer and in the predator populations, are given by

$$\dot{n}_1 = n_1(r - c(x_1)n_1 - a_2(x_1, x_2)n_2 - a_2(x_1, x_2')n_2'),$$
$$\dot{n}_2 = n_2(e_2 a_2(x_1, x_2)n_1 - d_2 - a_3(x_2, x_3)n_3),$$
$$\dot{n}_2' = n_2'(e_2 a_2(x_1, x_2')n_1 - d_2 - a_3(x_2', x_3)n_3),$$
$$\dot{n}_3 = n_3(e_3 a_3(x_2, x_3)n_2 + e_3 a_3(x_2', x_3)n_2' - d_3),$$

and

$$\dot{n}_1 = n_1(r - c(x_1)n_1 - a_2(x_1, x_2)n_2),$$
$$\dot{n}_2 = n_2(e_2 a_2(x_1, x_2)n_1 - d_2 - a_3(x_2, x_3)n_3 - a_3(x_2, x_3')n_3'),$$
$$\dot{n}_3 = n_3(e_3 a_3(x_2, x_3)n_2 - d_3),$$
$$\dot{n}_3' = n_3'(e_3 a_3(x_2, x_3')n_2 - d_3).$$

In line with Chapter 5 (see, in particular, Figure 5.1), resource intraspecific competition c is given by

$$c(x_1) = c_1 + c_2 (x_1 - c_0)^2,$$

where parameter c_0 is the *optimum resource trait* at which intraspecific competition is minimum, and the attack rates a_2 and a_3 are

$$a_2(x_1, x_2) = \exp\left(-\left(\frac{x_1 - a_{24}}{a_{21}}\right)^2 + 2a_{23}\frac{(x_1 - a_{24})(x_2 - a_{25})}{a_{21}a_{22}} - \left(\frac{x_2 - a_{25}}{a_{22}}\right)^2\right),$$

$$a_3(x_1, x_2) = \exp\left(-\left(\frac{x_1 - a_{34}}{a_{31}}\right)^2 + 2a_{33}\frac{(x_1 - a_{34})(x_2 - a_{35})}{a_{31}a_{32}} - \left(\frac{x_2 - a_{35}}{a_{32}}\right)^2\right),$$

where $a_{23} < 1$ and $a_{33} < 1$. If resource and consumer [consumer and predator] traits are tuned, i.e., if $x_1 = a_{24}$, $x_2 = a_{25}$ [$x_2 = a_{34}$, $x_3 = a_{35}$], the consumer [predator] attack rate is maximum. When resource and consumer [consumer and predator] traits are far from being tuned, the consumer [predator] attack rate vanishes.

At this point, the AD canonical equation can be derived since the resident demographic equilibrium (10.2) is explicitly known. For this, in accordance with the definition (3.19) in Chapter 3, the invasion fitnesses of the resource, consumer, and predator mutant populations are given by

$$\lambda_1(x_1, x_2, x_3, x_1') = r - c(x_1')\bar{n}_1(x_1, x_2, x_3) - a_2(x_1', x_2)\bar{n}_2(x_1, x_2, x_3),$$
$$\lambda_2(x_1, x_2, x_3, x_2') = e_2 a_2(x_1, x_2')\bar{n}_1(x_1, x_2, x_3) - d_2 - a_3(x_2', x_3)\bar{n}_3(x_1, x_2, x_3),$$
$$\lambda_3(x_1, x_2, x_3, x_3') = e_3 a_3(x_2, x_3')\bar{n}_2(x_1, x_2, x_3) - d_3,$$

where the equilibrium abundances $\bar{n}_i(x_1, x_2, x_3)$, $i = 1, 2, 3$, are given by (10.2). Then, the canonical equation is obtained as

$$\dot{x}_i = k_i \bar{n}_i(x_1, x_2, x_3) \left. \frac{\partial}{\partial x_i'} \lambda_i(x_1, x_2, x_3, x_i') \right|_{x_i'=x_i}, \quad i = 1, 2, 3, \qquad (10.4)$$

where $k_i = 1/2\mu_i\sigma_i^2$, $i = 1, 2, 3$, are constant mutational rates, proportional to the frequency and variance of mutations in the resource, consumer, and predator populations, respectively. The explicit expression of the canonical equation (10.4) is not reported because very long. However, it can be generated and handled by means of symbolic computation.

The canonical equation (10.4) could be studied through bifurcation analysis, as we have done throughout the book. However, this would require notions of bifurcation theory that we have not provided in Appendix A, which covers only the case of second-order systems. For this reason equation (10.4) will be used in the following only to carry out simulations, namely to determine the evolution of the three traits for a given parameter setting and for given ancestral conditions $x_i(0)$, $i = 1, 2, 3$. Of course, the ancestral conditions must guarantee the short-term coexistence of the three populations, i.e., $x_i(0)$, $i = 1, 2, 3$, must be such that condition (10.3) is satisfied. Condition (10.3) therefore defines the evolution set $\mathcal{X}$ of the community, which results in the bounded ovoid region shown in Figure 10.1A. The simulation of equation (10.4) starting from an initial point in $\mathcal{X}$ produces an evolutionary trajectory, which can either remain in $\mathcal{X}$ forever, thus converging toward an evolutionary attractor (an equilibrium in the example of Figure 10.1B), or reach the boundary of $\mathcal{X}$ in finite time, thus pointing out the evolutionary extinction of the predator population. This is actually a case of evolutionary murder (see Sections 1.8 and 3.6), since the boundary of $\mathcal{X}$ is a transcritical bifurcation of the resident model (10.1), at which the predator population abundance vanishes.

10.3 THE CHAOTIC EVOLUTIONARY ATTRACTOR

We now present a chaotic evolutionary attractor produced by the AD canonical equation (10.4). As far as we know, this is the first example of a chaotic attractor obtained through AD. For the moment we do not reveal how we were able to find a

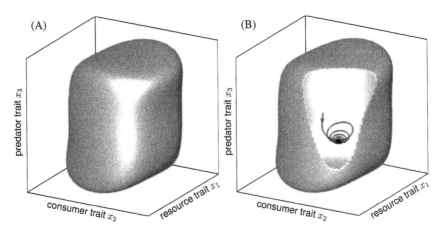

Figure 10.1 The evolution set $\mathcal{X}$ characterizing short-term stationary coexistence of the community at the demographic equilibrium (10.2) (A), and an evolutionary trajectory tending toward a stable equilibrium of the canonical equation (10.4) (B). Parameter values are $r = 0.5$, $d_2 = 0.05$, $d_3 = 0.02$, $e_2 = 0.14$, $e_3 = 0.14$, $c_1 = 0.5$, $c_2 = 3$, $c_0 = 0$, $a_{21} = 0.22$, $a_{22} = 0.25$, $a_{23} = 0.6$, $a_{24} = 0$, $a_{25} = 0.04$, $a_{31} = 0.22$, $a_{32} = 0.25$, $a_{33} = 0.6$, $a_{34} = 0$, $a_{35} = -0.04$, $k_1 = 0.15$, $k_2 = 1$, $k_3 = 1$. Ancestral conditions in B are $x_1(0) = -0.0411$, $x_2(0) = -0.0372$, $x_3(0) = 0.0075$.

parameter setting corresponding to chaotic evolutionary dynamics, but a discussion on this point is reported in the next section.

The chaotic attractor is shown in Figure 10.2. The left panel (Figure 10.2A) points out the attractor in the three-dimensional trait space, while the right panel (Figure 10.2B) shows three segments of the corresponding aperiodic time series $x_i(t)$, $i = 1, 2, 3$. Some features of the chaotic attractor can be identified through a visual inspection of Figure 10.2. Indeed, Figure 10.2B shows that the traits of the first two species vary almost periodically, while x_3 varies more irregularly. The intervals between successive peaks of the predator trait x_3 are almost constant (this property is called *coherence*), while the peaks of x_3 alternate irregularly, like in the classical two-band Rössler strange attractor (Rössler, 1976). Figure 10.2A shows that the attractor lies roughly on a Möbius strip and has therefore a fractal dimension very close to 2. Since chaotic attractors with fractal dimension close to 2 must have "peak-to-peak dynamics" (Candaten and Rinaldi, 2000), one should a priori expect that the so-called peak-to-peak plot, namely the set of all pairs of successive peaks of any trait, identifies a smooth curve. This is indeed the case, as shown in Figure 10.3, where each point represents a pair of successive peaks of the predator trait x_3 extracted from a long time series produced by model (10.4). The curve drawn in Figure 10.3, called the skeleton of the peak-to-peak plot, can be used to predict the value of the next peak of the predator trait from the value of the last peak. This is a rather intriguing property, since in a sense it shows that at least some predictions are possible even if the system is chaotic.

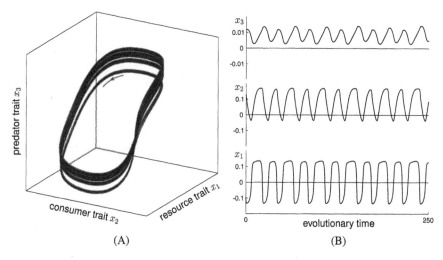

Figure 10.2 Chaotic evolutionary attractor (A) and corresponding time series of resource (x_1), consumer (x_2), and predator (x_3) traits (B). Parameter values as in Figure 10.1 except for $k_1 = 0.64$.

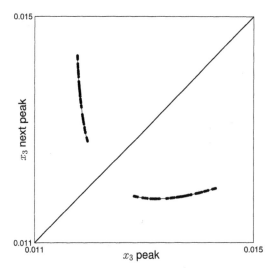

Figure 10.3 Peak-to-peak plot of the predator trait x_3. Each point represents a pair of consecutive peaks of an x_3 time series associated with the strange attractor of Figure 10.2. The curve drawn through the points can be used to forecast the next peak on the basis of the last peak.

Three so-called Lyapunov exponents, $L_1 > L_2 > L_3$, are associated with the chaotic attractor of Figure 10.2. Lyapunov exponents measure the mean exponential rates of initial divergence (if positive) or convergence (if negative) of nearby initial conditions along three independent directions suitably selected at each point of the attractor (see Ramasubramanian and Sriram, 2000, for computational issues). As anticipated in the Introduction, a chaotic attractor is characterized by both divergence (stretching) and convergence (folding) of nearby trajectories. Moreover, two close points on the same trajectory are neutral to the stretching and folding mechanisms (their distance initially evolve linearly with time), so that it must be $L_1 > 0$, $L_2 = 0$, $L_3 < 0$. In particular, the largest Lyapunov exponent L_1 measures the mean exponential rate of initial divergence of two generic nearby points on the chaotic attractor, i.e., the mean sensitivity to initial conditions. The estimates obtained with the so-called standard algorithm (see Ramasubramanian and Sriram, 2000) are

$$L_1 = 8.1321\ldots\times10^{-3}, \quad L_2 = -2.3923\ldots\times10^{-6}, \quad L_3 = -4.6270\ldots\times10^{-1},$$

and the corresponding attractor fractal dimension, obtained with the famous Kaplan-Yorke formula $(2 - L_1/L_3$; see, e.g., Alligood et al., 1996), is $2.0176\ldots$, which confirms the correctness of our visual interpretation of Figure 10.2A.

10.4 FEIGENBAUM CASCADE OF PERIOD-DOUBLING BIFURCATIONS

We now explain how we arrived at the strange evolutionary attractor described in the previous section. Hunting for strange attractors is a very peculiar game and the worst possible way to play this game is to do it randomly. One should use instead a mix of intuition and theory.

In the present case intuition was based on the results obtained in Chapter 5, where it was shown that one way of obtaining the most complex evolutionary dynamics (cyclic dynamics in that case) of a ditrophic food chain composed of resource and consumer was to increase the mutational rate of the resource. We could then retain that message, and hope, on a purely intuitive ground, that the resource mutational rate could be an effective control parameter for transforming simple (i.e., stationary) into complex (i.e., chaotic) evolutionary dynamics in tritrophic food chains. On the other hand, bifurcation theory of three-dimensional systems (not reported in Appendix A) is quite precise on this matter. It says that there are two common routes to chaos, namely two special sequences of bifurcations that characterize successive structural changes of the attractor, until the strange attractor appears. One route is marked by a particular global bifurcation, called Shil'nikov homoclinic bifurcation (briefly mentioned in Appendix A), and the other is the celebrated Feigenbaum cascade of period-doubling bifurcations (Feigenbaum, 1980).

Physically speaking, a *period doubling* (also called *flip*) bifurcation occurs when a small variation of a parameter p, from $p_1 - \epsilon$ to $p_1 + \epsilon$, transforms an attracting cycle of period T into an attractive cycle of period $2T$, as shown in Figure 10.4. A second period-doubling bifurcation at $p = p_2$ would transform the cycle of Figure 10.4B into a cycle with four loops, and so on. Of course, the period of the cycle at $p = p_2 - \epsilon$ might be radically different than $2T$, though in technical jargon one often says that the second period-doubling yields a "period-4" cycle. Thus, n successive period-doubling bifurcations $\{p_1, p_2, \ldots, p_n\}$ transform the period-1

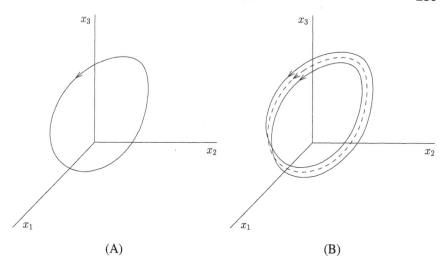

(A)   (B)

Figure 10.4 Period-doubling bifurcation: a stable cycle (solid trajectory in A) becomes unstable (dashed trajectory in B, a negative Floquet multiplier passes through -1
at the bifurcation) and a new stable cycle appears by tracing twice the bifur-
cating cycle (solid trajectory in B). Immediately before the bifurcation (A) the
attractor is a cycle with period T, while immediately after the bifurcation (B)
the attractor is a cycle with period $2T$.

cycle into a period-2^n cycle. The Feigenbaum cascade is an infinite sequence $\{p_i\}$
of period-doubling bifurcations in which the bifurcation values p_i accumulate at a
critical value p_∞ after which the attractor is a genuine strange attractor. Very often,
the strange attractor is coherent and its shape is similar to the period-1 cycle that
has originated the whole cascade.

With these elements in mind, we started from a parameter setting giving rise
to an evolutionary equilibrium, and tried to obtain an evolutionary cycle by varying some parameters. For doing this, we took into account the analysis performed
in Chapter 5, which suggests parameter settings giving rise to cyclic dynamics in
ditrophic food chains. Thus, once we had an evolutionary cycle we increased our
candidate control parameter (the resource mutational rate k_1) and after a few tri-
als we were able to detect a first period-doubling bifurcation, i.e., a clear warning
of a possible route to chaos. A further increase of the control parameter has confirmed the existence of a Feigenbaum cascade and has finally produced the strange
attractor of Figure 10.2. This route to chaos is visualized in Figure 10.5, where the
six panels A–F show the attractors corresponding to increasing values of the control parameter p, namely k_1: A, equilibrium; B, small cycle originated through a
Hopf bifurcation at $p = p_H$; C, large cycle; D, cycle after the first period-doubling
($p_1 < p < p_2$); E, cycle after the second period-doubling ($p_2 < p < p_3$); F,
strange attractor ($p > p_\infty$). Figure 10.6 reports the standard representation of the
Feigenbaum cascade, where the peaks of x_3 within the attractor are plotted for each
value of the control parameter.

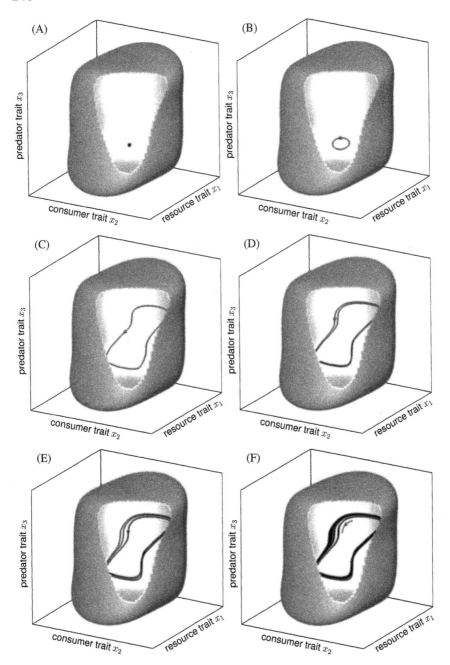

Figure 10.5 A sequence of evolutionary attractors obtained for increasing values of the resource mutational rate k_1: (A) $k_1 = 0.15$, equilibrium; (B) $k_1 = 0.2$, small
cycle; (C) $k_1 = 0.5$, big cycle; (D) $k_1 = 0.6$, period-2 cycle; (E) $k_1 = 0.635$,
period-4 cycle; (F) $k_1 = 0.64$, strange attractor. Other parameter values as in
Figure 10.1.

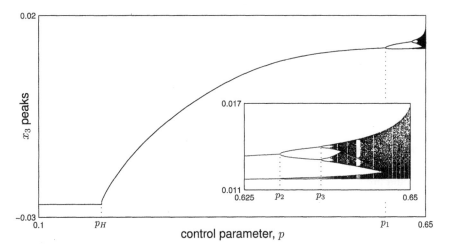

Figure 10.6 The standard representation of the Feigenbaum cascade obtained by plotting
the peaks of x_3 within the attractor corresponding to each value of the control
parameter p (i.e., k_1). Other parameter values as in Figure 10.1.

The shape of our strange attractor (see Figure 10.2) clearly reveals its Feigenbaum origin. Indeed, in the chaotic regime, the first two traits oscillate almost
periodically at the frequency $1/T$, while the third trait oscillates more irregularly
but still at the frequency $1/T$ (coherence). A physical image can be associated with
the chaotic evolutionary dynamics we have detected: one simply needs to stand on
a beach and observe the waves, which, indeed, arrive quite regularly on the beach
but with irregular (and often alternating) heights like the waves of the predator trait.

10.5 DISCUSSION AND CONCLUSIONS

We have shown in this chapter that the study of a Lotka-Volterra three-species food
chain reveals the possibility of chaotic evolutionary dynamics (Red Queen dynamics). This confirms that one of the long-term evolutionary scenarios predicted at
the end of Chapter 3 (a limited number of morphs per species which coevolve in an
apparently random fashion) is a possible consequence of mutation-selection processes.

The very special properties of our chaotic evolutionary attractor (coherence and
peak-to-peak dynamics) are certainly due to the extreme simplicity of the model
and would probably be lost with the addition of some extra realism. In other words,
we believe that the study of evolving systems with more complex structures would
reveal more complex chaotic regimes (so-called *hyperchaos*).

Although the results obtained in this chapter answer positively the basic question
raised in the Introduction (have evolutionary mechanisms involving a sufficiently
high number of traits the power of generating deterministic chaos?), they also bring
new and more subtle questions to our attention, such as:

1. Is it possible to identify other chaotic evolutionary attractors through mathematical models?

2. Is it possible to dig into field and laboratory evolutionary time series and detect the footprint of deterministic chaos?

The answer to the first question is certainly positive, though it might be hard, in practice, to discover new strange attractors. In a sense, we have already suggested how one could proceed. Good candidate models are those obtained by adding one species to any two-species interaction in which cyclic Red Queen dynamics have already been detected (e.g., the mutualistic interaction described in Chapter 6). Then one could proceed using the mix of intuition and theory described in the previous section, or, alternatively, perform the bifurcation analysis of the three-dimensional canonical equation, thus having higher chances to detect routes to chaos.

As for the second question, we suspect that scarcity of data, both in quantitative and qualitative terms, and exogenous randomness due, for example, to climatic variations, will prevent one from obtaining statistically significant answers. However, we believe it will be even more unlikely that one could prove the opposite, namely that there is no trace of endogenously produced evolutionary chaos in nature.

Appendix A

Second-order Dynamical Systems and Their Bifurcations

In this appendix we summarize the basic definitions and tools of analysis of dynamical systems, with particular emphasis on the asymptotic behavior of second-order continuous-time autonomous systems. In particular, the possible structural changes of the asymptotic behavior of the system under parameter variation, called bifurcations, are presented together with their analytical characterization and hints on their numerical analysis. The literature on dynamical systems is huge and we do not attempt to survey it here. Most of the results on bifurcations of second-order continuous-time systems are due to Andronov and Leontovich (see Andronov et al., 1973). More recent expositions for n-dimensional continuous- and discrete-time systems can be found in Guckenheimer and Holmes (1997) and Kuznetsov (2004), while less formal but didactically very effective treatments, rich in interesting examples and applications, are given in Strogatz (1994) and Alligood et al. (1996). Numerical aspects are well described in the fundamental papers by Keller (1977) and Doedel et al. (1991a,b), but see also Beyn et al. (2002) and the last chapter in Kuznetsov (2004).

A.1 DYNAMICAL SYSTEMS AND STATE PORTRAITS

The dynamical systems considered in this book are *continuous-time* and *finite-dimensional* dynamical systems described by n *autonomous* (i.e., time-independent) ordinary differential equations (ODEs) called *state equations*, i.e.,

$$\dot{x}_1(t) = f_1(x_1(t), x_2(t), \ldots, x_n(t)),$$
$$\dot{x}_2(t) = f_2(x_1(t), x_2(t), \ldots, x_n(t)),$$
$$\vdots$$
$$\dot{x}_n(t) = f_n(x_1(t), x_2(t), \ldots, x_n(t)),$$

where $x_i(t) \in \mathbf{R}$, $i = 1, 2, \ldots, n$, is the ith *state variable* at time $t \in \mathbf{R}$, $\dot{x}_i(t)$ is its time derivative, and functions $f_1, \ldots, f_n$ are assumed to be smooth.

In vector form, the state equations are

$$\dot{x}(t) = f(x(t)), \qquad (A.1)$$

where x and $\dot{x}$ are n-dimensional vectors (the *state vector* and its time derivative) and $f = [f_1, \ldots, f_n]^T$ (the T superscript denotes transposition).

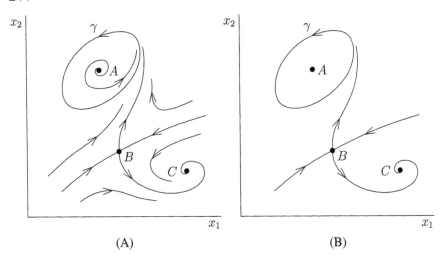

Figure A.1 Skeleton of the state portrait of a second-order system: (A) skeleton with 13
trajectories; (B) reduced skeleton (characteristic frame) with 8 trajectories (attractors, repellors, and saddles with stable and unstable manifolds).

Given the initial state (or condition) $x(0)$, the state equations uniquely define a
trajectory of the system, i.e., the state vector $x(t)$ for all $t \geq 0$. A trajectory is represented in state space by a curve starting from point $x(0)$, and vector $\dot{x}(t)$ is tangent
to the curve at point $x(t)$. Trajectories can be easily obtained numerically through
simulation (numerical integration) and the set of all trajectories (one for any $x(0)$)
is called the *state portrait*. If $n = 2$ (*second-order* or *planar* systems) the state
portrait is often represented by drawing a sort of qualitative skeleton, i.e., strategic trajectories (or finite segments of them), from which all other trajectories can
be intuitively inferred. For example, in Figure A.1A the skeleton is composed of
13 trajectories: three of them (A, B, C) are just points (corresponding to constant
solutions of (A.1)) and are called *equilibria*, while one (γ) is a closed trajectory
(corresponding to a periodic solution of (A.1)) called a *limit cycle*. The other trajectories allow one to conclude that A is a *repellor* (no trajectory starting close to A
tends or remains close to A), B is a *saddle* (almost all trajectories starting close to
B go away from B, but two trajectories tend to B and compose the so-called *stable
manifold*; the two trajectories emanating from B compose the *unstable manifold*
and both manifolds are also called *saddle separatrices*), while C and γ are *attractors* (all trajectories starting close to C [γ] tend to C [γ]). Attractors are said to be
(*asymptotically*) *stable* if all nearby trajectories remain close to them, *globally stable* if they attract all initial conditions (technically with the exclusion of sets with no
measure in state space), while saddles and repellors are *unstable*. Notice, however,
that attractors can also be unstable (see, e.g., the central panel of the forthcoming
Figure A.15, where the equilibrium *SN* attracts all nearby initial conditions, part
of which along trajectories going away from it). The skeleton of Figure A.1A also
identifies the *basin of attraction* of each attractor: in fact, all trajectories starting
above [below] the stable manifold of the saddle tend toward the limit cycle γ [the

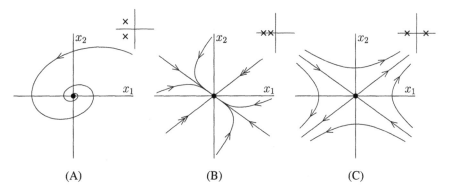

Figure A.2 Three state portraits of generic second-order continuous-time linear systems ($\lambda_1 \neq \lambda_2$, both with nonzero real part, see the complex plane associated with each panel): (A) (stable focus) and (B) (stable node) are attractors; the unstable focus (positive real part complex conjugate eigenvalues) and the unstable node (positive real eigenvalues) (repellors) are obtained by reversing all arrows in the state portraits (A) and (B), respectively; (C) is a saddle. Straight trajectories correspond to eigenvectors associated with real eigenvalues. Double arrows indicate the straight trajectories along which the state varies more rapidly.

equilibrium C]. Notice that the basins of attraction are open sets since their boundaries are the saddle and its stable manifold. Often, the full state portrait can be more easily imagined when the skeleton is reduced, as in Figure A.1B, to its basic elements, namely attractors, repellors, and saddles with their stable and unstable manifolds. From now on, the reduced skeleton is called the *characteristic frame*.

The asymptotic behaviors of continuous-time second-order systems are quite simple, because in the case $n = 2$ attractors can be equilibria (*stationary regimes*) or limit cycles (*cyclic* or *periodic regimes*). But in higher-dimensional systems, i.e., for $n \geq 3$, more complex behaviors are possible since attractors can also be *tori* (*quasi-periodic regimes*) or *strange attractors* (*chaotic regimes*).

In the simple but very important case of linear systems

$$\dot{x}(t) = Ax(t),$$

the state portrait can be immediately obtained from the eigenvalues and eigenvectors of the 2×2 matrix A (we recall that the eigenvalues of an $n \times n$ matrix A are the zeros $\lambda_1, \lambda_2, \ldots, \lambda_n$ of its characteristic polynomial $\det(\lambda I - A)$, where det denotes matrix determinant, and that the eigenvectors associated with an eigenvalue λ_i are nontrivial vectors $x^{(i)}$ satisfying the relationship $Ax^{(i)} = \lambda_i x^{(i)}$). There are five generic state portraits of second-order continuous-time linear systems: three of them are shown in Figure A.2 (the other two are obtained from cases A and B by reversing the sign of the eigenvalues and all arrows in the state portraits). When the two eigenvalues are complex (cases A), the trajectories spiral around the origin and tend to [diverge from] it if the real part of the eigenvalues is negative [positive]. By contrast, when the two eigenvalues are real (cases B and C), the trajectories do not spiral and there are actually special straight trajectories (corresponding to the eigenvectors) converging to [diverging from] the origin if the corresponding eigenvalue is negative [positive]. Along the straight trajectories both state variables vary in time as $\exp(\lambda_i t)$, while along all other trajectories they follow a more complex law of the kind $c_1 \exp(\lambda_1 t) + c_2 \exp(\lambda_2 t)$. Since in generic cases $\lambda_1 \neq \lambda_2$, one of the two exponential functions dominates the other for $t \to \pm\infty$ and all curved trajectories tend to align with one of the two straight trajectories. In particular, in the case of a stable node (characterized by $\lambda_2 < \lambda_1 < 0$, see Figure A.2B), both exponential functions tend to zero for $t \to +\infty$, but in the long run $\exp(\lambda_1 t) \gg \exp(\lambda_2 t)$ so that all trajectories, except the two straight trajectories corresponding to the second eigenvector $x^{(2)}$, tend to zero tangentially to the first eigenvector $x^{(1)}$.

Very similar definitions can be given for *discrete-time systems* described by n difference state equations of the form

$$x(t+1) = f(x(t)), \tag{A.2}$$

where the time t is an integer. In this case trajectories are sequences of points in state space and, again, asymptotic regimes can be stationary, cyclic, quasi-periodic, and chaotic. The major difference between continuous-time and discrete-time dynamical systems is that the former are always *reversible*, since under very general conditions system (A.1) has a unique solution for $t < 0$, while the latter can be *irreversible*. This implies that discrete-time systems can have quasi-periodic and chaotic regimes even if $n = 1$.

The equilibria of system (A.1) can be found by determining all solutions $\bar{x}$ of (A.1) with $\dot{x} = 0$. In second-order systems the equilibria are often determined graphically through the so-called *isoclines*, which are nothing but the lines in state space on which $f_1(x_1, x_2) = 0$ (x_1-isoclines) and $f_2(x_1, x_2) = 0$ (x_2-isoclines). Obviously, the equilibria are at the intersections of x_1- and x_2-isoclines. Moreover, all trajectories cross x_1- [x_2-] isoclines vertically [horizontally] because $\dot{x}_1$ [$\dot{x}_2$] is zero on x_1- [x_2-] isoclines. This property is often useful for devising qualitative geometric features of the state portrait.

The stability of an equilibrium $\bar{x}$ is not as easy to ascertain. However, it can very often be discussed through *linearization*, i.e., by approximating the behavior of the system in the vicinity of the equilibrium through a linear system. This can be done in the following way. Let

$$\delta x(t) = x(t) - \bar{x},$$

so that

$$\dot{\delta x}(t) = f(\bar{x} + \delta x(t)).$$

Under very general conditions, we can expand the function f in Taylor series, thus obtaining

$$\dot{\delta x}(t) = f(\bar{x}) + \left.\frac{\partial f}{\partial x}\right|_{x=\bar{x}} \delta x(t) + O(\|\delta x(t)\|^2),$$

where $\|\cdot\|$ is the standard norm in $\mathbf{R}^n$ and $O(\|\delta x(t)\|^2)$ stays for a term that vanishes as $\|\delta x(t)\|^2$ when $\delta x(t) \to 0$. Noticing that $f(\bar{x}) = 0$, since $\bar{x}$ is a constant solution of (A.1), we have

$$\dot{\delta x}(t) = \left.\frac{\partial f}{\partial x}\right|_{x=\bar{x}} \delta x(t) + O(\|\delta x(t)\|^2), \tag{A.3}$$

where the $n \times n$ constant matrix

$$
J = \left. \frac{\partial f}{\partial x} \right|_{x=\bar{x}} = \left[\begin{array}{ccc} \dfrac{\partial f_1}{\partial x_1} & \cdots & \dfrac{\partial f_1}{\partial x_n} \\ \vdots & & \vdots \\ \dfrac{\partial f_n}{\partial x_1} & \cdots & \dfrac{\partial f_n}{\partial x_n} \end{array} \right]_{x=\bar{x}}
\tag{A.4}
$$

is called the *Jacobian matrix* (or, more simply, *Jacobian*). One can easily imagine that, under suitable conditions, the behavior of system (A.3) (which is still system (A.1)) can be well approximated in the vicinity of $\bar{x}$, by the so-called *linearized system*, which, by definition, is

$$
\dot{\delta x}(t) = \left. \frac{\partial f}{\partial x} \right|_{x=\bar{x}} \delta x(t).
\tag{A.5}
$$

This is, indeed, the case. In particular, it can be shown that if the solution $\delta x(t)$ of (A.5) tends to 0 for all $\delta x(0) \neq 0$ (as in Figures A.2A and B), then the same is true for system (A.3) provided $\|\delta x(0)\|$ is sufficiently small. In other words, the stability of the linearized system implies the (local) stability of the equilibrium $\bar{x}$. This result is quite interesting because the stability of the linearized system can be numerically ascertained by checking if all eigenvalues λ_i, $i = 1, \ldots, n$, of the Jacobian matrix (A.4) have negative real part. A similar result holds also for the case of unstable equilibria. More precisely, if at least one eigenvalue λ_i of the Jacobian matrix has positive real part (as in Figure A.2C), then the equilibrium $\bar{x}$ is locally unstable (i.e., the solution of (A.3) diverges at least temporarily from zero for suitable $\delta x(0)$, no matter how small $\|\delta x(0)\|$ is). Similarly, the local stability of an equilibrium of a discrete-time system of the form (A.2) can be studied by simply looking at the module $|\lambda_i|$ of the n eigenvalues λ_i. In fact, if all $|\lambda_i| < 1$, the equilibrium is stable, while if at least one $|\lambda_i| > 1$, the equilibrium is unstable.

The study of the stability of limit cycles can also be carried out through linearization, following a very simple idea suggested by Poincaré (see Figure A.3). In the case of second-order systems the Poincaré method consists in cutting locally and transversally the limit cycle with a manifold $\mathcal{P}$, called the *Poincaré section*, and looking at the sequence $z(0), z(1), z(2), \ldots$ of points of return of the trajectory to $\mathcal{P}$. Since $\mathcal{P}$ is one dimensional, $z(t)$ is a scalar coordinate on $\mathcal{P}$ and the state equation (A.1) implicitly defines a first-order discrete-time system called the *Poincaré map*

$$
z(t+1) = P(z(t)).
\tag{A.6}
$$

The intersection $\bar{z}$ of the limit cycle γ with $\mathcal{P}$ is an equilibrium of the Poincaré map (since $\bar{z} = P(\bar{z})$) and γ is stable if and only if the equilibrium $\bar{z}$ of (A.6) is stable. One can therefore use the linearization technique, by taking into account that the single eigenvalue of the linearized Poincaré map, $\partial P / \partial z|_{z=\bar{z}}$ (called the *Floquet multiplier* of the cycle), cannot be negative, since trajectories cannot cross each other. Thus, a sufficient condition for the (local) stability of the limit cycle γ is

$$
\left. \frac{\partial P}{\partial z} \right|_{z=\bar{z}} < 1,
\tag{A.7}
$$

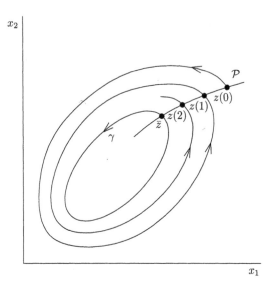

Figure A.3 A limit cycle γ, the Poincaré section $\mathcal{P}$, and the sequence $z(0), z(1), z(2), \ldots$ of return points. Since $\{z(t)\}$ tends to $\bar{z}$ as $t \to \infty$, the cycle γ is stable.

while the reverse inequality implies the instability of γ. It must be noticed, however, that condition (A.7) can only be verified numerically, since the cycle γ is, in general, not known analytically.

A.2 STRUCTURAL STABILITY

Structural stability is a key notion in the theory of dynamical systems, since it is needed to understand interesting phenomena like catastrophic transitions, bistability, hysteresis, frequency locking, synchronization, subharmonics, deterministic chaos, as well as many others. The final target of structural stability is the study of the asymptotic behavior of parameterized families of dynamical systems of the form

$$\dot{x}(t) = f(x(t), p), \tag{A.8}$$

for continuous-time systems, and

$$x(t + 1) = f(x(t), p), \tag{A.9}$$

for discrete-time systems, where p is a vector of constant *parameters*. Given the parameter vector p, all the definitions that we have seen in the previous section apply to the particular dynamical system of the family identified by p. Thus, all geometric and analytical properties of systems (A.8) and (A.9), e.g., trajectories, the state portrait, equilibria, limit cycles, their stability and associated Jacobian matrices and Poincaré maps, the basins of attraction, and, consequently, the asymptotic behavior of the system, now depend upon p.

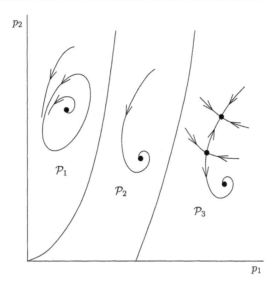

Figure A.4 Bifurcation diagram of a second-order system. The curves separating regions $\mathcal{P}_1$, $\mathcal{P}_2$, and $\mathcal{P}_3$ are bifurcation curves.

Structural stability allows one to rigorously explain why a small change in a parameter value can give rise to a radical change in the system behavior. More precisely, the aim is to find regions $\mathcal{P}_i$ in parameter space characterized by the same qualitative behavior of system (A.8), in the sense that all state portraits corresponding to values $p \in \mathcal{P}_i$ are topologically equivalent (i.e., they can be obtained one from the other through a simple deformation of the trajectories). Thus, varying $p \in \mathcal{P}_i$ the system conserves all the characteristic elements of the state portrait, namely its attractors, repellors, and saddles. In other words, when p is varied in $\mathcal{P}_i$, the characteristic frame varies but conserves its structure. Figure A.4 shows the typical result of a study of structural stability in the space (p_1, p_2) of two parameters of a second-order system. The parameter space is subdivided into three regions, $\mathcal{P}_1$, $\mathcal{P}_2$, and $\mathcal{P}_3$, and for all interior points of each one of these regions the state portrait is topologically equivalent to that sketched in the figure. In $\mathcal{P}_1$ the system is an oscillator, since it has a single attractor, which is a limit cycle. Also in $\mathcal{P}_2$ there is a single attractor, which is, however, an equilibrium. Finally, in $\mathcal{P}_3$ we have *bistability* since the system has two alternative attractors (two equilibria), each with its own basin of attraction delimited by the stable manifold of the saddle equilibrium.

If p is an interior point of a region $\mathcal{P}_i$, system (A.8) is said to be *structurally stable* at p since its state portrait is qualitatively the same as those of the systems obtained by slightly perturbing the parameters in all possible ways. By contrast, if p is on the boundary of a region $\mathcal{P}_i$ the system is not structurally stable since small perturbations can give rise to qualitatively different state portraits. The points of the boundaries of the regions $\mathcal{P}_i$ are called *bifurcation points*, and, in the case of two parameters, the boundaries are called *bifurcation curves*. Bifurcation points are therefore points of degeneracy. If they lie on a curve separating two distinct regions $\mathcal{P}_i$ and $\mathcal{P}_j$, $i \neq j$, they are called codimension-1 bifurcation points, while if they lie on the boundaries of three distinct regions they are called codimension-2 bifurcation points, and so on.

Notice that the simplest dynamical system, namely the first-order linear system $\dot{x}(t) = px(t)$, has a bifurcation at $p = p^* = 0$, i.e., when its eigenvalue p is equal to zero. In fact, such a system is stable for $p < 0$ and unstable for $p > 0$, while it is neutrally stable (i.e., the equilibrium $x = 0$ is not unstable but does not attract all nearby trajectories) for $p = p^* = 0$.

The discussion that follows is limited to continuous-time second-order systems, since only rarely are systems of different kinds mentioned in the book (the only relevant exception is Chapter 10). This will allow us to avoid difficult problems related to quasi-periodic and chaotic behaviors. Moreover, we will limit our review to codimension-1 bifurcations and give details on particular codimension-2 bifurcations only when needed.

A.3 BIFURCATIONS AS COLLISIONS

A generic element of the parameterized family of second-order dynamical systems (A.8) must be imagined to be structurally stable because if p is selected randomly it will be an interior point of a region $\mathcal{P}_i$ with probability 1. In generic conditions, attractors, repellors, saddles, and saddles stable and unstable manifolds are separated one from each other. Moreover, the eigenvalues of the Jacobian matrices associated with equilibria have nonzero real parts, while the eigenvalue of any linearized Poincaré map is different from 1. By continuity, small parametric variations will induce small variations of all attractors, repellors, saddles, and saddles stable and unstable manifolds which, however, will remain separated if the parametric variations are sufficiently small. The same holds for the eigenvalues of Jacobian matrices and linearized Poincaré maps, which, for sufficiently small parametric variations, will continue to be noncritical. Thus, in conclusion, starting from a generic condition, it is necessary to vary the parameters of a finite amount to obtain a bifurcation, which is generated by the collision of two or more elements of the characteristic frame, which then changes its structure at the bifurcation, thus involving a change of the state portrait of the system.

A bifurcation is called *local* when it involves the degeneracy of some eigenvalue of the Jacobians associated with equilibria or cycles. For example, the bifurcation described in Figure A.5, called *saddle-node bifurcation*, is a local bifurcation. Indeed, the bifurcation can be viewed as the collision, at $p = p^*$, of two equilibria: for $p < p^*$ the two equilibria (elements of the characteristic frame) are distinct and one is stable (the node N) while the other is unstable (the saddle S). Then, as p increases, the two equilibria approach one each other and finally collide when $p = p^*$ (and then disappear). Notice that the characteristic frame is degenerate at $p = p^*$ because it is composed of one element (an equilibrium), while there are two equilibria for $p < p^*$ and none for $p > p^*$. But the bifurcation can also be interpreted in terms of eigenvalue degeneracy. In fact, the eigenvalues of the Jacobian

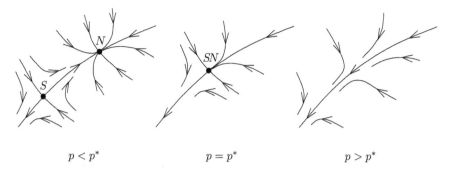

$$p < p^* \qquad\qquad p = p^* \qquad\qquad p > p^*$$

Figure A.5 Example of local bifurcation: saddle-node bifurcation.

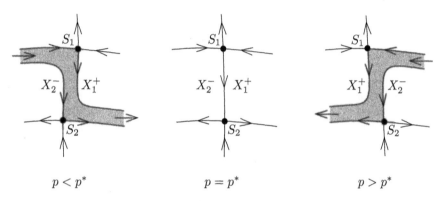

$$p < p^* \qquad\qquad p = p^* \qquad\qquad p > p^*$$

Figure A.6 Example of global bifurcation: heteroclinic bifurcation.

evaluated at the saddle are one positive and one negative, while the eigenvalues of the Jacobian evaluated at the node are both negative, so that when the two equilibria coincide, one of the two eigenvalues of the unique Jacobian matrix must be equal to zero.

By contrast, *global bifurcations* cannot be revealed by eigenvalue degeneracies. One example, known as *heteroclinic bifurcation*, is shown in Figure A.6, which presents the characteristic frames (two saddles and their stable and unstable manifolds) of a system for $p = p^*$ (bifurcation value) and for $p \neq p^*$. The characteristic frame for $p = p^*$ is structurally different from the others because it corresponds to the collision of the unstable manifold X_1^+ of the first saddle with the stable manifold X_2^- of the second saddle. However, the two Jacobian matrices associated with the two saddles do not degenerate at p^*, since their eigenvalues remain different from zero. In other words, the bifurcation cannot be revealed by the behavior of the system in the vicinity of an equilibrium, but is the result of the global behavior of the system.

When there is only one parameter p and there are various bifurcations at different values of the parameter, it is often advantageous to represent the dependence of the system behavior upon the parameter by drawing in the three-dimensional space (p, x_1, x_2), often called *control space*, the characteristic frame for all values

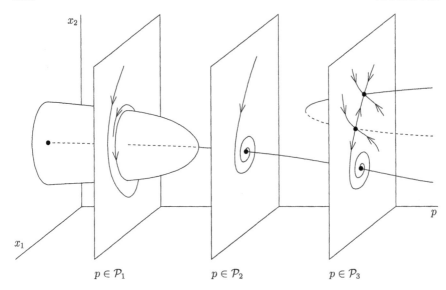

Figure A.7 Characteristic frame in the control space of a system with a Hopf and a saddle-
node bifurcation. Continuous lines represent trajectories in the three illustrated
state portraits and stable equilibria or limit cycles otherwise; dashed lines repre-
sent unstable equilibria. The symbols $\mathcal{P}_1$, $\mathcal{P}_2$, and $\mathcal{P}_3$ refer to Figure A.4.

of p. This is done, for example, in Figure A.7 for the same system described in
Figure A.4, with $p = p_1$ and constant p_2. Figure A.7 shows that for increasing
values of p a so-called *Hopf bifurcation* occurs, as the stable limit cycle shrinks to a
point, thus colliding with the unstable equilibrium that exists inside the cycle. This
is a local bifurcation, because the equilibrium is stable for higher values of p, so
that the bifurcation can be revealed by an eigenvalue degeneracy. The figure also
shows that a saddle-node bifurcation occurs at a higher value of the parameter, as
two equilibria, namely a stable node and a saddle, become closer and closer until
they collide and disappear. The Hopf and the saddle-node bifurcations are perhaps
the most popular local bifurcations of second-order systems and are discussed in
some detail in the next section.

A.4 LOCAL BIFURCATIONS

In this section we discuss the five most important local bifurcations of continuous-
time second-order systems. Three of them, called *transcritical*, saddle-node (already encountered above), and *pitchfork*, can be viewed as collisions of equilibria,
the fourth one, Hopf (also seen above), is the collision of an equilibrium with a
vanishing cycle, while the last one, called *tangent of limit cycles*, is the collision of
two limit cycles. Since the first three bifurcations can occur in first-order systems,
we present them in that context.

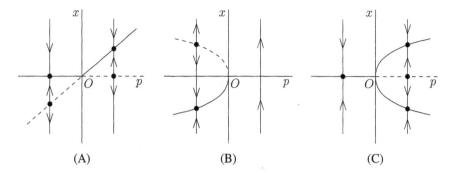

Figure A.8 Three local bifurcations viewed as collisions of equilibria: (A) transcritical; (B) saddle-node; (C) pitchfork.

Transcritical, Saddle-Node, and Pitchfork Bifurcations

Figure A.8 shows three different types of collisions of equilibria in first-order systems of the form (A.8). The state x and the parameter p have been normalized in such a way that the bifurcation occurs at $p^* = 0$ and that the corresponding equilibrium is zero. Continuous lines in the figure represent stable equilibria, while dashed lines indicate unstable equilibria. In Figure A.8A the collision is visible in both directions, while in Figures A.8B and C the collision is visible only from the left or from the right. The three bifurcations are called, respectively, transcritical, saddle-node, and pitchfork, and the three most simple state equations (called *normal forms*) giving rise to Figure A.8 are

$$\dot{x}(t) = px(t) - x^2(t), \quad \text{transcritical,} \qquad (A.10a)$$

$$\dot{x}(t) = p + x^2(t), \quad \text{saddle-node,} \qquad (A.10b)$$

$$\dot{x}(t) = px(t) - x^3(t), \quad \text{pitchfork.} \qquad (A.10c)$$

The first of these bifurcations is also called *exchange of stability* since the two equilibria exchange their stability at the bifurcation. The second is called saddle-node bifurcation because in second-order systems it corresponds to the collision of a saddle with a node, as shown in Figure A.5, but it is also known as *fold*, in view of the form of the graph of its equilibria. Due to the symmetry of the normal form, the pitchfork has three colliding equilibria, two stable and one unstable in the middle.

It is worth noticing that in changing the sign of the quadratic and cubic terms in the normal forms (A.10), three new normal forms are obtained, namely

$$\dot{x}(t) = px(t) + x^2(t), \quad \text{transcritical,} \qquad (A.11a)$$

$$\dot{x}(t) = p - x^2(t), \quad \text{saddle-node,} \qquad (A.11b)$$

$$\dot{x}(t) = px(t) + x^3(t), \quad \text{pitchfork,} \qquad (A.11c)$$

which have the bifurcation diagrams shown in Figure A.9. Comparing Figures A.8 and A.9, it is easy to verify that nothing changes from a phenomenological point of view in the first two cases. However, for the pitchfork bifurcation this is not true, since in case (A.10c) there is at least one attractor for each value of the parameter, while in case (A.11c), for $p > 0$, there is only a repellor. To distinguish the two

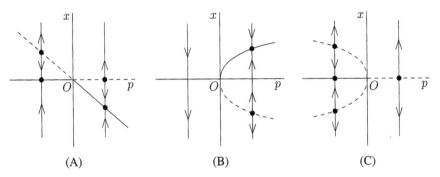

Figure A.9 Bifurcation diagrams corresponding to the normal forms (A.11).

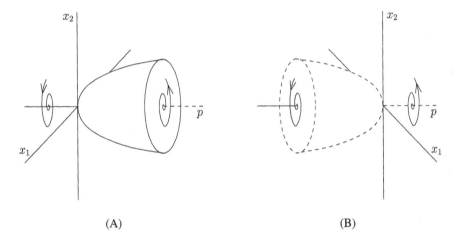

Figure A.10 Hopf bifurcation: (A) supercritical; (B) subcritical.

possibilities, the pitchfork (A.10c) is called *supercritical*, while the other is called *subcritical*.

Hopf Bifurcation

The Hopf bifurcation (actually discovered by A. A. Andronov for second-order systems; see Andronov et al., 1973, and Marsden and McCracken, 1976, for the English translation of Andronov and Hopf's original works) explains how a stationary regime can become cyclic as a consequence of a small variation of a parameter, a rather common phenomenon not only in physics but also in biology, economics, and life sciences. In terms of collisions, this bifurcation involves an equilibrium and a cycle which, however, shrinks to a point when the collision occurs. Figure A.10 shows the two possible cases, known as supercritical and subcritical Hopf bifurcations, respectively. In the supercritical case, a stable cycle has in its interior an unstable focus. When the parameter is varied the cycle shrinks until it collides with the equilibrium and after the collision only a stable equilibrium remains. By contrast, in the subcritical case the cycle is unstable and is the boundary of the basin of attraction of the stable equilibrium inside the cycle. Thus, after the collision there is only a repellor.

The normal form of the Hopf bifurcation is

$$\dot{x}_1(t) = \ \ \ px_1(t) + \omega x_2(t) + cx_1(t)\left(x_1^2(t) + x_2^2(t)\right),$$
$$\dot{x}_2(t) = -\omega x_1(t) + px_2(t) + cx_2(t)\left(x_1^2(t) + x_2^2(t)\right)),$$

which, in polar coordinates, becomes

$$\dot{\rho}(t) = p\rho(t) + c\rho^3(t),$$
$$\dot{\theta}(t) = \omega.$$

This last form shows that the trajectory spirals around the origin at constant angular velocity ω, while the distance from the origin varies in accordance with the first ODE, which is the normal form of the pitchfork. Thus, the stability of the cycle depends upon the sign of c, called the *Lyapunov coefficient*.

Taking into account Figures A.8C and A.9C, it is easy to check that the Hopf bifurcation is supercritical [subcritical] if $c < 0$ [$c > 0$] (in the case $c = 0$ the system is linear and for $p = p^* = 0$ the origin is neutrally stable and surrounded by an infinity of cycles). For $p = p^*$ the origin of the state space is stable in the supercritical case and unstable in the opposite case. The Jacobian of the normal form, evaluated at the origin, is

$$J = \begin{bmatrix} p & \omega \\ -\omega & p \end{bmatrix},$$

and its two eigenvalues $\lambda_{1,2} = p \pm i\omega$ cross the imaginary axis of the complex plane when $p = 0$. This is the property commonly used to detect Hopf bifurcations in second-order systems. In fact, denoting by $\bar{x}(p)$ an equilibrium of the system, the Jacobian evaluated at $\bar{x}(p)$ is

$$J = \begin{bmatrix} \dfrac{\partial f_1}{\partial x_1} & \dfrac{\partial f_1}{\partial x_2} \\[2ex] \dfrac{\partial f_2}{\partial x_1} & \dfrac{\partial f_2}{\partial x_2} \end{bmatrix}_{x=\bar{x}(p)},$$

and such a matrix has a pair of nontrivial and purely imaginary eigenvalues if and only if

$$\text{trace}(J) = \left.\frac{\partial f_1}{\partial x_1}\right|_{x=\bar{x}(p)} + \left.\frac{\partial f_2}{\partial x_2}\right|_{x=\bar{x}(p)} = 0,$$

$$\det(J) = \left.\frac{\partial f_1}{\partial x_1}\right|_{x=\bar{x}(p)} \left.\frac{\partial f_2}{\partial x_2}\right|_{x=\bar{x}(p)} - \left.\frac{\partial f_1}{\partial x_2}\right|_{x=\bar{x}(p)} \left.\frac{\partial f_2}{\partial x_1}\right|_{x=\bar{x}(p)} = 0.$$

In practice, one annihilates the trace of the Jacobian evaluated at the equilibrium and finds in this way the parameter values that are candidate Hopf bifurcations. Then, the test on the positivity of the determinant of J is used to select the true Hopf bifurcations among the candidates. Under suitable nondegeneracy conditions, the emerging cycle is unique and its frequency is $\omega = \sqrt{\det(J)}$, because $\sqrt{\det(J)} = \lambda_1 \lambda_2$, while its amplitude increases as $\sqrt{-c(p - p^*)}$.

Determining if a Hopf bifurcation is supercritical or subcritical is not easy. One can try to find out if the equilibrium is stable or unstable but this is quite difficult since linearization is unreliable at a bifurcation. Alternatively (but equivalently), one can determine the sign of the Lyapunov coefficient c by following a procedure that is often quite cumbersome (see, e.g., Guckenheimer and Holmes, 1997; Kuznetsov, 2004) and is therefore not reported here.

Example: Rosenzweig-MacArthur Resource-Consumer Model

The model most commonly used for describing the interactions between resources (prey) and consumers (predator) on the demographic timescale is the Rosenzweig and MacArthur (1963) model. The two state variables are the densities (or "numbers" of individuals) of the two populations, here indicated by n_1 and n_2 in accordance with the notation used throughout the book.

The state equations of the model are

$$\dot{n}_1(t) = rn_1(t)\left(1 - \frac{n_1(t)}{K}\right) - \frac{an_1(t)}{1 + a\tau n_1(t)}n_2(t), \qquad (A.12a)$$

$$\dot{n}_2(t) = e\frac{an_1(t)}{1 + a\tau n_1(t)}n_2(t) - dn_2(t), \qquad (A.12b)$$

where r and K are net growth rate and carrying capacity of the resource and a, τ, e, and d are attack rate, handling time, efficiency, and net death rate of the consumer. Model (A.12) is positive (as it should be), i.e., $n_i(0) \geq 0$, $i = 1, 2$, implies $n_i(t) \geq 0$, $i = 1, 2$, for all $t \geq 0$, so that the attention is restricted in the following to the positive quadrant of the state space.

Besides two trivial isoclines ($n_1 = 0$ for the resource and $n_2 = 0$ for the consumer) there are two nontrivial isoclines, namely the parabola

$$n_2 = \frac{r}{a}\left(1 - \frac{n_1}{K}\right)(1 + a\tau n_1)$$

for the resource and the vertical straight line

$$n_1 = \frac{d}{a(e - d\tau)}$$

for the consumer (see dotted lines in Figure A.11, where three state portraits of model (A.12) are shown for low, intermediate, and high consumer efficiency e). Thus, there are two trivial equilibria, namely

$$\bar{n}^{(1)} = \begin{bmatrix} 0 \\ 0 \end{bmatrix}, \quad \bar{n}^{(2)} = \begin{bmatrix} K \\ 0 \end{bmatrix},$$

and one positive equilibrium

$$\bar{n}^{(3)} = \begin{bmatrix} \dfrac{d}{a(e - d\tau)} \\ \dfrac{er\left(aK(e - d\tau) - d\right)}{a^2 K(e - d\tau)^2} \end{bmatrix},$$

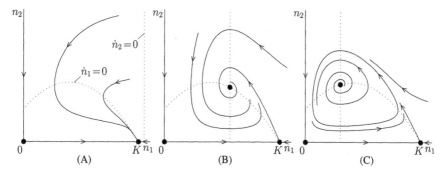

Figure A.11 Three state portraits of model (A.12): (A) consumer extinction ($e < e_1^*$); (B) stationary coexistence ($e_1^* < e < e_2^*$); (C) cyclic coexistence ($e > e_2^*$). Dotted lines: isoclines; filled circles: equilibria.

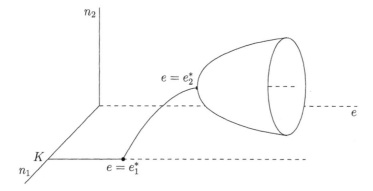

Figure A.12 Equilibria and limit cycles (characteristic frame) of model (A.12) in the control space (e, n_1, n_2).

provided (see Figures A.11B and C)

$$e > e_1^* = d\left(\tau + \frac{1}{aK}\right).$$

Notice that the third equilibrium collides with the second one for $e = e_1^*$, which is, therefore, a bifurcation. To determine which kind of bifurcation it is we can simply look at the graph of the equilibria in the control space (e, n_1, n_2) and recognize (see Figure A.12) that the bifurcation is a transcritical bifurcation. This can also be verified by computing the Jacobians $J^{(2)}$ and $J^{(3)}$ associated with the two colliding equilibria $\bar{n}^{(2)}$ and $\bar{n}^{(3)}$ and by checking that these equilibria exchange their stability at $e = e_1^*$ (see the transition between the state portraits A and B in Figure A.11). If $a\tau K > 1$ (this inequality holds if the vertex of the resource isocline (parabola) is in the positive quadrant and cannot be satisfied in a Lotka-Volterra model, where $\tau = 0$ and the resource isocline is a straight line) the model has also a Hopf bifurcation at

$$e = e_2^* = d\tau \frac{a\tau K + 1}{a\tau K - 1}.$$

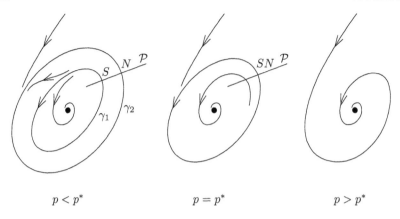

$$p < p^* \qquad\qquad p = p^* \qquad\qquad p > p^*$$

Figure A.13 Tangent bifurcation of limit cycles: two cycles γ_1 and γ_2 collide for $p = p^*$ and then disappear.

The expression of e_2^* ($> e_1^*$) can be determined by annihilating the trace of the Jacobian $J^{(3)}$ and by verifying that the determinant of $J^{(3)}$ for $e = e_2^*$ is positive (the computations require a substantial amount of algebra). Graphically, for $e > e_2^*$ the consumer isocline is on the left of the vertex of the resource isocline (see Figure A.11C). The Lyapunov coefficient of this Hopf bifurcation can be analytically computed (Kuznetsov, 2004) and turns out to be negative for all parameter values. Thus, the bifurcation is supercritical, as confirmed by the stability of the limit cycle in Figures A.11C and A.12.

The interpretation of Figure A.12 is straightforward. If consumers have a too low efficiency ($e < e_1^*$) they go extinct and only resources remain in the game. By contrast, if consumers are very efficient ($e > e_2^*$) the two populations coexist but fluctuate periodically, while for intermediate efficiencies ($e_1^* < e < e_2^*$) stationary coexistence is the only possible outcome. Moreover, the amplitude of the oscillations increases rapidly with the surplus of efficiency ($e - e_2^*$) and the period of the oscillations for e close to e_2^* is roughly equal to $2\pi/\sqrt{\det(J^{(3)})}$, which is the period of the emerging cycle.

Tangent Bifurcation of Limit Cycles

Other local bifurcations in second-order systems involve limit cycles and are somehow similar to transcritical, saddle-node, and pitchfork bifurcations of equilibria. In fact, the collision of two limit cycles can be studied as the collision of the two corresponding equilibria of the Poincaré map defined on a Poincaré section cutting both cycles. Thus, the transcritical, saddle-node, and pitchfork bifurcations of such equilibria correspond to analogous bifurcations of the colliding limit cycles.

The most common case is the saddle-node bifurcation of limit cycles, more often called fold or tangent bifurcation of limit cycles, where two cycles collide for $p = p^*$ and then disappear, as shown in Figure A.13. On the Poincaré section $\mathcal{P}$ the bifurcation is revealed by the collision of two equilibria of the Poincaré map,

S unstable and N stable, which then disappear. In terms of eigenvalue degeneracy, the eigenvalue of the linearized Poincaré map evaluated at S [N] is larger [smaller] than 1, so that when the two equilibria coincide, the eigenvalue of the unique linearized Poincaré map must be equal to 1.

Varying the parameter in the opposite direction, this bifurcation explains the sudden birth of a pair of cycles, one of which is stable. While in the case of the Hopf bifurcation the emerging cycle is degenerate (it has zero amplitude), in this case the emerging cycles are not degenerate.

A.5 GLOBAL BIFURCATIONS

As already said in Section A.3 global bifurcations cannot be detected through the analysis of the Jacobians associated with equilibria or cycles. However, they can still be viewed as structural changes of the characteristic frame.

Heteroclinic Bifurcation

In Figure A.6 we have already reported the bifurcation corresponding to the collision of a stable manifold of a saddle with the unstable manifold of another saddle. This bifurcation is called heteroclinic bifurcation, since the trajectory connecting the two saddles is called heteroclinic trajectory (or connection).

Homoclinic Bifurcation

A special but important global bifurcation is the so-called *homoclinic bifurcation*, characterized by the presence of a trajectory connecting an equilibrium with itself, called homoclinic trajectory (or connection).

There are two collisions that give rise to a homoclinic trajectory. The first and most common collision is that between the stable and unstable manifolds of the same saddle, as depicted in Figure A.14. The second collision, shown in Figure A.15, is that between a node and a saddle whose unstable manifold is connected to the node. The corresponding bifurcations are called *homoclinic bifurcation to standard saddle*, or simply homoclinic bifurcation, and *homoclinic bifurcation to saddle-node*.

Figure A.14 shows that the homoclinic bifurcation to standard saddle can also be viewed as the collision of a cycle $\gamma(p)$ with a saddle $S(p)$. When p approaches p^* the cycle $\gamma(p)$ gets closer and closer to the saddle $S(p)$, so that the period $T(p)$ of the cycle becomes longer and longer, since the state of the system moves very slowly when it is very close to the saddle. By contrast, Figure A.15 shows that the homoclinic bifurcation to saddle-node can be viewed as a saddle-node bifurcation on a cycle $\gamma(p)$, which therefore disappears. When p approaches p^* the system "feels" the forthcoming appearance of the two equilibria and therefore the state slows down close to the point where they are going to appear. Thus, in both cases, $T(p) \to \infty$ as $p \to p^*$ and this property is often used to detect homoclinic bifurcations through simulation. Another property used to detect homoclinic bifurcations

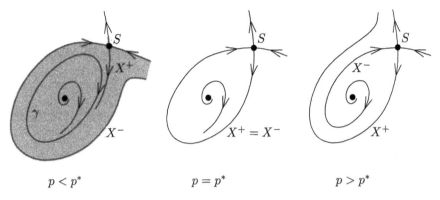

Figure A.14 Homoclinic bifurcation to standard saddle: for $p = p^*$ the stable manifold X^-
of the saddle S collides with the unstable manifold X^+ of the same saddle. The
bifurcation can also be viewed as the collision of the cycle γ with the saddle S.

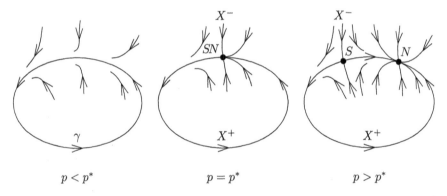

Figure A.15 Homoclinic bifurcation to saddle-node: for $p = p^*$ the unstable manifold X^+
of the saddle-node SN comes back to SN transversally to the stable manifold
X^-. The bifurcation can also be viewed as a saddle-node bifurcation on the
cycle γ.

to standard saddles through simulation is related to the form of the limit cycle which
becomes "pinched" close to the bifurcation, the angle of the pinch being the angle
between the stable and unstable manifolds of the saddle.

Looking at Figures A.14 and A.15 from the right to the left, we can recognize
that the homoclinic bifurcation explains the birth of a limit cycle. As in the case
of Hopf bifurcations, the emerging limit cycle is degenerate, but this time the degeneracy is not in the amplitude of the cycle but in its period, which is infinitely
long. The emerging limit cycles are stable in the figures (the gray region in Figure A.14 is the basin of attraction), but reversing the arrows of all trajectories the
same figures could be used to illustrate the cases of unstable emerging cycles. In
other words, homoclinic bifurcations in second-order systems are generically associated with a cycle emerging from the homoclinic trajectory existing at $p = p^*$
by suitably perturbing the parameter. It is interesting to note that the stability of

the emerging cycle can be easily predicted by looking at the sign of the so-called
saddle quantity σ, which is the sum of the two eigenvalues of the Jacobian matrix
associated with the saddle, i.e., the trace of the Jacobian (notice that one eigenvalue is equal to zero in the case of homoclinic bifurcation to saddle-node). More
precisely, if $\sigma < 0$ the cycle is stable, while if $\sigma > 0$ the cycle is unstable. As
proved by Andronov and Leontovich (see Andronov et al., 1973), this result holds
under a series of assumptions that essentially rule out a number of critical cases. A
very important and absolutely not simple extension of Andronov and Leontovich
theory is Shil'nikov theorem (Shil'nikov, 1968) concerning homoclinic bifurcations in three-dimensional systems. However, this extension is not reported here,
because it is not used in this book.

A.6 CATASTROPHES, HYSTERESIS, AND CUSP

We can now present a simple but comprehensive treatment of a delicate problem,
that of *catastrophic transitions* in dynamical systems. A lot has been said on this
topic in the last decades and the so-called *catastrophe theory* (Thom, 1972) has
often been invoked improperly, thus generating expectations that will never be satisfied. Reduced to its minimal terms, the problem of catastrophic transitions is the
following: assuming that a system is functioning in one of its asymptotic regimes,
is it possible that a microscopic variation of a parameter triggers a transient toward
a macroscopically different asymptotic regime? When this happens, we say that a
catastrophic transition occurs.

  To be more specific, assume that an instantaneous small perturbation from p to
$p + \Delta p$ occurs at time $t = 0$ when the system is on one of its attractors, say $\mathcal{A}(p)$,
or at a point $x(0)$ very close to $\mathcal{A}(p)$ in the basin of attraction $B\left(\mathcal{A}(p)\right)$. A first
possibility is that p and $p + \Delta p$ are not separated by any bifurcation. This implies
that the state portrait of the perturbed system $\dot{x} = f(x, p + \Delta p)$ can be obtained
by slightly deforming the state portrait of the original system $\dot{x} = f(x, p)$. In particular, if Δp is small, by continuity, the attractors $\mathcal{A}(p)$ and $\mathcal{A}(p + \Delta p)$, as well
as their basins of attraction $B\left(\mathcal{A}(p)\right)$ and $B\left(\mathcal{A}(p + \Delta p)\right)$, are almost coincident,
so that $x(0) \in B\left(\mathcal{A}(p + \Delta p)\right)$. This means that after the perturbation a transition
will occur from $\mathcal{A}(p)$ (or $x(0)$ close to $\mathcal{A}(p)$) to $\mathcal{A}(p + \Delta p)$. In conclusion, a microscopic variation of a parameter has generated a microscopic variation in system
behavior.

  The opposite possibility is that p and $p + \Delta p$ are separated by a bifurcation. In
such a case it can happen that the small parameter variation triggers a transient,
bringing the system toward a macroscopically different attractor. When this happens for all initial states $x(0)$ close to $\mathcal{A}(p)$, the bifurcation is called *catastrophic*.
By contrast, if the catastrophic transition is not possible, the bifurcation is called
noncatastrophic, while in all other cases the bifurcation is said to be *undetermined*.

  We can now revisit all bifurcations we have discussed in the previous sections.
Let us start with Figure A.8 and assume that p is small and negative, i.e., $p = -\epsilon$,
that $x(0)$ is different from zero but very small, i.e., close to the stable equilibrium,
and that $\Delta p = 2\epsilon$ so that, after the perturbation, $p = \epsilon$. In case A (transcritical

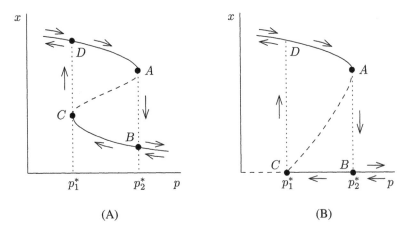

(A)   (B)

Figure A.16 Two systems with hysteresis generated by two saddle-node bifurcations (A),
and a saddle-node and a transcritical bifurcation (B).

bifurcation) $x(t) \to \epsilon$ if $x(0) > 0$ and $x(t) \to -\infty$ if $x(0) < 0$. Thus, this bifurcation is undetermined because it can, but does not always, give rise to a catastrophic
transition. In a case like this, the noise acting on the system has a fundamental role
since it determines the sign of $x(0)$, which is crucial for the behavior of the system after the parametric perturbation. We must notice, however, that in many cases
the sign of $x(0)$ is a priori fixed. For example, if the system is positive because
x represents the density of a population, then for physical reasons $x(0) > 0$ and
the bifurcation is therefore noncatastrophic. However, under the same conditions,
the transcritical bifurcation of Figure A.9A is catastrophic. Similarly, we can conclude that the saddle-node bifurcation of Figure A.8B is catastrophic, as well as
that of Figure A.9B, and that the pitchfork bifurcation can be noncatastrophic (as
in Figure A.8C) or catastrophic (as in Figure A.9C).

From Figure A.10 we can immediately conclude that the supercritical Hopf bifurcation is noncatastrophic, while the subcritical one is catastrophic. This is why
the two Hopf bifurcations are sometimes called catastrophic and noncatastrophic.
Finally, Figures A.13, A.14, and A.15 show that tangent and homoclinic bifurcations are catastrophic.

When a small parametric variation triggers a catastrophic transition from an at-
tractor $\mathcal{A}'$ to an attractor $\mathcal{A}''$ it is interesting to determine if it is possible to drive
the system back to the attractor $\mathcal{A}'$ by suitably varying the parameter. When this
is possible, the catastrophe is called *reversible*. The most simple case of reversible
catastrophes is the *hysteresis*, two examples of which (concerning first-order systems) are shown in Figure A.16. In case A the system has two saddle-node bifurcations, while in case B there is a transcritical bifurcation at p_1^* and a saddle-
node bifurcation at p_2^*. All bifurcations are catastrophic (because the transitions
$A \to B$ and $C \to D$ are macroscopic) and if p is varied back and forth between
$p_{\min} < p_1^*$ and $p_{\max} > p_2^*$ through a sequence of small steps with long time intervals between successive steps, the state of the system follows closely the cycle

$A \rightarrow B \rightarrow C \rightarrow D$ indicated in the figure and called a *hysteretic cycle* (or, briefly, hysteresis). The catastrophes are therefore reversible, but after a transition from $\mathcal{A}'$ to $\mathcal{A}''$ it is necessary to pass through a second catastrophe to come back to the attractor $\mathcal{A}'$. This simple type of hysteresis explains many phenomena in physics, chemistry, and electromechanics, but also in biology and social sciences. For example, the hysteresis of Figure A.16B was used by Noy-Meir (1975) to explain the possible collapse (saddle-node bifurcation) of an exploited population described by the equation

$$\dot{x} = rx \left(1 - \frac{x}{K}\right) - \frac{ax}{1 + a\tau x}p,$$

where x is resource density (e.g., density of grass) and p is the number of exploiters (e.g., number of cows). If p is increased step by step (e.g., by adding one extra cow every year) the resource declines smoothly until it collapses to zero when a threshold p_2^* is passed. To regenerate the resource, one is obliged to radically reduce the number of exploiters to $p < p_1^*$.

Hysteresis can be more complex than in Figure A.16 not only because the attractors involved in the hysteretic cycle can be more than two, but also because some of them can be cycles. To show the latter possibility, we reconsider the Rosenzweig-MacArthur model discussed in Section A.4 and simply add in the consumer equation (A.12b) a term representing the mortality rate due to a Holling type II predator with constant density p, i.e.,

$$\dot{n}_1 = rn_1 \left(1 - \frac{n_1}{K}\right) - \frac{a_2 n_1}{1 + a_2 \tau_2 n_1}n_2,$$
$$\dot{n}_2 = e_2 \frac{a_2 n_1}{1 + a_2 \tau_2 n_1}n_2 - dn_2 - \frac{a_3 n_2}{1 + a_3 \tau_3 n_2}p.$$

Thus, the model describes the dynamics of a tritrophic food chain in which, however, the top population is (or is kept) constant. Without entering into the details of the analysis of this second-order model (see Kuznetsov et al., 1995), we show in Figure A.17 the equilibria and the cycles of the system in the control space (p, n_1, n_2) for a specified value of all other parameters. The figure points out five bifurcations: a transcritical (TR), two homoclinic (h_1 and h_2), a supercritical Hopf (H), and a saddle-node (SN). Two of these bifurcations, namely the second homoclinic h_2 and the saddle-node SN, are catastrophic and irreversible. In fact, catastrophic transitions from h_2 and SN bring the system toward the trivial equilibrium $(K, 0)$ (extinction of the consumer population) and from this state it is not possible to return to h_2 or SN by varying p step by step. By contrast, the two other catastrophic bifurcations, namely the first homoclinic h_1 and the transcritical TR, are reversible and identify a hysteretic cycle obtained by varying back and forth the parameter p in an interval slightly larger than $[p_{TR}, p_{h_1}]$. On one extreme of the hysteresis we have a catastrophic transition from the equilibrium $(K, 0)$ to a resource-consumer limit cycle. Then, increasing p, the period of the limit cycle increases (and tends to infinity as $p \rightarrow p_{h_1}$), and on the other extreme of the hysteresis we have a catastrophic transition from a homoclinic cycle (in practice a cycle of very long period) to the equilibrium $(K, 0)$. Thus, if p is varied smoothly, slowly, and periodically from just below p_{TR} to just above p_{h_1}, one can expect that

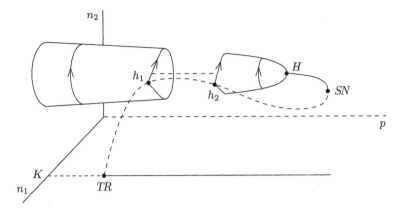

Figure A.17 Equilibria and limit cycles (characteristic frame) of the Rosenzweig-
MacArthur tritrophic food chain model with constant predator population p.

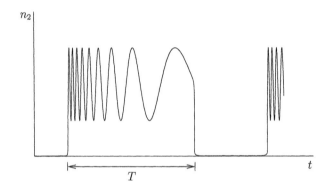

Figure A.18 Periodic variation of a consumer population induced by a periodic variation of
the predator.

the consumer population varies periodically in time, as shown in Figure A.18. In
conclusion, the consumer population remains very scarce for a long time and then
suddenly regenerates, giving rise to high-frequency resource-consumer oscillations
which, however, slow down before a crash of the consumer population occurs. Of
course, tritrophic food chains do not always have such wild dynamics. In fact,
many food chains have a bifurcation diagram with only two bifurcations, as in Figure A.12. In those cases, the food chain cannot have catastrophic transitions and
hysteresis.

An interesting variant of the hysteresis it the so-called *cusp*, described by the
normal form

$$\dot{x} = p_1 + p_2 x - x^3,$$

which is still a first-order system, but with two parameters. For $p_1 = 0$ the equation degenerates into the pitchfork normal form, while for $p_2 > 0$ the equation
points out a hysteresis with respect to p_1 with two saddle-nodes. The graph of

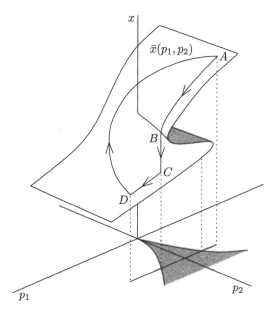

Figure A.19 Equilibria of the cusp normal form. The unstable equilibria are on the gray part of the surface, which corresponds to the gray cusp region in the parameter space.

the equilibria $\bar{x}(p_1, p_2)$ is reported in Figure A.19, which shows that for the parameters (p_1, p_2) belonging to the cusp region in parameter space, the system has three equilibria, two stable and one unstable (in the middle). In contrast with the hysteresis shown in Figure A.16, this time after a catastrophic transition from an attractor $\mathcal{A}'$ to an attractor $\mathcal{A}''$ (transition $B \to C$ in the figure), one can find the way to come back to $\mathcal{A}'$ without suffering a second catastrophic transition (path $C \to D \to A \to B$ in the figure).

A.7 EXTINCTION BIFURCATIONS

Until now we have considered second-order dynamical systems in which the state x was unconstrained. However, in the case of evolutionary dynamics, the state of the system must belong to an open set $\mathcal{X}$, called the *evolution set*. We assume that $\mathcal{X}$ is bounded and delimited by a smooth boundary. Outside the evolution set the system is not defined. In other words, the system persists if its state is inside $\mathcal{X}$ and is destroyed when the state hits the boundary of $\mathcal{X}$.

In systems of this kind we can have all the bifurcations we have discussed in the previous sections, which occur when the characteristic frame has a structural change not involving the boundary of the evolution set. But we can also have bifurcations involving the boundary of $\mathcal{X}$ (often called *border-collision bifurcations* in the dynamical system literature) when, for example, an equilibrium or a limit

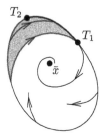

Figure A.20 An example of second-order system constrained in the evolution set $\mathcal{X}$. White
[gray] region: viable [unviable] set; thick segment $T_1 T_2$: extinction segment.

cycle hits the boundary.

To classify all possible bifurcations involving the boundary of a two-dimensional
evolution set, notice that $\mathcal{X}$ can be divided into two subsets, called viable and unviable (see Figure A.20). The *viable set* (white region) is the set of all initial conditions giving rise to trajectories that remain forever in $\mathcal{X}$. Conversely, the *unviable
set* (gray region) is the set of all initial conditions giving rise to trajectories that
reach the boundary of $\mathcal{X}$. Moreover, the boundary of $\mathcal{X}$ is reached only at points
belonging to parts of the boundary of the unviable set, called *extinction segments* in
the following (thick segments). At each point of an extinction segment the vector
$\dot{x}$ tangent to the trajectory points outside $\mathcal{X}$, so that extinction segments are generically delimited by so-called *tangent points*, where $\dot{x}$ is tangent to the boundary of
$\mathcal{X}$ (see points T_1 and T_2).

If we now include in the characteristic frame of the system all extinction segments and trajectories entering $\mathcal{X}$ from tangent points, we can still see bifurcations
as collisions of two or more elements of the characteristic frame and, in particular,
we call *extinction bifurcations* the collisions involving extinction segments. For example, in Figure A.20, the characteristic frame is composed of the equilibrium $\bar{x}$,
the extinction segment $T_1 T_2$, and the trajectory originating from T_1. The collision
of the equilibrium $\bar{x}$ with the boundary of $\mathcal{X}$ necessarily occurs at the tangent point
T_1 and is, therefore, an extinction bifurcation, called a *boundary equilibrium bifurcation* (see Figure A.21). Similarly, the collision of a limit cycle with the boundary
of $\mathcal{X}$ occurs at a tangent point and is called a *grazing bifurcation* (see Figure A.22).

If the system is on the attracting equilibrium (Figure A.21) or on the attracting
cycle (Figure A.22) for p slightly smaller than p^* and p is suddenly increased of a
small amount, so that after the perturbation p is slightly greater than p^*, the state
evolves until it reaches the boundary of $\mathcal{X}$, where the system is destroyed. Thus,
a microscopic variation of the parameter triggers a macroscopic variation of system behavior, so that both the boundary equilibrium and the grazing bifurcations
are catastrophic. While the first is a local bifurcation, Figure A.22 shows that for
$p < p^*$ and for $p > p^*$, the trajectory starting from the tangent point T does not
come back to T, while for $p = p^*$ it does. This property is of the same nature as
that characterizing homoclinic bifurcations and, indeed, the grazing bifurcation is
a global bifurcation.

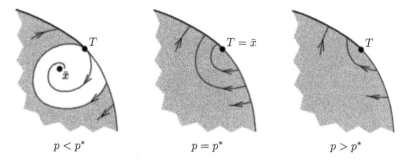

Figure A.21 Boundary equilibrium bifurcation: for $p < p^*$ the characteristic frame is composed of the equilibrium $\bar{x}$, the extinction (thick) segment, and the trajectory originating from the tangent point T, while for $p > p^*$ the characteristic frame is simply given by the extinction segment.

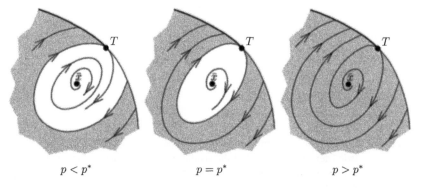

Figure A.22 Grazing bifurcation: for $p < p^*$ the characteristic frame is composed of the equilibrium $\bar{x}$, the limit cycle, the extinction (thick) segment, and the trajectory originating from the tangent point T, while for $p > p^*$ there is no cycle.

Other two possible extinction bifurcations are the collision of two tangent points, namely the appearance/disappearance of an extinction segment (local bifurcation) and the collision between the stable or unstable manifold of a saddle and a tangent point (global bifurcation) (see Figures A.23 and A.24, respectively). Finally, the last extinction bifurcation involves the collision between the trajectory emanating from a tangent point and a different tangent point (global bifurcation, not reported in figure because never encountered in our applications).

A.8 NUMERICAL METHODS AND SOFTWARE PACKAGES

All effective software packages for numerical bifurcation analysis are based on *continuation* (see, e.g., Keller, 1977; Doedel et al., 1991a,b; Beyn et al., 2002; Kuznetsov, 2004, Chapter 10), which is a general method for producing in $\mathbf{R}^q$ a

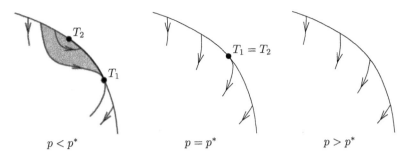

Figure A.23 Collision of tangent points: for $p < p^*$ the characteristic frame is composed of the extinction (thick) segment and the trajectory originating from the tangent point T_1, while for $p > p^*$ the characteristic frame is empty.

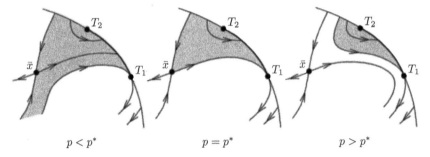

Figure A.24 Saddle-tangent-point connection: for both $p < p^*$ and $p > p^*$ the characteristic frame is composed of the saddle $\bar{x}$, its stable and unstable manifolds, the extinction (thick) segment $T_1 T_2$, and the trajectory originating from the tangent point T_1.

curve defined by $(q - 1)$ equations

$$F_1(w_1, w_2, \ldots, w_q) = 0,$$
$$F_2(w_1, w_2, \ldots, w_q) = 0,$$
$$\vdots$$
$$F_{q-1}(w_1, w_2, \ldots, w_q) = 0,$$

or, in compact form,

$$F(w) = 0, \quad w \in \mathbf{R}^q, \quad F : \mathbf{R}^q \to \mathbf{R}^{q-1}. \tag{A.13}$$

Given a point $w^{(0)}$ that is approximately on the curve, i.e., $F(w^{(0)}) \simeq 0$, the curve is produced by generating a sequence of points $w^{(i)}$, $i = 1, 2, \ldots$, that are approximately on the curve (i.e., $F(w^{(i)}) \simeq 0$), as shown in Figure A.25A. The ith iteration step, from $w^{(i)}$ to $w^{(i+1)}$, is a so-called prediction-correction procedure with adaptive step-size and is illustrated in Figure A.25B. The prediction $h v^{(i)}$ is taken along the direction tangent to the curve at $w^{(i)}$, where $v^{(i)}$ is computed as the vector of length 1 such that $\partial F / \partial w|_{w=w^{(i)}} v^{(i)} = 0$, the absolute value of h, called the step-size, is the prediction length, and the sign of h controls the direction of the

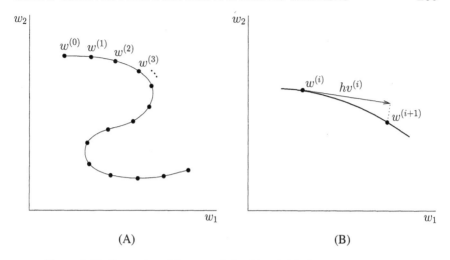

Figure A.25 Generation of the curve defined by (A.13) through continuation.

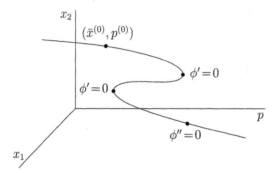

Figure A.26 The curve $\bar{x}(p)$ produced from $(\bar{x}^{(0)}, p^{(0)})$ through continuation in the three-dimensional control space (p, x_1, x_2) and three bifurcation points, detected through the annihilation of the bifurcation functions ϕ' and ϕ''.

continuation. Then, suitable corrections try to bring the predicted point back to the curve with the desired accuracy, thus determining $w^{(i+1)}$. If they fail, the step-size is reduced and the corrections are tried again until they succeed or the step-size goes below a minimum threshold at which the continuation halts with failure. By contrast, if corrections succeed at the first trial, the step-size is typically increased.

Given a second-order system $\dot{x} = f(x, p)$, where p is a single parameter, assume that an equilibrium $\bar{x}^{(0)}$ is known for $p = p^{(0)}$. Thus, starting from point $(\bar{x}^{(0)}, p^{(0)})$ in $\mathbf{R}^3$, the equilibria $\bar{x}(p)$ can be easily produced, as shown in Figure A.26, through continuation by considering (A.13) with

$$F(w) = f(x, p), \quad w = \begin{bmatrix} x \\ p \end{bmatrix}.$$

Moreover, at each step of the continuation, the Jacobian $J(\bar{x}(p), p)$ and its eigenvalues $\lambda_1(p)$ and $\lambda_2(p)$ are numerically estimated and a few indicators $\phi(\bar{x}(p), p)$,

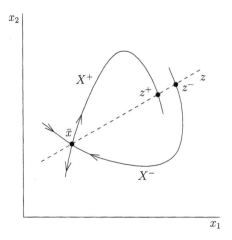

Figure A.27 The bifurcation function $\phi = z^+ - z^-$ is zero when there is a homoclinic bifurcation, i.e., when the stable and unstable manifolds X^- and X^+ of the saddle collide.

called *bifurcation functions*, are computed. These indicators annihilate at specific bifurcations, as shown in Figure A.26. For example, $\phi' = \det(J)$ is a bifurcation function of transcritical, saddle-node, and pitchfork bifurcations, since at these bifurcations one of the eigenvalues of the Jacobian matrix is zero and $\det(J) = \lambda_1 \lambda_2$. Similarly, $\phi'' = \text{trace}(J)$ is a Hopf bifurcation function (see Section A.4). Once a parameter value annihilating a bifurcation function has been found, a few simple tests are performed to check if the bifurcation is really present or to detect which is the true bifurcation within a set of potential ones. For example, as clearly pointed out by Figure A.8B, at a saddle-node bifurcation the p-component of the vector tangent to the curve $\bar{x}(p)$ annihilates. By contrast, at transcritical and pitchfork bifurcations (see Figures A.8A and C) two equilibrium curves, one of which is $\bar{x}(p)$, transversally cross each other, so that there are two tangent vectors at $p = p^*$, one with a vanishing p-component in the pitchfork case. Analogously, if $\phi''(\bar{x}(p^*), p^*) = 0$ one must first check that $\phi'(\bar{x}(p^*), p^*)$ is positive before concluding that $p = p^*$ is a Hopf bifurcation (see Section A.4).

Once a particular bifurcation has been detected through the annihilation of its bifurcation function ϕ, it can be continued by activating a second parameter. For this, (A.13) is written with

$$F(w) = \begin{bmatrix} f(x,p) \\ \phi(x,p) \end{bmatrix}, \quad w = \begin{bmatrix} x \\ p \end{bmatrix},$$

where w is now four dimensional since p is a vector of two parameters. If the curve obtained through continuation in $\mathbf{R}^4$ is projected on the two-dimensional parameter space, the desired bifurcation curve is obtained.

In the case of local bifurcations of limit cycles and global bifurcations, the functions ϕ are quite complex and their evaluation requires the solution of the ODEs $\dot{x} = f(x,p)$. Actually, a rigorous treatment of the problem brings one naturally to the formulation of two-boundary-value problems (Doedel et al., 1991b; Beyn et al.,

2002). For example, as shown in Figure A.27, homoclinic bifurcations can be detected by the function $\phi = z^+ - z^-$, where z^+ and z^- are the intersections of the unstable and stable manifolds of the saddle with an arbitrary axis z passing through the saddle. Thus, ϕ is zero if and only if the saddle has a homoclinic connection.

Many are the available software packages for bifurcation analysis, but the most interesting ones are AUTO (Doedel, 1981; Doedel et al., 1997, 2007), LOCBIF (Khibnik et al., 1993), CONTENT (Kuznetsov and Levitin, 1997), MATCONT (Dhooge et al., 2002), and SLIDECONT (Dercole and Kuznetsov, 2005). They can all be used to study systems with more than two state variables and, for this reason, they can detect and continue bifurcations that we did not mention in this appendix. AUTO is the most popular software for bifurcation analysis and is particularly suited for the analysis of global bifurcations. LOCBIF is more effective than AUTO for local bifurcations, since it can also continue codimension-2 bifurcations. However, LOCBIF runs only on MS-DOS and has therefore been reimplemented and improved in CONTENT, which runs on several software platforms. MATCONT, continuously updated, is aimed at encapsulating the best features of all previously mentioned software packages in a MATLAB environment. Finally, SLIDECONT is the only available software for detection and continuation of extinction bifurcations.

Appendix B

The Invasion Implies Substitution Theorem

In this appendix we prove the *invasion implies substitution theorem* presented in Section 3.4. Following the notation introduced in Sections 3.1, 3.2, and 3.4, the theorem is the following.

Given (x, X) in the evolution set $\mathcal{X}$, if

$$\left. \frac{\partial}{\partial x'} \lambda(x, x', X) \right|_{x'=x} (x' - x) > 0 \tag{B.1}$$

and $|x' - x|$ and

$$\|(n(0) + n'(0) - \bar{n}(x, X), N(0) - \bar{N}(x, X))\| \tag{B.2}$$

are sufficiently small, then the trajectory $(n(t), n'(t), N(t))$ of the resident-mutant model (3.2) tends for $t \to \infty$ toward the equilibrium (3.17), i.e., toward

$$(n, n', N) = (0, \bar{n}(x', X), \bar{N}(x', X)). \tag{B.3}$$

To prove the above statement, we introduce a new notation for the per-capita growth rate of the resident and mutant populations. More precisely, we consider a second (virtual) mutant population, with abundance n'' and trait value x'', and write the demographic dynamics of the resulting community as

$$\dot{n} = n f''(n, n', n'', N, x, x', x'', X), \tag{B.4a}$$

$$\dot{n}' = n' f''(n', n, n'', N, x', x, x'', X), \tag{B.4b}$$

$$\dot{n}'' = n'' f''(n'', n, n', N, x'', x, x', X), \tag{B.4c}$$

$$\dot{N} = F''(n, n', n'', N, x, x', x'', X), \tag{B.4d}$$

where the functions f'' and F'' satisfy properties of the kind (3.3) and (3.4), i.e.,

$$f''(n, 0, n'', N, x, x', x'', X) = f''(n, 0, n'', N, x, \cdot, x'', X), \tag{B.5a}$$

$$f''(n, n', n'', N, x, x, x'', X) = f''((1 - \phi)(n + n'), \phi(n + n'), n'', N, x, x, x'', X), \tag{B.5b}$$

$$f''(n, n', n'', N, x, x', x', X) = f''(n, (1 - \phi)(n' + n''), \phi(n' + n''), N, x, x', x', X), \tag{B.5c}$$

$$f''(n, n', n'', N, x, x', x'', X) = f''(n, n'', n', N, x, x'', x', X), \tag{B.5d}$$

and

$$F''(n, 0, n'', N, x, x', x'', X) = F''(n, 0, n'', N, x, \cdot, x'', X), \tag{B.6a}$$

$$F''(n, n', n'', N, x, x, x'', X) = F''((1 - \phi)(n + n'), \phi(n + n'), n'', N, x, x, x'', X), \tag{B.6b}$$

$$F''(n, n', n'', N, x, x', x'', X) = F''(n', n, n'', N, x', x, x'', X), \quad \text{(B.6c)}$$

$$F''(n, n', n'', N, x, x', x'', X) = F''(n'', n', n, N, x'', x', x, X), \quad \text{(B.6d)}$$

$0 \leq \phi \leq 1$. Moreover, model (B.4) degenerates into the resident-mutant model (3.2) when $n'' = 0$, i.e.,

$$f''(n, n', 0, N, x, x', \cdot, X) = f(n, n', N, x, x', X), \quad \text{(B.7a)}$$

$$F''(n, n', 0, N, x, x', \cdot, X) = F(n, n', N, x, x', X). \quad \text{(B.7b)}$$

Let

$$g(n, n', N, x, x', X, x'') = f''(0, n, n', N, x'', x, x', X) \quad \text{(B.8)}$$

be the per-capita growth rate of the newly introduced population when absent (see (B.4c) with $n'' = 0$). Thus, g is the initial per-capita growth rate of a second mutant population, which appears when the first mutant population is already part of the community. Since we consider mutations as rare events on the demographic timescale, we do not need to consider two mutant populations present at the same time in the community. However, we now show that the virtual presence of a second mutant population allows us to rewrite the resident-mutant model (3.2) in terms of function g, and this new formulation will be helpful in the proof of the theorem.

First, function g inherits properties (B.5) from function f'', i.e.,

$$g(0, n', N, x, x', X, x'') = g(0, n', N, \cdot, x', X, x''), \quad \text{(B.9a)}$$

$$g(n, n', N, x, x, X, x'') = g((1 - \phi)(n + n'), \phi(n + n'), N, x, x, X, x''), \quad \text{(B.9b)}$$

$$g(n, n', N, x, x', X, x'') = g(n', n, N, x', x, X, x''), \quad \text{(B.9c)}$$

$0 \leq \phi \leq 1$. Second, function g can be used to obtain the per-capita growth rates of the resident and of the real mutant populations by simply imagining that x'' coincides with x and x', respectively. In fact, from (B.5), (B.7a), and (B.8) it follows that

$$g(n, n', N, x, x', X, x) \overset{\text{(B.8)}}{=} f''(0, n, n', N, x, x, x', X)$$

$$\overset{\text{(B.5b)}}{=} f''(n, 0, n', N, x, x, x', X) \overset{\text{(B.5a)}}{=} f''(n, 0, n', N, x, \cdot, x', X)$$

$$\overset{\text{(B.5d)}}{=} f''(n, n', 0, N, x, x', \cdot, X) \overset{\text{(B.7a)}}{=} f(n, n', N, x, x', X)$$

and

$$g(n, n', N, x, x', X, x') \overset{\text{(B.8)}}{=} f''(0, n, n', N, x', x, x', X)$$

$$\overset{\text{(B.5d)}}{=} f''(0, n', n, N, x', x', x, X) \overset{\text{(B.5b)}}{=} f''(n', 0, n, N, x', x', x, X)$$

$$\overset{\text{(B.5a)}}{=} f''(n', 0, n, N, x', \cdot, x, X) \overset{\text{(B.5d)}}{=} f''(n', n, 0, N, x', x, \cdot, X)$$

$$\overset{\text{(B.7a)}}{=} f(n', n, N, x', x, X)$$

(where each equality holds in view of the relationship indicated above the equality sign; $\phi = 0$ in (B.5b)), so that we can rewrite the resident-mutant model (3.2) as

$$\dot{n} = ng(n, n', N, x, x', X, x), \quad \text{(B.10a)}$$

$$\dot{n}' = n'g(n, n', N, x, x', X, x'), \quad \text{(B.10b)}$$

$$\dot{N} = F(n, n', N, x, x', X). \quad \text{(B.10c)}$$

The function g is sometimes called the *generating fitness function* or *g-function* (see Brown and Vincent, 1987b; Vincent and Brown, 2005), because it generates the demographic dynamics of the resident and mutant populations by simply placing the trait of the virtual population at the corresponding resident and mutant values. In terms of the g-function, the invasion fitness of the mutant population reads

$$\lambda(x, x', X) = g(\bar{n}(x, X), 0, \bar{N}(x, X), x, \cdot, X, x'). \tag{B.11}$$

Let us now consider the change of variables

$$s = n + n', \quad r = \frac{n'}{n + n'}, \tag{B.12}$$

which is invertible through the transformation

$$n = s(1 - r), \quad n' = sr. \tag{B.13}$$

Then, by suitably expanding functions g and F in Taylor series with respect to x' around x and denoting by Δx the mutational step $(x' - x)$, the resident-mutant model (B.10) becomes

$$
\begin{aligned}
\dot{s} &= \dot{n} + \dot{n}' = ng(n, n', N, x, x', X, x) + n'g(n, n', N, x, x', X, x') \\
&= ng(n, n', N, x, x, X, x) + n'g(n, n', N, x, x, X, x) + O(\Delta x) \\
&\stackrel{(\text{B.9b})}{=} (n + n')g(n + n', 0, N, x, x, X, x) + O(\Delta x) \\
&\stackrel{(\text{B.9a, c})}{=} sg(s, 0, N, x, \cdot, X, x) + O(\Delta x),
\end{aligned}
$$

$$
\begin{aligned}
\dot{r} &= \frac{\dot{n}'(n + n') - n'(\dot{n} + \dot{n}')}{(n + n')^2} = \frac{n}{(n + n')^2}\dot{n}' - \frac{n'}{(n + n')^2}\dot{n} \\
&= \frac{nn'}{(n + n')^2}g(n, n', N, x, x', X, x') - \frac{n'n}{(n + n')^2}g(n, n', N, x, x', X, x) \\
&= r(1-r)\left(\frac{\partial}{\partial \Delta x}g(n, n', N, x, x + \Delta x, X, x + \Delta x)\Big|_{\Delta x = 0}\Delta x \right. \\
&\quad \left. - \frac{\partial}{\partial \Delta x}g(n, n', N, x, x + \Delta x, X, x)\Big|_{\Delta x = 0}\Delta x\right) + O(\Delta x^2) \\
&\stackrel{(*)}{=} r(1-r)\frac{\partial}{\partial \Delta x}g(n, n', N, x, x, X, x + \Delta x)\Big|_{\Delta x = 0}\Delta x + O(\Delta x^2) \\
&\stackrel{(\text{B.9b})}{=} r(1-r)\frac{\partial}{\partial \Delta x}g(n + n', 0, N, x, x, X, x + \Delta x)\Big|_{\Delta x = 0}\Delta x + O(\Delta x^2) \\
&\stackrel{(\text{B.9a, c})}{=} r(1-r)\frac{\partial}{\partial \Delta x}g(s, 0, N, x, \cdot, X, x + \Delta x)\Big|_{\Delta x = 0}\Delta x + O(\Delta x^2),
\end{aligned}
$$

$$\dot{N} = F(n, n', N, x, x, X) + O(\Delta x) \stackrel{(\text{3.4a, b})}{=} F(s, 0, N, x, \cdot, X) + O(\Delta x).$$

Notice that the advantage of the g-function formulation is evident at the equality marked by $(*)$. In fact, the influence of Δx on g is split between the fifth and the seventh arguments of g, i.e., between the traits of the mutant and virtual populations, and this leads to the simplification at $(*)$. In other words, the g-function formulation formally reveals the slow-fast nature of the resident-mutant dynamics, where r is the slow variable, whose rate of change $\dot{r}$ is proportional to the small mutational step Δx, while s and N are the fast variables. The timescaling $\tau = |\Delta x|t$ yields the standard slow-fast formulation

$$|\Delta x|\frac{ds}{d\tau} = sg(s, 0, N, x, \cdot, X, x) + O(\Delta x),\tag{B.14a}$$

$$\frac{dr}{d\tau} = r(1-r)\left.\frac{\partial}{\partial \Delta x}g(s, 0, N, x, \cdot, X, x+\Delta x)\right|_{\Delta x=0}\frac{\Delta x}{|\Delta x|} + O(\Delta x),\tag{B.14b}$$

$$|\Delta x|\frac{dN}{d\tau} = F(s, 0, N, x, \cdot, X) + O(\Delta x),\tag{B.14c}$$

where condition (B.1) implies

$$\frac{\Delta x}{|\Delta x|} = \begin{cases} 1 & \text{if } \left.\frac{\partial}{\partial x'}\lambda(x, x', X)\right|_{x'=x} > 0 \\[2mm] -1 & \text{if } \left.\frac{\partial}{\partial x'}\lambda(x, x', X)\right|_{x'=x} < 0, \end{cases}$$

i.e.,

$$\left.\frac{\partial}{\partial \Delta x}g(s, 0, N, x, \cdot, X, x + \Delta x)\right|_{\Delta x=0}\frac{\Delta x}{|\Delta x|}$$
$$= \left|\left.\frac{\partial}{\partial \Delta x}g(s, 0, N, x, \cdot, X, x + \Delta x)\right|_{\Delta x=0}\right|$$

(see (B.11)).

The proof of the theorem follows by applying *singular perturbation theory* (Tikhonov, 1952; Hoppensteadt, 1966) to model (B.14). Singular perturbation theory gives conditions under which the solutions of the "perturbed" system (B.14), characterized by a small parameter Δx perturbed around its nominal ("singular") value $\Delta x = 0$, converge as $\Delta x \to 0$ to the *singular solutions*, namely to the solutions of the differential-algebraic system obtained from system (B.14) by setting $\Delta x = 0$ (*singular system*), i.e.,

$$0 = sg(s, 0, N, x, \cdot, X, x),\tag{B.15a}$$

$$\frac{dr}{d\tau} = r(1-r)\left|\left.\frac{\partial}{\partial \Delta x}g(s, 0, N, x, \cdot, X, x + \Delta x)\right|_{\Delta x=0}\right|,\tag{B.15b}$$

$$0 = F(s, 0, N, x, \cdot, X).\tag{B.15c}$$

More precisely, if for any fixed value $0 \le r \le 1$ of the slow variable the fast variables (s, N) converge to an equilibrium, then the singular solutions are given by the dynamics of the slow variable, which track the corresponding singular equilibrium of the fast variables (i.e., the equilibrium obtained for $\Delta x = 0$). In particular, the result by Hoppensteadt (1966) says that under suitable regularity conditions, which we assume to hold for $(x, X) \in \mathcal{X}$, for sufficiently small Δx, and for initial values of the fast variables in the basin of attraction of the equilibrium corresponding to the initial value of the slow variable, the solution of system (B.14) exists for all $\tau > 0$ and converges as $\Delta x \to 0$, uniformly on any closed interval $[0, \tau]$, to the singular solution.

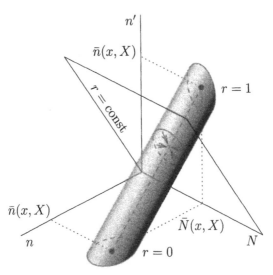

Figure B.1 Dynamics of system (B.14) close to segment (B.17) (dashed line) in the limit $\Delta x \to 0$.

To apply Hoppensteadt's result to system (B.14), notice that since
$$g(\bar{n}(x, X), 0, \bar{N}(x, X), x, \cdot, X, x) = 0,$$
$$F(\bar{n}(x, X), 0, \bar{N}(x, X), x, \cdot, X) = 0,$$
by definition of the resident demographic equilibrium (3.6), then
$$(s, N) = (\bar{n}(x, X), \bar{N}(x, X)) \tag{B.16}$$
is the singular equilibrium of the fast variables for any fixed $0 \le r \le 1$. Thus, as sketched in Figure B.1, in the limit $\Delta x \to 0$, the trajectories of system (B.14) originating close enough to segment
$$(n, n', N) = ((1 - \phi)\bar{n}(x, X), \phi\bar{n}(x, X), \bar{N}(x, X)) \tag{B.17}$$
(dashed line), quickly converge to it at constant r (i.e., by remaining on a plane with constant n'/n ratio, see (B.12)). Such trajectories are characterized by a small (B.2), i.e., they originate in an open tube with segment (B.17) as the axis, which becomes the invariant tube of Figure B.1 for sufficiently small Δx. The existence of the invariant tube, into which resident-mutant dynamics are trapped for sufficiently small Δx, has been proved by Geritz et al. (2002) under much more general conditions (physiologically structured resident communities coexisting at generic attractors) and is called the "tube theorem."

Once the fast variables are at their singular equilibrium (B.16), the equation (B.15b) of the slow variable says that $dr/d\tau$ is positive for any $0 < r < 1$, i.e., at any point of segment (B.17), so that the singular solution moves along segment (B.17) and converges as $\tau \to \infty$ to the equilibrium
$$(s, r, N) = (\bar{n}(x, X), 1, \bar{N}(x, X)),$$
i.e., to
$$(n, n', N) = (0, \bar{n}(x, X), \bar{N}(x, X)).$$
This completes the proof of the theorem.

Appendix C

The Probability of Escaping Accidental Extinction

In this appendix we show that the probability $P_{1,I}$ that a mutant population initially composed of a single individual is after some time composed of I individuals is given by (3.34), i.e.,

$$P_{1,I} = \frac{1 - \dfrac{\lambda_d(x, x', X)}{\lambda_b(x, x', X)}}{1 - \left(\dfrac{\lambda_d(x, x', X)}{\lambda_b(x, x', X)}\right)^I}, \tag{C.1}$$

where $\lambda_b(x, x', X)$ and $\lambda_d(x, x', X)$, introduced in Section 3.5, approximate the per-capita birth and death rates of the mutant population close to the equilibrium (3.15) of the resident-mutant model (3.2).

As discussed in Section 3.5, as long as the number of individuals of the mutant population is not very large, the abundances (n, n', N) of the resident-mutant model (3.2) remain close to equilibrium (3.15), since deterministic demographic models can be used only when the actual number of individuals in each population is large enough to avoid accidental extinction. Moreover, birth and death processes are assumed to be independent Markov processes, so that the probabilities that a mutant individual gives rise to a birth or dies in a small demographic time interval $[t, t + dt]$ are given by

$$\lambda_b(x, x', X)dt, \tag{C.2a}$$

$$\lambda_d(x, x', X)dt \tag{C.2b}$$

(independently of the actual number of mutant individuals present at time t), and the probability of more than one birth or death in the same time interval is $O(dt^2)$ and gives no contribution to $P_{1,I}$, in the limit $dt \to 0$. Thus, just after the appearance of the first individual of the mutant population, the resident-mutant demographic dynamics are described by the resident populations at their demographic equilibrium (3.6) and by the stochastic dynamics of the mutant population governed by the Markov chain of Figure C.1 (where the probabilities (C.2) are abbreviated by $\lambda_b dt$ and $\lambda_d dt$, since (x, x', X) are frozen on the demographic timescale).

Notice that the extinction state 0 is the only node with no outward edges, i.e., it is a so-called *absorbing state*. Since all realizations of the mutant demographic dynamics starting from state 1 and never passing through state I necessarily have to end in state 0, then $1 - P_{1,I}$ is the probability of reaching state 0 from state 1 without ever passing through state I. In other words, all and only those realizations of the mutant demographic dynamics that contribute to $P_{1,I}$ start from state 1 and reach state I, regardless of whether they eventually end in state 0 or not. Thus, $P_{1,I}$

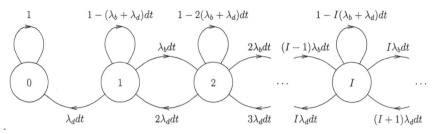

Figure C.1 Markov chain describing the demographic dynamics of the mutant population after the appearance of the first mutant. Numbers in nodes represent the number of individuals in the population (i.e., the possible states of the population), while edges represent $O(1)$ and $O(dt)$ probabilities of state transitions in the time interval $[t,\, t + dt]$.

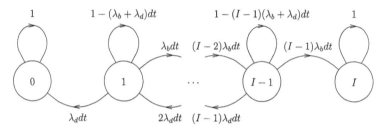

Figure C.2 Markov chain for the computation of $P_{1,I}$. The number of states is finite, so that all realizations eventually end in one of the two absorbing states 0 and I.

can be computed as the probability of reaching state I from state 1 in the Markov chain of Figure C.2, where state I is assumed to be a second absorbing state.

For brevity, we define $q = \lambda_d/\lambda_b$, so that (C.1) becomes

$$P_{1,I} = \frac{1-q}{1-q^I}. \tag{C.3}$$

Moreover, recalling the well-known result about geometric series

$$\sum_{i=0}^{I-1} q^i = \frac{1-q^I}{1-q}, \tag{C.4}$$

we can write (C.3) as

$$P_{1,I} = \left(\sum_{i=0}^{I-1} q^i\right)^{-1}. \tag{C.5}$$

To prove (C.5), we proceed by induction. For $I = 1$ (C.5) is readably satisfied, since the probability of reaching state 1 starting from state 1 itself is obviously equal to 1. Thus, all we have to prove is

$$P_{1,I+1} = \left(\sum_{i=0}^{I} q^i\right)^{-1}, \tag{C.6}$$

giving (C.5) as granted.

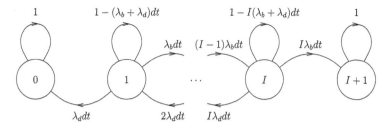

Figure C.3 Markov chain for the computation of $P_{1,I+1}$.

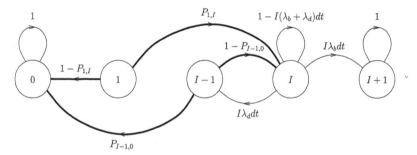

Figure C.4 Markov chain equivalent to that of Figure C.3 under hypothesis (C.5). Thick edges represent state transitions occurring in finite time.

For this, we must consider the Markov chain of Figure C.3, where states 0 and $(I + 1)$ are absorbing states, and compute $P_{1,I+1}$ as the probability of reaching state $(I + 1)$ from state 1. However, by virtue of (C.5), we can transform the Markov chain of Figure C.3 into that of Figure C.4, where thick edges represent state transitions occurring in finite time. In fact, from (C.5), the probability $P_{1,I}$ of reaching state I at some time in the future starting from state 1 is nonzero (see thick edge $(1, I)$ in Figure C.4). On the other hand, if state I is not reached in finite time, then the state of the mutant population is bounded between 0 and $(I - 1)$, so that the extinction state 0 remains the only absorbing state and will be certainly reached (see thick edge $(1, 0)$). Moreover, denoting by $P_{I-1,0}$ the probability of reaching the extinction state 0 starting from state $(I - 1)$ without ever passing through state I (see thick edge $(I - 1, 0)$), then $1 - P_{I-1,0}$ is the probability of going in finite time from state $(I - 1)$ to state I (see thick edge $(I - 1, I)$).

Notice that all realizations starting from state $(I - 1)$ and passing through state I give no contribution to $P_{I-1,0}$, which can therefore be computed as the probability of reaching state 0 from state $(I - 1)$ in the Markov chain of Figure C.2, where state I is an absorbing state. In Figure C.2, state $(I - 1)$ is separated from state 0 by $(I - 1)$ state transitions which occur with probabilities respectively obtained from those of the transitions separating state 1 from state I by reversing the order and exchanging λ_b with λ_d, i.e., by replacing q with $1/q$. Thus, under hypothesis (C.5), it must be

$$P_{I-1,0} = \left(\sum_{i=0}^{I-1} \left(\frac{1}{q} \right)^i \right)^{-1} = q^{I-1} \left(\sum_{i=0}^{I-1} q^i \right)^{-1}. \tag{C.7}$$

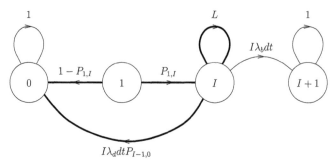

Figure C.5 Markov chain equivalent to that of Figure C.4 (see (C.8) for the computation of probability L). Thick edges represent state transitions occurring in finite time.

The Markov chain of Figure C.4 can be further transformed into that of Figure C.5, by removing state $(I-1)$ and accordingly computing the probabilities of the transitions $(I, 0)$ and (I, I). The first is the probability of reaching the extinction state 0 starting from state I without ever repassing through state I and is given by the product of the probabilities of transitions $(I, I-1)$ and $(I-1, 0)$ in Figure C.4. The second, denoted by L in Figure C.5, is the probability of going from state I back to state I without ever passing through state $(I+1)$ and is obtained from Figure C.4 as the product of the probabilities of transitions $(I, I-1)$ and $(I-1, I)$ plus the probability of transition (I, I), i.e.,

$$L = I\lambda_d dt(1 - P_{I-1,0}) + 1 - I(\lambda_b + \lambda_d)dt$$
$$= 1 - I\lambda_d dt P_{I-1,0} - I\lambda_b dt. \tag{C.8}$$

The probability $P_{1,I+1}$ can now be easily computed. In fact, $P_{1,I+1}$ is given by the sum of the probabilities of all realizations going from state 1 to state $(I+1)$ in Figure C.5, i.e.,

$$P_{1,I+1} = P_{1,I}\left(1 + L + L^2 + \cdots\right) I\lambda_b dt$$
$$= P_{1,I}\left(\sum_{i=0}^{\infty} L^i\right) I\lambda_b dt = \frac{P_{1,I} I\lambda_b dt}{1 - L}, \tag{C.9}$$

where the limit of the geometric series (C.4) (with $I \to \infty$ and L instead of q, $0 < L < 1$) has been taken into account. Substituting (C.5), (C.7), and (C.8) into (C.9), and recalling that $q = \lambda_d/\lambda_b$, we finally get

$$P_{1,I+1} = \frac{I\lambda_b dt}{\sum_{i=0}^{I-1} q^i \left(I\lambda_d dt\, q^{I-1} \left(\sum_{i=0}^{I-1} q^i\right)^{-1} + I\lambda_b dt \right)}$$
$$= \left(qq^{I-1} + \sum_{i=0}^{I-1} q^i\right)^{-1} = \left(\sum_{i=0}^{I} q^i\right)^{-1},$$

i.e., (C.6), as we wanted to show.

Appendix D

The Branching Conditions

In this appendix we prove that a stable evolutionary equilibrium $(\bar{x}, \bar{X})$ of the AD canonical equation (3.24) is a branching point with respect to trait x if conditions (3.41a) and (3.41b) are satisfied.

In Section 3.7 we showed (see, in particular, Figure 3.8) that, given $X = \bar{X}$, a small neighborhood of point $(\bar{x}, \bar{x})$ in the (x, x') plane is generically partitioned into four regions ①–④ corresponding to different stability configurations of the equilibria (3.15) and (3.17) of the resident-mutant model (3.2): ① both unstable; ② (3.15) unstable, (3.17) stable; ③ (3.15) stable, (3.17) unstable; ④ both stable (see bottom-right panel of Figure 3.8). Moreover, the invasion implies substitution theorem (Appendix B) guarantees that the state portraits of the resident-mutant model corresponding to regions ② and ③ are those reported in Figure 3.9. Thus, the diagonal $x' = x$ is a degenerate transcritical bifurcation of the resident-mutant model, at which both equilibria (3.15) and (3.17) change stability (with the presence, for $x' = x$, of a continuum of strictly positive equilibria lying on the segment connecting them). By contrast, the boundaries separating regions ① and ④ from regions ② and ③ are generic transcritical bifurcations. In fact, on such boundaries, only one of the two invasion eigenvalues $\lambda(x, x', \bar{X})$ and $\lambda(x', x, \bar{X})$ vanishes, while the corresponding equilibrium ((3.15) or (3.17)) exists on both sides of the boundary. More precisely, on boundaries separating region ① from region ③ [②] and region ④ from region ② [③], equilibrium (3.15) [(3.17)] collides and exchanges stability with a strictly positive equilibrium, which is present in regions ① (stable) and ④ (unstable). Generically, there are no other bifurcations of the resident-mutant model for (x, x') in a small neighborhood of $(\bar{x}, \bar{x})$ (see Meszéna et al., 2005, for a formal justification), so that the state portraits ① and ④ of Figure 3.9 and the resident-mutant coexistence condition (3.41a) are now justified.

What remains to be shown is that condition (3.41b) implies that two trait values x_1 and x_2, both initially close to $\bar{x}$ and characterizing two similar resident populations with other traits initially at $\bar{X}$, diverge one from the other according to the higher-dimensional canonical equation (3.49). For this we use the notation introduced in Section 3.7 and assume $x_1 < x_2$.

The resident community is settled at the demographic equilibrium (3.48) and is characterized by traits $(x_1, x_2, X^{(1)}, X^{(2)}, X^{(r)})$ in the evolution set $\mathcal{X}^{(1,2,r)}$. In particular, we can be more precise about the fitness functions λ_1 and λ_2 in (3.49). In fact, $X^{(1)} = X^{(2)}$ (the two similar resident populations differ only in one trait), so that $\lambda_1(x_1, x_2, X^{(1)}, X^{(1)}, X^{(r)}, x_1')$ and $\lambda_2(x_1, x_2, X^{(1)}, X^{(1)}, X^{(r)}, x_2')$ give the initial exponential rate of growth of two mutant populations also differing in a single trait, which takes value x_1' in one population and x_2' in the other. Thus, if we denote by

$$\lambda_{1,2}(x_1, x_2, X^{(1)}, X^{(r)}, x')$$ (D.1)

the fitness of a mutant population characterized by traits $(x', X^{(1)})$, we can write

$$\lambda_i(x_1, x_2, X^{(1)}, X^{(1)}, X^{(r)}, x'_i) = \lambda_{1,2}(x_1, x_2, X^{(1)}, X^{(r)}, x'_i), \quad i = 1, 2. \quad (D.2)$$

We first show that point (3.51), i.e.,

$$(x_1, x_2, X^{(1)}, X^{(2)}, X^{(r)}) = (\bar{x}, \bar{x}, \bar{X}^{(1)}, \bar{X}^{(1)}, \bar{X}^{(r)}), \quad (D.3)$$

is an evolutionary equilibrium of the higher-dimensional canonical equation (3.49). Obviously, if populations 1 and 2 are identical, i.e., if $x_1 = x_2 = x$ and $X^{(1)} = X^{(2)}$, they form a single population characterized by traits x and $X^{(1)}$ and the resident demographic equilibrium (3.48) is not uniquely defined, since

$$(n_1, n_2, N^{(r)}) = (\phi\bar{n}(x, X), (1 - \phi)\bar{n}(x, X), \bar{N}(x, X))$$

is an equilibrium of the resident model for any $0 \le \phi \le 1$. Moreover, since the individuals of a mutant population cannot distinguish between individuals of populations 1 and 2 if they are identical, for any $0 \le \phi \le 1$ it must be

$$\lambda_{1,2}(x, x, X^{(1)}, X^{(r)}, x') = \lambda(x, x', X), \quad (D.4)$$

so that

$$\left.\frac{\partial}{\partial x'}\lambda_{1,2}(\bar{x}, \bar{x}, \bar{X}^{(1)}, \bar{X}^{(r)}, x')\right|_{x'=\bar{x}} = \left.\frac{\partial}{\partial x'}\lambda(\bar{x}, x', \bar{X})\right|_{x'=\bar{x}} = 0 \quad (D.5)$$

(recall (3.39)). Similarly, the fitness of a mutant population originated by a mutation of one of the traits $X^{(1)}$, $X^{(2)}$, $X^{(r)}$ is unaffected by the virtual splitting of a resident population into two identical subpopulations, so that all selection derivatives annihilate at point (D.3).

The stability of equilibrium (D.3) cannot be assessed via linearization, since its associated eigenvalues depend on ϕ, as one can easily verify. However, in the following we show that, independently of ϕ, the evolutionary rates of change $\dot{x}_1$ and $\dot{x}_2$ (given by (3.49a) and (3.49b)) have opposite sign in $\mathcal{X}^{(1,2,r)}$ close to equilibrium (D.3) and that $\dot{x}_2$ and the second derivative (3.52), i.e.,

$$\left.\frac{\partial^2}{\partial x'^2}\lambda(\bar{x}, x', \bar{X})\right|_{x'=\bar{x}}, \quad (D.6)$$

have the same sign (we closely follow Geritz et al., 1998, Appendix 1). This completes the proof (recall that $x_1 < x_2$, so that we want to show that condition (3.41b) implies $\dot{x}_1 < 0$ and $\dot{x}_2 > 0$).

Let us consider points in $\mathcal{X}^{(1,2,r)}$ close to equilibrium (D.3) at which $X^{(1)} = X^{(2)}$. Then, by a continuity argument, the result also holds if $X^{(1)}$ and $X^{(2)}$ are slightly different, but close to $\bar{X}^{(1)}$. Second-order Taylor expansion of the fitness function (D.1) around

$$(x_1, x_2, X^{(1)}, X^{(r)}, x') = (\bar{x}, \bar{x}, \bar{X}^{(1)}, \bar{X}^{(r)}, \bar{x})$$

yields

$$\lambda_{1,2}(x_1, x_2, X^{(1)}, X^{(r)}, x') = \lambda_0 + \lambda_1(x_1 - \bar{x}) + \lambda_2(x_2 - \bar{x})$$
$$+ \lambda_3(X^{(1)} - \bar{X}^{(1)}) + \lambda_4(X^{(r)} - \bar{X}^{(r)}) + \lambda_5(x' - \bar{x}) + \frac{1}{2}\lambda_{11}(x_1 - \bar{x})^2$$
$$+ \lambda_{12}(x_1 - \bar{x})(x_2 - \bar{x}) + \lambda_{13}(x_1 - \bar{x})(X^{(1)} - \bar{X}^{(1)})$$
$$+ \lambda_{14}(x_1 - \bar{x})(X^{(r)} - \bar{X}^{(r)}) + \lambda_{15}(x_1 - \bar{x})(x' - \bar{x})$$
$$+ \frac{1}{2}\lambda_{22}(x_2 - \bar{x})^2 + \lambda_{23}(x_2 - \bar{x})(X^{(1)} - \bar{X}^{(1)}) + \lambda_{24}(x_2 - \bar{x})(X^{(r)} - \bar{X}^{(r)})$$
$$+ \lambda_{25}(x_2 - \bar{x})(x' - \bar{x}) + \frac{1}{2}(X^{(1)} - \bar{X}^{(1)})^T \lambda_{33}(X^{(1)} - \bar{X}^{(1)})$$
$$+ (X^{(1)} - \bar{X}^{(1)})^T \lambda_{34}(X^{(r)} - \bar{X}^{(r)}) + (X^{(1)} - \bar{X}^{(1)})^T \lambda_{35}(x' - \bar{x})$$
$$+ \frac{1}{2}(X^{(r)} - \bar{X}^{(r)})^T \lambda_{44}(X^{(r)} - \bar{X}^{(r)}) + (X^{(r)} - \bar{X}^{(r)})^T \lambda_{45}(x' - \bar{x})$$
$$+ \frac{1}{2}\lambda_{55}(x' - \bar{x})^2 + O(\|(x_1 - \bar{x}, x_2 - \bar{x}, X^{(1)} - \bar{X}^{(1)}, X^{(r)} - \bar{X}^{(r)}, x' - \bar{x})\|^3),$$

$$\text{(D.7)}$$

where

$$\lambda_0 = \lambda_{1,2}(\bar{x}, \bar{x}, \bar{X}^{(1)}, \bar{X}^{(r)}, \bar{x}),$$

while λ_i and λ_{ij}, $i, j = 1, \ldots, 5$, denote first- and second-order derivatives of the fitness function (D.1) with respect to the ith and jth arguments evaluated at $(\bar{x}, \bar{x}, \bar{X}^{(1)}, \bar{X}^{(r)}, \bar{x})$. In (D.7) $X^{(1)}$ and $X^{(r)}$ are column vectors, λ_i and λ_{ij} are scalar quantities, row or column vectors, or suitable matrices, depending on i and j (λ_i has one row and as many columns as the number of components of the ith argument of the fitness function (D.1); λ_{ij} has one row [column] for each component of the ith [jth] argument), and $\lambda_{ij} = \lambda_{ji}^T$, where the T superscript denotes transposition.

Notice that the fitness function (D.1) is symmetric with respect to its first two arguments, i.e.,

$$\lambda_{1,2}(x_1, x_2, X^{(1)}, X^{(r)}, x') = \lambda_{1,2}(x_2, x_1, X^{(1)}, X^{(r)}, x'), \qquad \text{(D.8)}$$

since both sides of (D.8) are the fitness of a mutant population characterized by traits $(x', X^{(1)})$ and interacting with populations 1, 2, regardless of their order, and with all other resident populations. Moreover, the fitness function (D.1) must vanish whenever x' coincides with x_1 or x_2, i.e.,

$$\lambda_{1,2}(x_1, x_2, X^{(1)}, X^{(r)}, x_1) = \lambda_{1,2}(x_1, x_2, X^{(1)}, X^{(r)}, x_2) = 0,$$

which we split, for convenience, into

$$\lambda_{1,2}(x_1, x_2, X^{(1)}, X^{(r)}, x_1) = \lambda_{1,2}(x_1, x_2, X^{(1)}, X^{(r)}, x_2) \qquad \text{(D.9a)}$$

and

$$\lambda_{1,2}(x_1, x_2, X^{(1)}, X^{(r)}, x_1) = 0. \qquad \text{(D.9b)}$$

Expanding both sides of (D.8) by means of (D.7) we get, after some algebra, the following first- and second-order terms:

$$\lambda_1(x_1 - \bar{x}) + \lambda_2(x_2 - \bar{x}) + \frac{1}{2}\lambda_{11}(x_1 - \bar{x})^2 + \lambda_{13}(x_1 - \bar{x})(X^{(1)} - \bar{X}^{(1)})$$
$$+ \lambda_{14}(x_1 - \bar{x})(X^{(r)} - \bar{X}^{(r)}) + \lambda_{15}(x_1 - \bar{x})(x' - \bar{x})$$
$$+ \frac{1}{2}\lambda_{22}(x_2 - \bar{x})^2 + \lambda_{23}(x_2 - \bar{x})(X^{(1)} - \bar{X}^{(1)})$$
$$+ \lambda_{24}(x_2 - \bar{x})(X^{(r)} - \bar{X}^{(r)}) + \lambda_{25}(x_2 - \bar{x})(x' - \bar{x})$$
$$= \lambda_1(x_2 - \bar{x}) + \lambda_2(x_1 - \bar{x}) + \frac{1}{2}\lambda_{11}(x_2 - \bar{x})^2 + \lambda_{13}(x_2 - \bar{x})(X^{(1)} - \bar{X}^{(1)})$$
$$+ \lambda_{14}(x_2 - \bar{x})(X^{(r)} - \bar{X}^{(r)}) + \lambda_{15}(x_2 - \bar{x})(x' - \bar{x})$$
$$+ \frac{1}{2}\lambda_{22}(x_1 - \bar{x})^2 + \lambda_{23}(x_1 - \bar{x})(X^{(1)} - \bar{X}^{(1)})$$
$$+ \lambda_{24}(x_1 - \bar{x})(X^{(r)} - \bar{X}^{(r)}) + \lambda_{25}(x_1 - \bar{x})(x' - \bar{x}),$$

so that balancing common powers of $(x_1 - \bar{x})$, $(x_2 - \bar{x})$, $(X^{(1)} - \bar{X}^{(1)})$, $(X^{(r)} - \bar{X}^{(r)})$, and $(x' - \bar{x})$, we can conclude that

$$\lambda_1 = \lambda_2, \tag{D.10a}$$
$$\lambda_{11} = \lambda_{22}, \tag{D.10b}$$
$$\lambda_{13} = \lambda_{23}, \tag{D.10c}$$
$$\lambda_{14} = \lambda_{24}, \tag{D.10d}$$
$$\lambda_{15} = \lambda_{25}. \tag{D.10e}$$

Analogously, expanding both sides of (D.9a) by means of (D.7) we get

$$\lambda_5(x_1 - \bar{x}) + \lambda_{15}(x_1 - \bar{x})^2 + \lambda_{25}(x_2 - \bar{x})(x_1 - \bar{x})$$
$$+ (X^{(1)} - \bar{X}^{(1)})^T \lambda_{35}(x_1 - \bar{x}) + (X^{(r)} - \bar{X}^{(r)})^T \lambda_{45}(x_1 - \bar{x})$$
$$+ \frac{1}{2}\lambda_{55}(x_1 - \bar{x})^2$$
$$= \lambda_5(x_2 - \bar{x}) + \lambda_{15}(x_1 - \bar{x})(x_2 - \bar{x}) + \lambda_{25}(x_2 - \bar{x})^2$$
$$+ (X^{(1)} - \bar{X}^{(1)})^T \lambda_{35}(x_2 - \bar{x}) + (X^{(r)} - \bar{X}^{(r)})^T \lambda_{45}(x_2 - \bar{x})$$
$$+ \frac{1}{2}\lambda_{55}(x_2 - \bar{x})^2,$$

so that balancing common powers we obtain

$$\lambda_5 = 0, \tag{D.11a}$$
$$\lambda_{15} + \frac{1}{2}\lambda_{55} = 0, \tag{D.11b}$$
$$\lambda_{15} = \lambda_{25}, \tag{D.11c}$$
$$\lambda_{35} = 0, \tag{D.11d}$$
$$\lambda_{45} = 0, \tag{D.11e}$$
$$\lambda_{25} + \frac{1}{2}\lambda_{55} = 0, \tag{D.11f}$$

where (D.11c) coincides with (D.10e).

Applying the same procedure to (D.9b) and taking (D.11) into account, we get

$$\lambda_0 + \lambda_1(x_1 - \bar{x}) + \lambda_2(x_2 - \bar{x}) + \lambda_3(X^{(1)} - \bar{X}^{(1)}) + \lambda_4(X^{(r)} - \bar{X}^{(r)})$$

$$+ \frac{1}{2}\lambda_{11}(x_1 - \bar{x})^2 + \lambda_{12}(x_1 - \bar{x})(x_2 - \bar{x}) + \lambda_{13}(x_1 - \bar{x})(X^{(1)} - \bar{X}^{(1)})$$

$$+ \lambda_{14}(x_1 - \bar{x})(X^{(r)} - \bar{X}^{(r)}) + \lambda_{15}(x_1 - \bar{x})^2 + \frac{1}{2}\lambda_{22}(x_2 - \bar{x})^2$$

$$+ \lambda_{23}(x_2 - \bar{x})(X^{(1)} - \bar{X}^{(1)}) + \lambda_{24}(x_2 - \bar{x})(X^{(r)} - \bar{X}^{(r)})$$

$$+ \lambda_{25}(x_2 - \bar{x})(x_1 - \bar{x}) + \frac{1}{2}(X^{(1)} - \bar{X}^{(1)})^T \lambda_{33}(X^{(1)} - \bar{X}^{(1)})$$

$$+ (X^{(1)} - \bar{X}^{(1)})^T \lambda_{34}(X^{(r)} - \bar{X}^{(r)}) + \frac{1}{2}(X^{(r)} - \bar{X}^{(r)})^T \lambda_{44}(X^{(r)} - \bar{X}^{(r)})$$

$$+ \frac{1}{2}\lambda_{55}(x_1 - \bar{x})^2 = 0,$$

so that balancing common powers we obtain

$$\lambda_0 = 0, \tag{D.12a}$$
$$\lambda_1 = 0, \tag{D.12b}$$
$$\lambda_2 = 0, \tag{D.12c}$$
$$\lambda_3 = 0, \tag{D.12d}$$
$$\lambda_4 = 0, \tag{D.12e}$$
$$\frac{1}{2}\lambda_{11} + \lambda_{15} + \frac{1}{2}\lambda_{55} = 0, \tag{D.12f}$$
$$\lambda_{12} + \lambda_{25} = 0, \tag{D.12g}$$
$$\lambda_{13} = 0, \tag{D.12h}$$
$$\lambda_{14} = 0, \tag{D.12i}$$
$$\lambda_{22} = 0, \tag{D.12j}$$
$$\lambda_{23} = 0, \tag{D.12k}$$
$$\lambda_{24} = 0, \tag{D.12l}$$
$$\lambda_{33} = 0, \tag{D.12m}$$
$$\lambda_{34} = 0, \tag{D.12n}$$
$$\lambda_{44} = 0. \tag{D.12o}$$

Notice that (D.12f) and (D.11b) imply $\lambda_{11} = 0$.

Taking (D.10)–(D.12) into account (see, in particular, (D.11b, c, f) and (D.12g), which allow one to express λ_{12}, λ_{15}, and λ_{25} as functions of λ_{55}) the expansion (D.7) reduces to

$$\lambda_{1,2}(x_1, x_2, X^{(1)}, X^{(r)}, x') = \frac{1}{2}\lambda_{55}(x_1 - \bar{x})(x_2 - \bar{x})$$

$$- \frac{1}{2}\lambda_{55}(x_1 - \bar{x})(x' - \bar{x}) - \frac{1}{2}\lambda_{55}(x_2 - \bar{x})(x' - \bar{x})$$

$$+ \frac{1}{2}\lambda_{55}(x' - \bar{x})^2 + O(\|\cdots\|^3)$$

$$= \frac{1}{2}\lambda_{55}(x' - x_1)(x' - x_2) + O(\|\cdots\|^3). \tag{D.13}$$

Moreover, evaluating (D.13) at $(\bar{x}, \bar{x}, \bar{X}^{(1)}, \bar{X}^{(r)}, x')$, taking (D.4) and (D.5) into account, and recalling that $\lambda(x, x, X) = 0$, we obtain

$$\lambda_{1,2}(\bar{x}, \bar{x}, \bar{X}^{(1)}, \bar{X}^{(r)}, x') = \frac{1}{2}\lambda_{55}(x' - \bar{x})^2 + O((x' - \bar{x})^3)$$

$$= \lambda(\bar{x}, x', \bar{X}) = \frac{1}{2}\frac{\partial^2}{\partial x'^2}\lambda(\bar{x}, x', \bar{X})\bigg|_{x'=\bar{x}} (x' - \bar{x})^2 + O((x' - \bar{x})^3),$$

so that

$$\lambda_{55} = \frac{\partial^2}{\partial x'^2}\lambda(\bar{x}, x', \bar{X})\bigg|_{x'=\bar{x}}.$$

Thus, from (D.13) we finally obtain

$$\frac{\partial}{\partial x'}\lambda_{1,2}(x_1, x_2, X^{(1)}, X^{(r)}, x')\bigg|_{x'=x_i}$$

$$= \frac{1}{2}\frac{\partial^2}{\partial x'^2}\lambda(\bar{x}, x', \bar{X})\bigg|_{x'=\bar{x}} (2x_i - x_1 - x_2) + O(\|\cdots\|^2), \quad i = 1, 2,$$

i.e., $\dot{x}_1$ and $\dot{x}_2$ have, according to (3.49a), (3.49b), and (D.2), opposite sign and $\dot{x}_2$ has the sign of (D.6) (recall that $x_1 < x_2$).

Bibliography

Abarbanel, H. D. I. (1996) *Analysis of Observed Chaotic Data*. Springer-Verlag, New York.

Abrams, P. A. (1992a) Adaptive foraging by predators as a cause of predator-prey cycles. *Evol. Ecol.* **6**, 56–72.

Abrams, P. A. (1992b) Predators that benefit prey and prey that harm predators: Unusual effects of interacting forgaing adaptations. *Am. Nat.* **140**, 573–600.

Abrams, P. A. (2000) The evolution of predator-prey interactions: Theory and evidence. *Annu. Rev. Ecol. Syst.* **31**, 79–105.

Abrams, P. A. (2001) Modelling the adaptive dynamics of traits involved in inter and intraspecific interactions: An assessment of three methods. *Ecol. Lett.* **4**, 166–175.

Abrams, P. A., Harada, Y., and Matsuda, H. (1993a) On the relationship between quantitative genetics and ESS models. *Evolution* **47**, 982–985.

Abrams, P. A. and Matsuda, H. (1997) Prey evolution as a cause of predator-prey cycles. *Evolution* **51**, 1742–1750.

Abrams, P. A., Matsuda, H., and Harada, Y. (1993b) Evolutionary unstable fitness maxima and stable fitness minima of continuous traits. *Evol. Ecol.* **7**, 465–487.

Abrams, P. A. and Shen, L. (1989) Population dynamics of systems with consumers that maintain a constant ratio of intake rates of two resources. *Theor. Popul. Biol.* **35**, 51–89.

Abrams, P. A. and Walters, C. J. (1996) Invulnerable prey and the paradox of enrichment. *Ecology* **77**, 1125–1133.

Addicott, J. F. (1985) Competition in mutualistic systems. In *The Biology of Mutualism*, ed. D. H. Boucher, pp. 217–247, Croom Helm, London.

Addicott, J. F. (1996) Cheaters in yucca/moth mutualism. *Nature* **380**, 114–115.

Addicott, J. F. and Bao, T. (1999) Limiting the costs of mutualism: Multiple modes of interaction between yuccas and yucca moths. *Proc. R. Soc. Lond.* B **266**, 197–202.

Aghion, P. and Howitt, P. (1992) A model of growth through creative destruction. *Econometrica* **60**, 323–351.

Albertson, R. C., Markert, J. A., Danley, P. D., and Kocher, T. D. (1999) Phylogeny of a rapidly evolving clade: The cichlid fishes of Lake Malawi, East Africa. *Proc. Natl. Acad. Sci.* **96**, 5107–5110.

Allee, W. C. (1931) *Animal Aggregations: A study in General Sociology*. The University of Chicago Press, Chicago.

Alligood, K. T., Sauer, T. D., and Yorke, J. A. (1996) *Chaos: An Introduction to Dynamical Systems*. Springer-Verlag, New York.

Alvarez, L. W., Alvarez, W., Asaro, F., and Michel, H. V. (1980) Extraterrestrial cause for the Cretaceous-Tertiary extinction. *Science* **208**, 1095–1108.

Andronov, A. A., Leontovich, E. A., Gordon, I. J., and Maier, A. G. (1973) *Theory of Bifurcations of Dynamical Systems on a Plane*. Israel Program for Scientific Translations, Jerusalem.

Anstett, M. C., Gibernau, M., and Hossaert-McKey, M. (1998) Partial avoidance of female inflorescences of a dioecious fig by their mutualistic pollinator wasps. *Proc. R. Soc. Lond.* B **265**, 45–50.

Arthur, W. B. (1990) Silicon valley's locational clusters: When increasing returns imply monopoly. *Math. Soc. Sci.* **19**, 235–251.

Ascher, U. M. and Spiteri, R. J. (1994) Collocation software for boundary value differential-algebric equations. *SIAM J. Sci. Comput.* **15**, 938–952.

Ausubel, J. H. (1990) Hydrogen and the green wave. *The Bridge* **20**, 17–22.

Ausubel, J. H. (1996) Can technology spare the Earth? *Am. Sci.* **84**, 166–178.

Ausubel, J. H., Marchetti, C., and Meyer, P. (1998) Toward green mobility: The evolution of transport. *Europ. Rev.* **6**, 137–156.

Axelrod, R. and Hamilton, W. D. (1981) The evolution of cooperation. *Science* **211**, 1390–1396.

Ayres, R. U. (1988) Complexity, reliability, and design: Manufacturing implications. *Manuf. Rev.* **1**, 26–35.

Bailey, N. T. J. (1964) *The Elements of Stochastic Processes with Applications to the Natural Sciences*. Wiley, New York.

Bakker, R. T. (1983) The deer flees, the wolf pursues: Incongruencies in predator-prey coevolution. In *Coevolution*, eds. D. J. Futuyma and M. Slatkin, pp. 350–382, Sinauer Associates, Sunderland, MA.

Baltensweiler, W. (1971) The relevance of changes in the composition of larch bud moth populations for the dynamics of its number. In *Dynamics of Populations*, eds. P. J. den Boer and G. R. Gradwell, pp. 208–219, Center for Agricultural Publishing and Documentation, Wageningen, The Netherlands.

Barluenga, M., Stölting, K. N., Salzburger, W., Muschick, M., and Meyer, A. (2006) Sympatric speciation in Nicaraguan crater lake cichlid fish. *Nature* **439**, 719–723.

Bartlett, M. S., ed. (1960) *Stochastic Population Models in Ecology and Epidemiology*. Methuen, London.

Baumgartner, T. R., Soutar, A., and Ferreira-Bartrina, V. (1992) Reconstruction of the history of Pacific sardine and northern anchovy populations over the past two millennia from sediments of the Santa Barbara basin, California. *Cal. Coop. Ocean. Fish.* **33**, 24–40.

Bell, G. (1982) *The Masterpiece of Nature: The Evolution and Genetics of Sexuality*. University of California Press, Berkeley.

Benkman, C. W. (1999) The selection mosaic and diversifying coevolution between crossbills and lodgepole pine. *Am. Nat.* **154**, S75–S91.

Benkman, C. W. (2003) Divergent selection drives the adaptive radiation of crossbills. *Evolution* **57**, 1176–1181.

Benton, M. J. and Pearson, P. N. (2001) Speciation in the fossil record. *Trends Ecol. Evol.* **16**, 405–411.

Bernard, J., Canter, U., Hnausch, H., and Westerman, G. (1994) Detecting technological performance and variety. EUNETIC Conference, October, Strasbourg.

Beyn, W.-J., Champneys, A. R., Doedel, E. J., Govaerts, W., Kuznetsov, Yu. A., and Sandstede, B. (2002) Numerical continuation, and computation of normal forms. In *Handbook of Dynamical Systems*, vol. 2, ed. B. Fiedler, pp. 149–219, Elsevier Science, Burlington, MA.

Bishop, B. E. (1996) Mendel's opposition to evolution and to Darwin. *J. Hered.* **87**, 205–213.

Blackburn, T. M. and Lawton, J. H. (1994) Population abundance and body size in animal assemblages. *Philos. T. Roy. Soc.* B **343**, 33–39.

Bolker, B. and Pacala, S. W. (1997) Using moment equations to understand stochastically driven spatial pattern formation in ecological systems. *Theor. Popul. Biol.* **52**, 179–197.

Bomze, I. and Bürger, R. (1995) Stability by mutation in evolutionary games. *Games Econ. Behav.* **11**, 146–172.

Bomze, I. and Pötscher, B. (1989) *Game Theoretical Foundations of Evolutionary Stability*. Springer-Verlag, Berlin.

Boucher, D. H., James, S., and Keeler, K. H. (1992) The ecology of mutualism. *Annu. Rev. Ecol. Syst.* **13**, 315–347.

Boyd, R. and Richerson, P. J., eds. (1985) *Culture and the Evolutionary Process*. The University of Chicago Press, Chicago.

Boyd, R. and Richerson, P. J. (1992) Punishment allows the evolution of cooperation (or anything else) in sizable groups. *Ethol. Sociobiol.* **13**, 171–195.

Bradshaw, H. D., Jr. and Schemske, D. W. (2003) Allele substitution at a flower colour locus produces a pollinator shift in monkeyflowers. *Nature* **426**, 176–178.

Bronstein, J. L. (2001a) The exploitation of mutualisms. *Ecol. Lett.* **4**, 277–287.

Bronstein, J. L. (2001b) Mutualisms. In *Evolutionary Ecology, Perspectives and Synthesis*, eds. C. Fox, D. Fairbairn, and D. Roff, pp. 149–162, Oxford University Press, New York.

Brooks, H. (1980) Technology, evolution, and purpose. *Daedalus* **109**, 65–81.

Brooks, J. L. and Dodson, S. I. (1965) Predation, body size, and composition of plankton. *Science* **150**, 28–35.

Brown, J. H., ed. (1995) *Macroecology*. The University of Chicago Press, Chicago.

Brown, J. S. and Pavlovic, N. B. (1992) Evolution in heterogeneous environments: Effects of migration on habitat specialization. *Evol. Ecol.* **6**, 360–382.

Brown, J. S. and Vincent, T. L. (1987a) Coevolution as an evolutionary game. *Evolution* **41**, 66–79.

Brown, J. S. and Vincent, T. L. (1987b) A theory for the evolutionary game. *Theor. Popul. Biol.* **31**, 140–166.

Brown, J. S. and Vincent, T. L. (1992) Organization of predator-prey communities as an evolutionary game. *Evolution* **46**, 1269–1283.

Brues, A. M. (1954) Stochastic tests of selection in the ABO blood groups. *Am. J. Phys. Anthropol.* **21**, 287–299.

Buerkle, A. C., Morris, R. J., Asmussen, M. A., and Rieseberg, L. H. (2000) The likelihood of homoploid hybrid speciation. *Heredity* **84**, 441–451.

Bull, J. J. and Rice, W. R. (1991) Distinguishing mechanisms for the evolution of co-operation. *J. Theor. Biol.* **149**, 63–74.

Bulmer, M. G. (1980) *The Mathematical Theory of Quantitative Genetics*. Oxford University Press, New York.

Bultman, T. L., Welch, A. M., Boning, R. A., and Bowdish, T. I. (2000) The cost of mutualism in a fly-fungus interaction. *Oecologia* **124**, 85–90.

Burda, M. and Wyplosz, Ch. (1997) *Macroeconomics: A European Text*. Oxford University Press, Oxford, UK.

Bürger, R. (2000) *The Mathematical Theory of Selection, Recombination, and Mutation*. Wiley, New York.

Byström, P. and Garcia-Berthóu, E. (1999) Density dependent growth and size specific competitive interactions in young fish. *Oikos* **86**, 217–232.

Calder, III., W. A. (1996) *Size, Function, and Life History*. Dover, Mineola, NY.

Callaway, R. M. and Walker, L. R. (1997) Competition and facilitation: A synthetic approach to interactions in plant communities. *Ecology* **78**, 1958–1965.

Candaten, M. and Rinaldi, S. (2000) Peak-to-peak dynamics: A critical survey. *Int. J. Bifurcat. Chaos* **10**, 1805–1819.

Carroll, L. (1871) *Through the Looking-Glass and What Alice Found There*. MacMillan, London.

Case, T. J. (1995) Surprising behavior from a familiar model and implications for competition theory. *Am. Nat.* **146**, 961–966.

Cavalli-Sforza, L. L. and Feldman, M. W. (1981) *Cultural Transmission and Evolution: A Quantitative Approach*. Princeton University Press, Princeton, NJ.

Cavalli-Sforza, L. L. and Zei, G. (1967) Experiments with an artificial population. In *Proceedings of the Third International Congress of Human Genetics*, eds. J. F. Crow and J. V. Neel, pp. 473–478, The Johns Hopkins University Press, Baltimore, MD.

Champagnat, N., Ferrière, R., and Ben Arous, G. (2001) The canonical equation of adaptive dynamics: A mathematical view. *Selection* **2**, 73–83.

Champagnat, N., Ferrière, R., and Méléard, S. (2006) Unifying evolutionary dynamics: From individual stochastic processes to macroscopic models. *Theor. Popul. Biol.* **69**, 297–321.

Charlesworth, B. and Charlesworth, D. (2003) *Evolution: A Very Short Introduction*. Oxford University Press, New York.

Chesson, J. (1983) The estimation and analysis of preference and its relationship to foraging models. *Ecology* **64**, 1297–1304.

Chown, S. L. and Smith, V. R. (1993) Climate change and the short-term impact of feral house mice at the sub-Antarctic Prince Edwards Islands. *Oecologia* **96**, 508–516.

Christaller, W. (1933) *Central Places in Southern Germany*. Prentice-Hall, Engle-wood Cliffs, NJ.

Christensen, K. M., Whitham, T. G., and Balda, R. P. (1991) Discrimination among pinyon pine trees by clark's nutcrackers: Effects of cone crop size and cone characters. *Oecologia* **86**, 402–407.

Christiansen, F. B. (1991) On conditions for evolutionary stability for a continu-ously varying character. *Am. Nat.* **138**, 37–50.

Christiansen, F. B. (2000) *Population Genetics of Multiple Loci*. Wiley, New York.

Claessen, D. and de Roos, A. M. (2003) Bistability in a size-structured population model of cannibalistic fish–a continuation study. *Theor. Popul. Biol.* **64**, 49–65.

Claessen, D., de Roos, A. M., and Persson, L. (2000) Dwarfs and giants: Cannibal-ism and competition in size-structured populations. *Am. Nat.* **155**, 219–237.

Claessen, D., de Roos, A. M., and Persson, L. (2004) Population dynamic theory of size-dependent cannibalism. *Proc. R. Soc. Lond.* B **271**, 333–340.

Cohen, D. (1966) Optimizing reproduction in a randomly varying environment. *J. Theor. Biol.* **12**, 119–129.

Cohen, J. E., Pimm, S. L., Yodzis, P., and Saldan, J. (1993) Body sizes of animal predator and animal prey in food webs. *J. Anim. Ecol.* **62**, 67–78.

Colas, B., Olivieri, I., and Riba, M. (1997) *Centaurea corymbosa*, a cliff-dwelling species tottering on the brink of extinction. a demographic and genetic study. *Proc. Natl. Acad. Sci.* **94**, 3471–3476.

Cole, L. C. (1954) The population consequences of life history phenomena. *Q. Rev. Biol.* **29**, 103–137.

Connell, J. H. and Sousa, W. P. (1983) On the evidence needed to judge ecological stability or persistence. *Am. Nat.* **121**, 789–824.

Courchamp, F., Clutton-Brock, T., and Grenfell, B. (1999) Inverse density depen-dence and the Allee effect. *Trends Ecol. Evol.* **14**, 405–410.

Coyne, J. A. and Orr, A. H. (2004) *Speciation*. Sinauer Associates, Sunderland, MA.

Crawley, M. J., ed. (1992) *Natural Enemies*. Blackwell Scientific, Oxford, UK.

Cressman, R. (1990) Strong stability and density-dependent evolutionarily stable strategies. *J. Theor. Biol.* **145**, 319–330.

Cressman, R. (1992) *The Stability Concept of Evolutionary Game Theory*. Springer-Verlag, Berlin.

Crick, F. and Watson, J. (1953) A structure of deoxyribonucleic acid. *Nature* **171**, 737–738.

Crow, J. F. and Kimura, M. (1970) *An Introduction to Population Genetics Theory.* Harper & Row, New York.

Currie, C. R., Scott, J. A., Summerbell, R. C., and Malloch, D. (1999) Fungus-growing ants use antibiotic-producing bacteria to control garden parasites. *Nature* **398**, 701–704.

Cury, P. (1988) Pressions sélectives et nouveautés évolutives: une hypothèse pour comprendre certains aspects des fluctuations à long terme des poissons pélagiques côtiers. *Can. J. Fish. Aquat. Sci.* **45**, 1099–1107 (in French).

Cushing, J. M. (1991) A simple model of cannibalism. *Math. Biosci.* **107**, 47–71.

Dalle, J. M. (1998) Local interaction structures, heterogeneity and diffusion of technological innovations. In *Advances in Self-organization and Evolutionary Economics*, eds. J. Lesourne and A. Orléan, pp. 240–264, Economica, London.

Darwin, C. (1839) *Journal of Researches into the Geology and Natural History of the Various Countries Visited by H. M. S. Beagle.* Colburn, London.

Darwin, C. (1858) On the variation of organic beings in a state of nature; on the natural means of selection; on the comparison of domestic races and true species. *Journal of the Proceedings of the Linnean Society, Zoology* **3**, 46–50.

Darwin, C. (1859) *The Origin of Species by Means of Natural Selection, or The Preservation of Favoured Races in the Struggle for Life.* John Murray, London.

Darwin, C. (1871) *The Descent of Man, and Selection in Relation to Sex.* John Murray, London.

Dawkins, R. (1976) *The Selfish Gene.* Oxford University Press, Oxford, UK.

Dawkins, R. (1982) *The Extended Phenotype: The Long Reach of the Gene.* Oxford University Press, Oxford, UK.

Dawkins, R. (1986) *The Blind Watchmaker: Why the Evidence of Evolution Reveals a Universe Without Design.* W. W. Norton, New York.

Dawkins, R. (1989) The evolution of evolvability. In *Artificial Life*, ed. C. Langton, pp. 201–220, Addison-Wesley, Reading, MA.

Day, T. (2005) Modelling the ecological context of evolutionary change: Déjà vu or something new? In *Ecological Paradigms Lost*, eds. K. Cuddington and B. E. Beisner, pp. 273–309, Elsevier, Burlington, MA.

De Angelis, D. L. and Gross, L. J., eds. (1992) *Individual-Based Models and Approaches in Ecology: Populations, Communities and Ecosystems.* Chapman & Hall, New York.

De Feo, O. and Ferrière, R. (2000) Bifurcation analysis of population invasion: On-off intermittency and basin riddling. *Int. J. Bifurcat. Chaos* **10**, 443–452.

De Feo, O. and Rinaldi, S. (1997) Yield and dynamics of tritrophic food chains. *Am. Nat.* **150**, 328–345.

De Feo, O. and Rinaldi, S. (1998) Singular homoclinic bifurcations in tritrophic food chains. *Math. Biosci.* **148**, 7–20.

De Palma, A., Kilani, K., and Lesourne, J. (1998) How network externalities affect product variety. In *Advances in Self-organization and Evolutionary Economics*, eds. J. Lesourne and A. Orléan, pp. 57–76, Economica, London.

De Vries, T. J. and Pearcy, W. G. (1982) Fish debris in sediments of the upwelling zone off central Peru: A late Quaternary record. *Deep-Sea Res.* **28**, 87–109.

Deng, B. (2001) Food chain chaos due to junction-fold point. *Chaos* **11**, 514–525.

Dennis, B. (1989) Allee effects: Population growth, critical density, and the chance of extinction. *Nat. Resour. Modelling* **3**, 481–538.

Dercole, F. (2002) *Evolutionary Dynamics through Bifurcation Analysis: Methods and Applications*. Thesis, Department of Electronics and Information, Politecnico di Milano, Milano, Italy.

Dercole, F. (2003) Remarks on branching-extinction evolutionary cycles. *J. Math. Biol.* **47**, 569–580.

Dercole, F. (2005) Border collision bifurcations in the evolution of mutualistic interactions. *Int. J. Bifurcat. Chaos* **15**, 2179–2190.

Dercole, F., Dieckmann, U., Obersteiner, M., and Rinaldi, S. (2008) Adaptive dynamics and technological change. *Technovation* (to appear).

Dercole, F., Ferrière, R., and Rinaldi, S. (2002a) Ecological bistability and evolutionary reversals under asymmetrical competition. *Evolution* **56**, 1081–1090.

Dercole, F., Gragnani, A., Ferrière, R., and Rinaldi, S. (2006) Coevolution of slow-fast populations: An application to prey-predator systems. *Proc. R. Soc. Lond. B* **273**, 983–990.

Dercole, F., Gragnani, A., and Rinaldi, S. (2002b) Sliding bifurcations in relay control systems: An application to natural resources management. In *Proceedings 15th IFAC World Congress,* Barcelona.

Dercole, F., Gragnani, A., and Rinaldi, S. (2007a) Bifurcation analysis of Filippov's ecological models. *Theor. Popul. Biol.* **72**, 186–196.

Dercole, F., Irisson, J.-O., and Rinaldi, S. (2003) Bifurcation analysis of a prey-predator coevolution model. *SIAM J. App. Math.* **63**, 1378–1391.

Dercole, F. and Kuznetsov, Yu. A. (2005) Slidecont: An Auto97 driver for bifurcation analysis of Filippov systems. *ACM T.Math. Software* **31**, 95–119.

Dercole, F., Loiacono, D., and Rinaldi, S. (2007b) Synchronization in population networks: A byproduct of darwinian evolution? *Int. J. Bifurcat. Chaos* **7**, 2435–2446.

Dercole, F. and Rinaldi, S. (2002) Evolution of cannibalism: Scenarios derived from adaptive dynamics. *Theor. Popul. Biol.* **62**, 365–374.

Dercole, F. and Rinaldi, S. (2008) Evolutionary dynamics can be chaotic: A first example. *Int. J. Bifurcat. Chaos* (to appear).

Després, L. and Jaeger, N. (1999) Evolution of oviposition strategies and speciation in the globeflower flies *Chiastocheta* spp. (Anthomyiidae). *J. Evol. Biol.* **12**, 822–831.

Dhooge, A., Govaerts, W., and Kuznetsov, Yu. A. (2002) MATCONT: A MATLAB package for numerical bifurcation analysis of ODEs. *ACM T.Math. Software* **29**, 141–164.

Dieckmann, U. (1994) Coevolutionary dynamics of stochastic replicator systems. Report n. 3018, Research Center Juelich, Germany.

Dieckmann, U. (1997) Can adaptive dynamics invade? *Trends Ecol. Evol.* **12**, 128–131.

Dieckmann, U. and Doebeli, M. (1999) On the origin of species by sympatric speciation. *Nature* **400**, 354–357.

Dieckmann, U., Doebeli, M., Metz, J. A. J., and Tautz, D., eds. (2004) *Adaptive Speciation*. Cambridge University Press, Cambridge, UK.

Dieckmann, U. and Ferrière, R. (2004) Adaptive dynamics and evolving biodiversity. In *Evolutionary Conservation Biology*, eds. R. Ferrière, U. Dieckmann, and D. Couvet, pp. 188–224, Cambridge University Press, Cambridge, UK.

Dieckmann, U., Heino, M., and Parvinen, K. (2006) The adaptive dynamics of function-valued traits. *J. Theor. Biol.* **241**, 370–389.

Dieckmann, U. and Law, R. (1996) The dynamical theory of coevolution: A derivation from stochastic ecological processes. *J. Math. Biol.* **34**, 579–612.

Dieckmann, U. and Law, R. (2000) Relaxation projections and the method of moments. In *The Geometry of Ecological Interactions*, eds. U. Dieckmann, R. Law, and J. A. J. Metz, pp. 412–455, Cambridge University Press, Cambridge, UK.

Dieckmann, U., Law, R., and Metz, J. A. J., eds. (2000) *The Geometry of Ecological Interactions*. Cambridge University Press, Cambridge, UK.

Dieckmann, U., Marrow, U., and Law, R. (1995) Evolutionary cycling in predator-prey interactions: Population dynamics and the Red Queen. *J. Theor. Biol.* **176**, 91–102.

Dieckmann, U., Metz, J. A. J., Sabelis, M. W., and Sigmund, K., eds. (2002) *Adaptive Dynamics of Infectious Diseases: In Pursuit of Virulence Management.* Cambridge University Press, Cambridge, UK.

Diekmann, O. (2004) A beginner's guide to adaptive dynamics. Banach Center Publications n. 63, 47–86, Banach International Mathematical Center, Warsaw, Poland.

Diekmann, O., Jabin, P.-E., Mischler, S., and Perthame, B. (2005) The dynamics of adaptation: An illuminating example and a Hamilton-Jacobi approach. *Theor. Popul. Biol.* **67**, 257–271.

Diekmann, O., Mylius, S. D., and ten Donkelaar, J. R. (1999) Saumon à la kaitala et getz, sauce hollandaise. *Evol. Ecol. Res.* **1**, 261–275.

Diekmann, O., Nisbet, R. M., Gurney, W. S. C., and van den Bosch, F. (1986) Simple mathematical models for cannibalism: A critique and a new approach. *Math. Biosci.* **78**, 21–46.

Dobzanski, T. (1937) *Genetics and the Origin of Species.* Columbia University Press, New York.

Doebeli, M. (1996a) A quantitative genetic competition model for sympatric speciation. *J. Evol. Biol.* **9**, 893–909.

Doebeli, M. (1996b) Quantitative genetics and population dynamics. *Evolution* **50**, 532–546.

Doebeli, M. (1998) Invasion of rare mutants does not imply their evolutionary success: A counterexample from metapopulation theory. *J. Evol. Biol.* **11**, 389–401.

Doebeli, M. and Dieckmann, U. (2000) Evolutionary branching and sympatric speciation caused by different types of ecological interactions. *Am. Nat.* **156**, 77–101.

Doebeli, M. and Dieckmann, U. (2003) Speciation along environmental gradients. *Nature* **421**, 259–264.

Doebeli, M., Dieckmann, U., Metz, J. A. J., and Tautz, D. (2005) What we have also learned: Adaptive speciation is theoretically plausible. *Evolution* **59**, 691–695.

Doebeli, M., Hauert, C., and Killingback, T. (2004) The evolutionary origin of cooperators and defectors. *Science* **306**, 859–862.

Doebeli, M. and Knowlton, N. (1998) The evolution of interspecific mutualisms. *Proc. Natl. Acad. Sci.* **95**, 8676–8680.

Doebeli, M. and Koella, J. C. (1995) Evolution of simple population dynamics. *Proc. R. Soc. Lond.* B **260**, 119–125.

Doebeli, M. and Ruxton, G. D. (1997) Evolution of dispersal rates in metapopulation models: Branching and cyclic dynamics in phenotype space. *Evolution* **51**, 1730–1741.

Doedel, E. J., Champneys, A. R., Fairgrieve, T. F., Kuznetsov, Yu. A., Sandstede, B., and Wang, X. J. (1997) AUTO97: Continuation and bifurcation software for ordinary differential equations. Department of Computer Science, Concordia University, Montreal, QC.

Doedel, E. J., Champneys, A. R., Fairgrieve, T. F., Kuznetsov, Yu. A., Oldeman, B., Paffenroth, R. C., Sandstede, B., Wang, X. J., and Zhang, C. H. (2007) AUTO-07p: Continuation and bifurcation software for ordinary differential equations. Department of Computer Science, Concordia University, Montreal, QC.

Doedel, E. J. (1981) AUTO, a program for the automatic bifurcation analysis of autonomous systems. *Cong. Numer.* **30**, 265–384.

Doedel, E. J., Keller, H. B., and Kernévez, J.-P. (1991a) Numerical analysis and control of bifurcation problems (I): Bifurcation in finite dimensions. *Int. J. Bifurcat. Chaos* **1**, 493–520.

Doedel, E. J., Keller, H. B., and Kernévez, J.-P. (1991b) Numerical analysis and control of bifurcation problems (II): Bifurcation in infinite dimensions. *Int. J. Bifurcat. Chaos* **1**, 745–772.

Dunning, J. H. (2000) Regions, globalization, and the knowledge economy: The issues stated. In *Regions, Globalization, and the Knowledge Economy*, ed. J. Dunning, pp. 1–29, Oxford University Press, Oxford, UK.

Durinx, M., Metz, J. A. J., and Meszéna, G. (2007) Adaptive dynamics for physiologically structured population models. *J. Math. Biol.* (to appear).

Dybdahl, M. F. and Storfer, A. (2003) Parasite local adaptation: Red Queen versus Suicide King. *Trends Ecol. Evol.* **18**, 523–530.

Ebenman, M., Johanson, A., Jonsson, T., and Wennergren, U. (1996) Evolution of stable population dynamics through natural selection. *Proc. R. Soc. Lond.* B **263**, 1145–1151.

Ehrlich, P. R. and Levin, S. A. (2005) The evolution of norms. *PLoS Biology* **3**, e194.

Ellner, S. P. and Turchin, P. (1995) Chaos in a noisy world: New methods and evidence from time series analysis. *Am. Nat.* **145**, 343–375.

Emlen, J. M. (1987) Evolutionary ecology and the optimality assumption. In *The Latest on the Best*, ed. J. Dupre, pp. 163–177, MIT Press, Cambridge, MA.

Enard, W., Przeworski, M., Fisher, S. E., Lai, C. S., Wiebe, V., Kitano, T., Monaco, A. P., and Pääbo, S. (2002) Molecular evolution of FOXP2, a gene involved in speech and language. *Nature* **418**, 869–872.

Engländer, O. (1926) *Kritisches und Positives zu einer allgemeinen reinen Lehre von Standort*. Zeitschrift für Volkswirtschaft und Sozialpolitik, Neue Folge 5 (in German).

Ernande, B. and Dieckmann, U. (2004) The evolution of phenotypic plasticity in spatially structured environments: Implications of intraspecific competition, plasticity costs and environmental characteristics. *J. Evol. Biol.* **17**, 613–628.

Ernande, B., Dieckmann, U., and Heino, M. (2004) Adaptive changes in harvested populations: Plasticity and evolution of age and size at maturation. *Proc. R. Soc. Lond.* B **271**, 415–423.

Eshel, I. (1983) Evolutionary and continuous stability. *J. Theor. Biol.* **103**, 99–111.

Eshel, I. (1996) On the changing concept of evolutionary population stability as a reflection of a changing point of view in the quantitative theory of evolution. *J. Math. Biol.* **34**, 485–510.

Eshel, I. and Feldman, M. W. (1984) Initial increase of new mutants and some continuity properties of ESS in two-locus systems. *Am. Nat.* **124**, 631–640.

Eshel, I., Feldman, M. W., and Bergman, A. (1998) Long-term evolution, short-term evolution, and population genetic theory. *J. Theor. Biol.* **191**, 391–396.

Eshel, I. and Motro, U. (1981) Kin selection and strong stability of mutual help. *Theor. Popul. Biol.* **19**, 420–433.

Eshel, I., Motro, U., and Sansone, E. (1997) Continuous stability and evolutionary convergence. *J. Theor. Biol.* **185**, 333–343.

Ewens, W. J. (1979) *Mathematical Population Genetics*. Springer-Verlag, Berlin.

Falconer, D. S. (1989) *Introduction to Quantitative Genetics*. Longman, Harlow.

Feder, J. L. (1995) The effects of parasitoids on sympatric host races of *Rhagoletis pomonella* (diptera: Tephritidae). *Ecology* **76**, 801–813.

Fehr, E. and Gächter, S. (1998) Reciprocity and economics: The economic implications of homo reciprocans. *Europ. Econ. Rev.* **42**, 845–859.

Fehr, E. and Gächter, S. (2000) Cooperation and punishment in public goods experiments. *Am. Econ. Rev.* **90**, 980–994.

Fehr, E. and Gächter, S. (2002) Altruistic punishment in humans. *Nature* **415**, 137–140.

Feigenbaum, M. J. (1980) The metric universal properties of period doubling bi-furcations and the spectrum for a route to turbulence. *Ann. NY Acad. Sci.* **357**, 330–336.

Felsenstein, J. (1981) Skepticism towards Santa Rosalia, or why are there so few kinds of animals? *Evolution* **35**, 124–138.

Ferdy, J.-B. and Godelle, B. (2005) Diversification of transmission modes and the evolution of mutualism. *Am. Nat.* **166**, 613–627.

Ferrière, R. (2000) Adaptive responses to environmental threats: Evolutionary sui-cide, insurance and rescue. Options Spring 2000, 12–16, International Institute for Applied Systems Analysis, Laxenburg, Austria.

Ferrière, R., Bronstein, J. L., Rinaldi, S., Law, R., and Gauduchon, M. (2002) Cheating and the evolutionary stability of mutualisms. *Proc. R. Soc. Lond.* B **269**, 773–780.

Ferrière, R. and Cazelles, C. (1999) Universal power laws govern intermittent rarity in communities of interacting species. *Ecology* **80**, 1505–1521.

Ferrière, R., Dieckmann, U., and Couvet, D., eds. (2004a) *Evolutionary Conserva-tion Biology*. Cambridge University Press, Cambridge, UK.

Ferrière, R., Dieckmann, U., and Couvet, D. (2004b) Introduction. In *Evolutionary Conservation Biology*, eds. R. Ferrière, U. Dieckmann, and D. Couvet, pp. 1–14, Cambridge University Press, Cambridge, UK.

Ferrière, R. and Fox, G. A. (1995) Chaos and evolution. *Trends Ecol. Evol.* **10**, 480–485.

Ferrière, R. and Gatto, M. (1993) Chaotic dynamics can result from natural selec-tion. *Proc. R. Soc. Lond.* B **251**, 33–38.

Ferrière, R. and Gatto, M. (1995) Lyapunov exponents and the mathematics of invasion in oscillatory or chaotic populations. *Theor. Popul. Biol.* **48**, 126–171.

Filippov, A. F. (1988) *Differential Equations with Discontinuous Righthand Sides*. Kluwer Academic, Dordrecht, The Netherlands.

Fischbacher, U., Gächter, S., and Fehr, E. (2001) Are people conditionally cooper-ative? Evidence from a public goods experiment. *Econ. Lett.* **71**, 397–404.

Fisher, J. C. and Pry, R. H. (1971) A simple substitution model of technological change. *Technol. Forecast. Soc.* **3**, 75–88.

Fisher, R. A. (1930) *The Genetical Theory of Natural Selection*. Clarendon Press, Oxford, UK.

Ford, E. B. (1949) *Mendelism and Evolution*. Methuen, London.

Foster, M. S. and Delay, L. S. (1998) Dispersal of mimetic seeds of three species of *Ormosia* (Leguminosae). *J. Trop. Ecol.* **14**, 389–411.

Fox, L. R. (1975) Cannibalism in natural populations. *Annu. Rev. Ecol. Syst.* **6**, 87–106.

Frank, S. A. (1994) The origin of synergistic symbiosis. *J. Theor. Biol.* **176**, 403–410.

Frank, S. A. (1996) Models of parasite virulence. *Q. Rev. Biol.* **71**, 37–78.

Fudenberg, D. and Levine, K. (1998) *The Theory of Learning in Games*. MIT Press, Cambridge, MA.

Futuyma, D. J. (1998) *Evolutionary Biology*, 3rd ed., Sinauer Associates, Sunderland, MA.

Futuyma, D. J. and Slatkin, M., eds. (1983) *Coevolution*. Sinauer Associates, Sunderland, MA.

Galis, F., van Alphen, J. J. M., and Metz, J. A. J. (2001) Why five fingers? Evolutionary constraints on digit numbers. *Trends Ecol. Evol.* **16**, 637–646.

Gaston, K. J. and Kunin, W. E. (1997) Rare-common differences: An overview. In *The Biology of Rarity*, eds. W. E. Kunin and K. J. Gaston, pp. 12–29, Chapman & Hall, London.

Gatto, M. (1990) A general minimum principle for competing populations: Some ecological and evolutionary consequences. *Theor. Popul. Biol.* **37**, 369–388.

Gatto, M. (1993) The evolutionary optimality of oscillatory and chaotic dynamics in simple population models. *Theor. Popul. Biol.* **43**, 310–336.

Gavrilets, S. (1997) Coevolutionary chase in exploiter-victim systems with polygemic characters. *J. Theor. Biol.* **186**, 527–534.

Gavrilets, S. (2003) Models of speciation: What have we learned in 40 years? *Evolution* **57**, 2197–2215.

Gavrilets, S. (2004) *Fitness Landscapes and the Origin of Species*. Princeton University Press, Princeton, NJ.

Genkai-Kato, M. and Yamamura, N. (1999) Unpalatable prey resolves the paradox of enrichment. *Proc. R. Soc. Lond.* B **266**, 1215–1219.

Geritz, S. A. H. (2005) Resident-invader dynamics and the coexistence of similar strategies. *J. Math. Biol.* **50**, 67–82.

Geritz, S. A. H., Gyllenberg, M., Jacobs, F. J. A., and Parvinen, K. (2002) Invasion dynamics and attractor inheritance. *J. Math. Biol.* **44**, 548–560.

Geritz, S. A. H. and Kisdi, E. (2000) Adaptive dynamics in diploid, sexual populations and the evolution of reproductive isolation. *Proc. R. Soc. Lond.* B **267**, 1671–1678.

Geritz, S. A. H., Kisdi, E., Meszéna, G., and Metz, J. A. J. (1998) Evolutionarily singular strategies and the adaptive growth and branching of the evolutionary tree. *Evol. Ecol.* **12**, 35–57.

Geritz, S. A. H., Kisdi, E., van der Meijden, E., and Metz, J. A. J. (1999) Evolutionarily dynamics of seed size and seedling competitive ability. *Theor. Popul. Biol.* **55**, 324–343.

Geritz, S. A. H., Metz, J. A. J., Kisdi, E., and Meszéna, G. (1997) The dynamics of adaptation and evolutionary branching. *Phys. Rev. Lett.* **78**, 2024–2027.

Getto, P., Diekmann, O., and de Roos, A. M. (2005) On the (dis) advantages of cannibalism. *J. Math. Biol.* **51**, 695–712.

Gibbons, A. (1996) The species problem. *Science* **273**, 1501.

Gillespie, D. T. (1976) A general method for numerically simulating the stochastic time evolution of coupled chemical reactions. *J. Comp. Phys.* **22**, 403–434.

Goldberg, D. E. (1989) *Genetic Algorithms in Search, Optimization, and Machine Learning*. Addison-Wesley, Reading, MA.

Grafen, A. (1984) Natural selection, kin selection and group selection. In *Behavioral Ecology: An Evolutionary Approach*, eds. J. R. Krebs and N. B. Davies, pp. 62–86, Blackwell Scientific, Oxford, UK.

Grant, P. R. (1986) *Ecology and Evolution of Darwin's Finches*. Princeton University Press, Princeton, NJ.

Grant, P. R. and Grant, B. R. (2002) Unpredictable evolution in a 30-year study of Darwin's finches. *Science* **296**, 707–711.

Grant, P. R., Grant, B. R., Smith, J. N. M., Abbott, I. J., and Abbott, L. K. (1976) Darwin's finches: Population variation and natural selection. *Proc. Natl. Acad. Sci.* **73**, 257–261.

Grant, P. R., Schluter, D., Curry, R. L., and Abbott, L. K. (1985) Variation in the size and shape of Darwin's finches. *Biol. J. Linn. Soc.* **25**, 1–39.

Greif, A. (1994) Cultural beliefs and the organization of society: A historical and theoretical reflection on collectivist and individualist societies. *J. Polit. Econ.* **102**, 912–950.

Grimm, V. and Railsback, S. F. (2005) *Individual-based Modeling and Ecology*. Princeton University Press, Princeton, NJ.

Grossman, G. M. and Helpman, E. (1991) *Innovation and Growth in the Global Economy*. MIT Press, Cambridge, MA.

Grübler, A. (1990a) *The Rise and Fall of Infrastructure: Dynamics of Evolution and Technological Change in Transport*. Physica-Verlag, Heidelberg.

Grübler, A. (1990b) Technological diffusion in a long wave context: The case of the steel and coal industries. In *Life Cycles and Long Waves*, eds. T. Vasko, R. Ayres, and L. Fontvieille, pp. 117–146, Springer-Verlag, Berlin.

Grübler, A. (1998) *Technology and Global Change*. The Press Syndicate of the University of Cambridge, Cambridge, UK.

Guckenheimer, J. and Holmes, P. (1997) *Nonlinear Oscillations, Dynamical Systems and Bifurcations of Vector Fields*, 5th ed., Springer-Verlag, New York.

Gyllenberg, M. and Parvinen, K. (2001) Necessary and sufficient conditions for evolutionary suicide. *Bull. Math. Biol.* **63**, 981–993.

Gyllenberg, M., Parvinen, K., and Dieckmann, U. (2002) Evolutionary suicide and evolution of dispersal in structured metapopulations. *J. Math. Biol.* **45**, 79–105.

Haccou, P., Jagers, P., and Vatutin, V. A. (2005) *Branching Processes: Variation, Growth, and Extinction of Populations*. Cambridge University Press, Cambridge, UK.

Hadeler, K. P. (1981) Stable polymorphisms in a selection model with mutation. *SIAM J. App. Math.* **41**, 1–7.

Haldane, J. B. S. (1932) *The Causes of Evolution*. Longmans Green, London.

Haldane, J. B. S. and Jayakar, S. D. (1963) Polymorphism due to selection depending on the composition of a population. *J. Genet.* **58**, 318–323.

Hamilton, W. D. (1963) The evolution of altruistic behaviour. *Am. Nat.* **97**, 354–356.

Hamilton, W. D. (1964a) The genetical evolution of social behaviour, I. *J. Theor. Biol.* **7**, 1–16.

Hamilton, W. D. (1964b) The genetical evolution of social behaviour, II. *J. Theor. Biol.* **7**, 17–52.

Hamilton, W. D. (1967) Extraordinary sex ratios. *Science* **156**, 477–488.

Hammerstein, P. (1996a) Darwinian adaptation, population genetics and the streetcar theory of evolution. *J. Math. Biol.* **34**, 511–532.

Hammerstein, P. (1996b) Streetcar theory and long-term evolution. *Science* **23**, 1029–1032.

Hanski, I. (1985) Single-species spatial dynamics may contribute to long-term rarity and commonness. *Ecology* **66**, 335–343.

Harberger, A. C. (1998) A vision of the growth process. *Am. Econ. Rev.* **88**, 1–32.

Hardin, G. (1960) The competitive exclusion principle. *Science* **131**, 1292–1298.

Hardin, G. (1968) The tragedy of the commons. *Science* **162**, 1243–1248.

Hardy, G. H. (1908) Mendelian proportions in a mixed population. *Science* **28**, 49–50.

Hartl, D. L. and Clark, A. G. (1997) *Principles of Population Genetics*, 3rd ed., Sinauer Associates, Sunderland, MA.

Hastings, A. and Powell, T. (1991) Chaos in a three species food chain. *Ecology* **72**, 896–903.

Hauert, C., De Monte, S., Hofbauer, J., and Sigmund, K. (2002) Volunteering as Red Queen mechanism for cooperation in public goods games. *Science* **296**, 1129–1132.

Hauert, C. and Doebeli, M. (2004) Spatial structure often inhibits the evolution of cooperation in the snowdrift game. *Nature* **428**, 643–646.

Hauser, M. D. (1996) *The Evolution of Communication*. MIT Press, Cambridge, MA.

Hauser, M. D., Chomsky, N., and Fitch, W. T. (2002) The faculty of language: What is it, who has it, and how did it evolve? *Science* **298**, 1569–1579.

Hayashi, C., Ueda, Y., Akamatsu, N., and Itakura, H. (1970) On the behavior of self-oscillatory systems with external force. *Electronics & Communication in Japan* **53-A**, 150–158 (in Japanese).

Hayek, F. A. (1967) *Studies in Philosophy, Politics, and Economics*. Routledge & Kegan Paul, London.

Heino, M., Metz, J. A. J., and Kaitala, V. (1998) The enigma of frequency-dependent selection. *Trends Ecol. Evol.* **13**, 367–370.

Henson, S. M. (1997) Cannibalism can be beneficial even when its mean yield is less than one. *Theor. Popul. Biol.* **51**, 107–117.

Herre, E. A., Knowlton, N., Mueller, U. G., and Rehmer, S. A. (1999) The evolution of mutualisms: Explaining the paths between conflict and cooperation. *Trends Ecol. Evol.* **14**, 49–53.

Hibbett, D. S., Gilbert, L. B., and Donoghue, M. J. (2000) Evolutionary instability of ectomycorrhizal symbioses in basidiomycetes. *Nature* **407**, 506–508.

Higashi, M., Takimoto, G., and Yamamura, N. (1999) Sympatric speciation by sexual selection. *Nature* **402**, 523–526.

Hodgson, G. M. (1997) The evolutionary and non-Darwinian economics of Joseph Schumpeter. *J. Evol. Econ.* **7**, 131–146.

Hofbauer, J., Schuster, P., and Sigmund, K. (1979) A note on evolutionarily stable strategies and game dynamics. *J. Theor. Biol.* **81**, 609–612.

Hofbauer, J. and Sigmund, K. (1990) Adaptive dynamics and evolutionary stability. *Appl. Math. Lett.* **3**, 75–79.

Hofbauer, J. and Sigmund, K. (1998) *Evolutionary Games and Population Dynamics*. Cambridge University Press, Cambridge, UK.

Hofbauer, J. and Sigmund, K. (2003) Evolutionary games dynamics. *Bull. Am. Math. Soc.* **40**, 479–519.

Holcik, J. (1977) Changes in fish community of Kikava Reservoir with particular reference to Eurasian perch. *Fish. Res. Bd. Can.* **34**, 1734–1747.

Holland, J. H. (1975) *Adaptation in Natural and Artificial Systems*. University of Michigan Press, Ann Arbor.

Holling, C. S. (1965) The functional response of predators to prey density and its role in mimicry and population regulation. *Mem. Entomol. Soc. Can.* **45**, 5–60.

Holling, C. S. (1973) Resilience and stability of ecological systems. *Annu. Rev. Ecol. Syst.* **4**, 1–23.

Holt, R. D. (1997) Rarity and evolution: Some theoretical considerations. In *The Biology of Rarity*, eds. W. E. Kunin and K. J. Gaston, pp. 209–234, Chapman & Hall, London.

Hoppensteadt, F. (1966) Singular perturbations on the infinite interval. *T. Am. Math. Soc.* **123**, 521–535.

Hori, M. (1993) Frequency-dependent natural selection in the handedness of scale-eating cichlid fish. *Science* **260**, 216–219.

Hsu, S. B., Hubbel, S. P., and Waltman, P. A. (1978) A contribution to the theory of competing predators. *Ecol. Monographs* **48**, 337–349.

Huey, R. B. and Ward, P. D. (2005) Hypoxia, global warming, and terrestrial late permian extinctions. *Science* **308**, 398–401.

Hugill, P. J. (1993) *World Trade since 1431: Geography, Technology, and Capitalism*. The Johns Hopkins University Press, Baltimore, MD.

Hutchinson, G. E. (1959) Homage to Santa Rosalia or why are there so many kinds of animals? *Am. Nat.* **93**, 145–159.

Irwin, R. E. and Brody, A. K. (1998) Nectar robbing in *Ipomopsis aggregata*: Effects on pollinator behavior and plant fitness. *Oecologia* **116**, 519–527.

Iwasa, Y., de Jong, T. J., and Klinkhamer, P. G. L. (1995) Why pollinators visit only a fraction of the open flowers on a plant: The plant's point of view. *J. Evol. Biol.* **8**, 439–453.

Iwasa, Y. and Pomiankowski, A. (1995) Continual change in mate preferences. *Nature* **377**, 420–422.

Iwasa, Y. and Pomiankowski, A. (1999) Good parent and good genes models of handicap evolution. *J. Theor. Biol.* **200**, 97–109.

Iwasa, Y., Pomiankowski, A., and Nee, S. (1991) The evolution of costly mate preferences, II: The "handicap" principle. *Evolution* **45**, 1431–1442.

Johnson, N. C., Graham, J. H., and Smith, F. A. (1997) Functioning of mycorrhizal associations along the mutualism—parasitism continuum. *New Phytol.* **135**, 575–585.

Johnson, P. A., Hoppenstaedt, F. C., Smith, J. J., and Bush, G. L. (1996) Conditions for sympatric speciation: A diploid model incorporating habitat fidelity and non-habitat assortative mating. *Evol. Ecol.* **10**, 187–205.

Jonard, N. and Yildizoglu, M. (1998a) Interaction between local interactions: Localized learning and network externalities. In *The Economics of Networks: Interaction and Behaviours*, eds. P. Cohendet, P. Llerena, H. Stahn, and G. Umbhauer, pp. 189–204, Springer-Verlag, Berlin.

Jonard, N. and Yildizoglu, M. (1998b) Technological diversity in an evolutionary model with localized learning and network externalities. *Struct. Change Econ. Dynam.* **9**, 35–55.

Jonard, N. and Yildizoglu, M. (1999) Sources of technological diversity. Cahiers de l'innovation n. 99030, CNRS, Paris.

Karban, R. (1986) Interspecific competition between folivorous insects on *Erigeron glaucus*. *Ecology* **67**, 1063–1072.

Kawecki, T. (1996) Sympatric speciation driven by beneficial mutations. *Proc. R. Soc. Lond.* B **263**, 1515–1520.

Keller, H. B. (1977) Numerical solution of bifurcation and nonlinear eigenvalue problems. In *Applications of Bifurcation Theory*, ed. P. H. Rabinowitz, pp. 359–384, Academic Press, New York.

Kelm, M. (1997) Schumpeter's theory of economic evolution: A Darwinian interpretation. *J. Evol. Econ.* **7**, 97–130.

Khibnik, A. I. and Kondrashov, A. S. (1997) Three mechanisms of Red Queen dynamics. *Proc. R. Soc. Lond.* B **264**, 1049–1056.

Khibnik, A. I., Kuznetsov, Yu. A., Levitin, V. V., and Nikolaev, E. V. (1993) Continuation techniques and interactive software for bifurcation analysis of ODEs and iterated maps. *Physica* D **62**, 360–370.

Kiester, A. R., Lande, R., and Schemske, D. W. (1984) Models of coevolution and speciation in plants and their pollinators. *Am. Nat.* **124**, 220–243.

Kimura, M. (1983) *The Neutral Theory of Molecular Evolution.* Cambridge University Press, Cambridge, UK.

Kisdi, E. (1999) Evolutionary branching under asymmetric competition. *J. Theor. Biol.* **197**, 149–162.

Kisdi, E. (2002) Dispersal: Risk spreading versus local adaptation. *Am. Nat.* **159**, 579–596.

Kisdi, E. and Geritz, S. A. H. (1999) Adaptive dynamics in allele space: Evolution of genetic polymorphism by small mutations in a heterogeneous environment. *Evolution* **53**, 993–1008.

Kisdi, E. and Geritz, S. A. H. (2001) Evolutionary disarmament in interspecific competition. *Proc. R. Soc. Lond.* B **268**, 2589–2594.

Kisdi, E., Jacobs, F. J. A., and Geritz, S. A. H. (2001) Red Queen evolution by cycles of evolutionary branching and extinction. *Selection* **2**, 161–176.

Knox, E. B. and Palmer, J. D. (1995) Chloroplast DNA variation and the recent radiation of the giant senecios (*Asteraceae*) on the tall mountains of eastern africa. *Proc. Natl. Acad. Sci.* **92**, 10349–10353.

Koch, A. L. (1974) Competitive coexistence of two predators utilizing the same prey under constant environmental conditions. *J. Theor. Biol.* **44**, 387–395.

Koella, J. C. and Doebeli, M. (1999) Population dynamics and the evolution of virulence in epidemiological models with discrete host generations. *J. Theor. Biol.* **198**, 461–475.

Kondô, K. (1986) Relationships between long term fluctuations in the Japanese sardine (*Sardinops melanosticus*) and oceanographic conditions. International Symposium on Long term Changes in Marine Fish Populations, November, Vigo.

Kondrashov, A. S. and Mina, M. V. (1986) Sympatric speciation: When is it possible? *Biol. J. Linn. Soc.* **27**, 201–223.

Kortum, S. (1997) Research, patenting, and technological change. *Econometrica* **65**, 1389–1419.

Křivan, V. (1998) Effects of optimal antipredator behavior of prey on predator-prey dynamics: The role of refuges. *Theor. Popul. Biol.* **53**, 131–142.

Křivan, V. and Sikder, A. (1999) Optimal foraging and predator-prey dynamics, II. *Theor. Popul. Biol.* **55**, 111–126.

Kuznetsov, Yu. A. (2004) *Elements of Applied Bifurcation Theory*, 3rd ed., Springer-Verlag, Berlin.

Kuznetsov, Yu. A. and Levitin, V. V. (1997) CONTENT: A multiplatform environment for analyzing dynamical systems. Dynamical Systems Laboratory, Centrum voor Wiskunde en Informatica, Amsterdam, The Netherlands (ftp.cwi.nl/pub/CONTENT).

Kuznetsov, Yu. A., Muratori, S., and Rinaldi, S. (1995) Homoclinic bifurcations in slow-fast second-order systems. *Nonlinear Anal.* **25**, 747–762.

Kuznetsov, Yu. A., Rinaldi, S., and Gragnani, A. (2003) One parameter bifurcations in planar Filippov systems. *Int. J. Bifurcat. Chaos* **13**, 2157–2188.

Lai, C. S., Fisher, S. E., Hurst, J. A., Vargha-Khadem, F., and Monaco, A. P. (2001) A forkhead-domain gene is mutated in a severe speech and language disorder. *Nature* **413**, 519–523.

Lande, R. (1976) Natural selection and random genetic drift in phenotypic evolution. *Evolution* **30**, 314–334.

Lande, R. (1979) Quantitative genetic analysis of multivariate evolution, applied to brain: Body size allometry. *Evolution* **33**, 402–416.

Law, R. (1985) Evolution in a mutualistic environment. In *The Biology of Mutualism*, ed. D. H. Boucher, pp. 145–170, Croom Helm, London.

Law, R. and Dieckmann, U. (1997) Symbiosis through exploitation and the merger of lineages in evolution. *Proc. R. Soc. Lond.* B **265**, 1245–1253.

Lawton, J. H. (1989) What is the relationship between population density and body size in animals? *Oikos* **55**, 429–434.

Lawton, J. H. and Hassell, M. P. (1981) Asymmetrical competition in insects. *Nature* **289**, 793–795.

Le Galliard, J.-F., Ferrière, R., and Dieckmann, U. (2003) The adaptive dynamics of altruism in spatially heterogeneous populations. *Evolution* **57**, 1–17.

Le Galliard, J.-F., Ferrière, R., and Dieckmann, U. (2005) Adaptive evolution of social traits: Origin, trajectories, and correlations of altruism and mobility. *Am. Nat.* **165**, 206–225.

Legendre, S. (2002) ZEN: Eco-evolutionary software. Tech. rep., Laboratoire d'écologie, Ecole Normale Supérieure, Paris, France (www.biologie.ens.fr/legendre/zen/zen.html).

Lenski, R. E. and Bennett, A. F. (1993) Evolutionary response of *Escherichia coli* to thermal stress. *Am. Nat.* **142**, S47–S64.

Lenski, R. E., Rose, M. R., Simpson, S. C., and Tadler, S. C. (1991) Long-term experimental evolution in *Escherichia coli*, I: Adaptation and divergence during 2,000 generations. *Am. Nat.* **138**, 1315–1341.

Leon, J. A. (1974) Selection in contexts of interspecific competition. *Am. Nat.* **108**, 739–757.

Lessard, S. (1984) Evolutionary dynamics in frequency-dependent two-phenotype models. *Theor. Popul. Biol.* **25**, 210–234.

Lessard, S. (2006) ESS theory now. *Theor. Popul. Biol.* **69**, 231–233.

Levin, S. A. and Muller-Landau, H. C. (2000) The evolution of dispersal and seed size in plant communities. *Evol. Ecol. Res.* **2**, 409–435.

Levins, R. (1962a) Theory of fitness in a heterogeneous environment, I: The fitness set and adaptive function. *Am. Nat.* **96**, 361–373.

Levins, R. (1962b) Theory of fitness in a heterogeneous environment, II: Developmental flexibility and niche selection. *Am. Nat.* **97**, 74–90.

Levins, R. (1968) *Evolution in Changing Environments*. Princeton University Press, Princeton, NJ.

Lewontin, R. C. (1958) A general method for investigating the equilibrium of gene frequencies in a population. *Genetics* **43**, 419–434.

Lewontin, R. C. (1974) *The Genetic Basis of Evolutionary Change*. Columbia University Press, New York.

Lewontin, R. C. (1979) Fitness, survival, and optimality. In *Analysis of Ecological Systems*, eds. D. J. Horn, G. R. Stairs, and R. D. Mitchell, pp. 3–22, Ohio State University Press, Columbus, OH.

Lewontin, R. C. (1982) Keeping it clean. *Nature* **300**, 113–114.

Lewontin, R. C. (1983) Gene, organism and environment. In *Evolution from Molecules to Men*, ed. D. S. Bendall, pp. 273–285, Cambridge University Press, Cambridge, UK.

Lewontin, R. C. (1987) The shape of optimality. In *The Latest on the Best*, ed. J. Dupre, pp. 151–159, MIT Press, Cambridge, MA.

Li, C. C. (1955) *Population Genetics*. The University of Chicago Press, Chicago.

Lively, C. M. (1993) Rapid evolution by biological enemies. *Trends Ecol. Evol.* **8**, 345–346.

Lively, C. M. (1996) Host-parasite coevolution and sex: Do interactions between biological enemies maintain genetic variation and cross-fertilization? *Bioscience* **46**, 107–114.

Loesch, A. (1941) *The Economics of Location*. Yale University Press, New Haven, CT.

Lorenz, E. N. (1963) Deterministic nonperiodic flow. *J. Atmos. sci.* **20**, 130–141.

Lotka, A. J. (1920) Undamped oscillations derived from the law of mass action. *J. Am. Chem. Soc.* **42**, 1595–1599.

Lowe, E. D. and Lowe, J. W. (1990) Velocity of the fashion process in women's formal evening dress, 1789–1980. *Clothing and Textiles Research Journal* **9**, 50–58.

Ludwig, D., Jones, D. D., and Holling, C. S. (1978) Qualitative analysis of insect outbreak systems: The spruce budworm and forest. *J. Anim. Ecol.* **47**, 315–332.

Lythgoe, K. A. and Read, A. F. (1998) Catching the Red Queen? The advice of the Rose. *Trends Ecol. Evol.* **13**, 473–474.

MacArthur, R. H. (1969) Species packing, and what interspecies competition minimizes. *Proc. Natl. Acad. Sci.* **64**, 1369–1371.

MacArthur, R. H. (1970) Species packing and competitive equilibrium for many species. *Theor. Popul. Biol.* **1**, 1–11.

Machado, C. A., Herre, E. A., McCafferty, S., and Bermingham, E. (1996) Molecular phylogenies of fig pollinating and non-pollinating wasps and the implications for the origin and evolution of the fig-fig wasp mutualism. *J. Biogeogr.* **23**, 521–530.

Madan, R., ed. (1993) *Chua's Circuit: A Paradigm for Chaos*. World Scientific, Singapore.

Malmgren, B. A., Berggren, W. A., and Lohmann, G. P. (1983) Evidence for punctuated gradualism in the late Neogene *Globorotalia tumida* lineage of planktonic foraminifera. *Paleobiology* **9**, 377–389.

Malthus, T. (1798) *An Essay on the Principle of Population*. Printed for J. Johnson in St. Paul's Church-Yard, London.

Mandelbrot, B. B. (1977) *The Fractal Geometry of Nature*. Freeman, New York.

Marchetti, C. (1979) Energy systems: The broader context. *Technol. Forecast. Soc.* **14**, 191–203.

Marquet, P. A., Navarette, S. A., and Castilla, J. C. (1990) Scaling population density to body size in rocky intertidal communities. *Science* **250**, 1125–1127.

Marrow, P. and Cannings, C. (1993) Evolutionary instability in predator-prey systems. *J. Theor. Biol.* **160**, 135–150.

Marrow, P., Dieckmann, U., and Law, R. (1996) Evolutionary dynamics of predator-prey systems: An ecological perspective. *J. Math. Biol.* **34**, 556–578.

Marrow, P. and Johnstone, R. A. (1996) Riding the evolutionary streetcar: Where population genetics and game theory meet. *Trends Ecol. Evol.* **11**, 445–446.

Marrow, P., Law, R., and Cannings, C. (1992) The coevolution of predator-prey interactions: ESSs and Red Queen dynamics. *Proc. R. Soc. Lond.* B **250**, 133–141.

Marsden, J. and McCracken, M. (1976) *Hopf Bifurcation and its Applications.* Springer-Verlag, New York.

Marshall, A. (1920) *Principles of Economics*, 8th ed., MacMillan, London.

Matessi, C. and Di Pasquale, C. (1996) Long-term evolution of multilocus traits. *J. Math. Biol.* **34**, 613–653.

Matessi, C. and Eshel, I. (1992) Sex-ratio in the social hymenoptera. A population genetics study of long term evolution. *Am. Nat.* **139**, 276–312.

Matessi, C. and Gimelfarb, A. (2006) Discrete polymorphisms due to disruptive selection on a continuous trait, I: The one-locus case. *Theor. Popul. Biol.* **69**, 283–295.

Matessi, C., Gimelfarb, A., and Gavrilets, S. (2001) Long term buildup of reproductive isolation promoted by disruptive selection: How far does it go? *Selection* **2**, 41–64.

Mathias, A. and Kisdi, E. (2002) Adaptive diversification of germination strategies. *Proc. R. Soc. Lond.* B **269**, 151–156.

Matsuda, H. and Abrams, P. A. (1994a) Runaway evolution to self-extinction under asymmetrical competition. *Evolution* **48**, 1764–1772.

Matsuda, H. and Abrams, P. A. (1994b) Timid consumers: Self extinction due to adaptive change in foraging and anti-predator effort. *Theor. Popul. Biol.* **45**, 76–91.

Matsuyama, K. (1995) Complementarities and cumulative processes in models of monopolistic competition. *J. Econ. Lit.* **33**, 701–729.

Matthew, P. (1831) *On Naval Timber and Arboriculture.* Adam Black, Edinburgh.

May, R. M. (1977) Thresholds and breakpoints in ecosystems with a multiplicity of stable states. *Nature* **269**, 471–477.

May, R. M. and Anderson, R. M. (1983) Epidemiology and genetics in the coevolution of parasites and hosts. *Proc. R. Soc. Lond.* B **219**, 281–313.

Maynard Smith, J. (1966) Sympatric speciation. *Am. Nat.* **100**, 637–650.

Maynard Smith, J. (1974) The theory of games and the evolution of animal conflicts. *J. Theor. Biol.* **47**, 209–221.

Maynard Smith, J. (1982) *Evolution and the Theory of Games*. Cambridge University Press, Cambridge, UK.

Maynard Smith, J. (1989) *Evolutionary Genetics*. Oxford University Press, Oxford, UK.

Maynard Smith, J. (1992) Byte-sized evolution. *Nature* **335**, 772–773.

Maynard Smith, J. (1993) *The Theory of Evolution*, 3rd ed., Cambridge University Press, Cambridge, UK.

Maynard Smith, J. (1996) The games lizards play. *Nature* **380**, 198–199.

Maynard Smith, J. and Price, J. (1973) The logic of animal conflicts. *Nature* **246**, 15–18.

Maynard Smith, J. and Szathmary, E. (1995) *The Major Transitions in Evolution*. Freeman, Oxford, UK.

Mayr, E. (1942) *Systematics and the Origin of Species*. Columbia University Press, New York.

Mayr, E. (1963) *Animal Species and Evolution*. Harvard University Press, Cambridge, MA.

Mayr, E. (1982) *The Growth of Biological Thought*. Harvard University Press, Cambridge, MA.

Mayr, E. (2001) Introduction. In *The Origin of Species, a Facsimile of the First Edition* by C. Darwin, 17th ed., pp. vii–xxvii, Harvard University Press, Cambridge, MA.

McCauley, E., Nisbet, R. M., Murdoch, W. W., de Roos, A. M., and Gurney, W. S. C. (1999) Large-amplitude cycles of *Daphnia* and its algal prey in enriched environments. *Nature* **402**, 653–656.

McGill, B. J. and Brown, J. S. (2007) Evolutionary game theory and adaptive dynamics of continuous traits. *Annu. Rev. Ecol. Evol. Syst.* **38**, 403–435.

McHenry, H. M. (1994) Tempo and mode in human evolution. *Proc. Natl. Acad. Sci.* **91**, 6780–6786.

McKinnon, J. S. and Rundle, H. D. (2002) Speciation in nature: The threespine stickleback model systems. *Trends Ecol. Evol.* **17**, 480–488.

Mendel, G. J. (1865) Experiments in plant hybridization. *Verhandlungen des naturforschenden Vereines in Brünn, Abhandlungen* **4**, 3–47 (in German).

Meszéna, G., Gyllenberg, M., Jacobs, F. J., and Metz, J. A. J. (2005) Link between population dynamics and dynamics of Darwinian evolution. *Phys. Rev. Lett.* **95**, 078105.

Meszéna, G., Kisdi, E., Dieckmann, U., Geritz, S. A. H., and Metz, J. A. J. (2001) Evolutionary optimization models and matrix games in the unified perspective of adaptive dynamics. *Selection* **2**, 193–210.

Metcalfe, J. S. (1988) The diffusion of innovation: An interpretative survey. In *Technical Change and Economic Theory*, eds. G. Dosi, C. Freeman, R. N. G. Silverberg, and L. Soete, pp. 560–589, Francis Pinter, London.

Metz, J. A. J., Geritz, S. A. H., Meszéna, G., Jacobs, F. J. A., and van Heerwaarden, J. S. (1996) Adaptive dynamics: A geometrical study of the consequences of nearly faithful reproduction. In *Stochastic and Spatial Structures of Dynamical Systems*, eds. S. J. van Strien and S. M. Verduyn Lunel, pp. 183–231, Elsevier Science, Burlington, MA.

Metz, J. A. J., Nisbet, R. M., and Geritz, S. A. H. (1992) How should we define fitness for general ecological scenarios? *Trends Ecol. Evol.* **7**, 198–202.

Meyer, A., Kocher, T. D., Basasibwaki, P., and Wilson, A. C. (1990) Monophyletic origin of Lake Victoria cichlid fishes suggested by mitochondrial DNA sequences. *Nature* **347**, 550–553.

Mitchell, R. J. (1994) Effects of floral traits, pollinator visitation and plant size on *Ipomopsis aggregata* fruit production. *Am. Nat.* **143**, 870–889.

Moyà-Solà, S., Köhler, M., Alba, D. M., Casanovas-Vilar, I., and Galindo, J. (2004) *Pierolapithecus catalaunicus*, a new middle Miocene great ape from Spain. *Science* **306**, 1339–1344.

Mueller, L. D. (1997) Theoretical and empirical examination of density-dependent selection. *Annu. Rev. Ecol. Syst.* **28**, 269–288.

Muratori, S. and Rinaldi, S. (1989) Remarks on competitive coexistence. *SIAM J. App. Math.* **49**, 1462–1472.

Muratori, S. and Rinaldi, S. (1991) A separation condition for the existence of limit cycles in slow-fast systems. *Appl. Math. Modelling* **61**, 312–318.

Muratori, S. and Rinaldi, S. (1992) Low- and high-frequency oscillations in three-dimensional food chain systems. *SIAM J. App. Math.* **52**, 1688–1706.

Murdoch, W. W. (1969) Switching in general predators: Experiments on prey specificity and stability of prey populations. *Ecol. Monographs* **39**, 335–354.

Murdoch, W. W., Briggs, C. J., and Nisbet, R. M. (2003) *Consumer-Resource Dynamics*. Monographs in Population Biology, 36. Princeton University Press, Princeton, NJ.

Murdoch, W. W., Nisbet, R. M., McCauley, E., de Roos, A. M., and Gurney, W. S. C. (1998) Plankton abundance and dynamics across nutrient levels: Test of hypotheses. *Ecology* **79**, 1339–1356.

Mylius, S. D. and Diekmann, O. (2001) The resident strikes back: On the evolutionary jumping between population dynamical attractors. *J. Theor. Biol.* **211**, 297–311.

Nagel, L. and Schluter, D. (1998) Body size, natural selection, and speciation in sticklebacks. *Evolution* **52**, 209–218.

Nakicenovic, N. (1998) Dynamics and replacement of U.S. transport infrastructures. In *Cities and Their Vital Systems: Infrastructures Past, Present, and Future*, eds. J. H. Ausubel and R. H. Herman, pp. 175–221, The National Academies Press, Washington, DC.

Nakicenovic, N., Grübler, A., Ishitani, H., Johansson, T., Marland, G., Moreira, J. R., and Rogner, H.-H. (1996) Energy primer. In *Climate Change 1995, Impacts, Adaptation and Mitigation of Climate Change: Scientific-Technical Analyses*, eds. R. T. Watson, M. C. Zinyowera, and R. H. Moss, pp. 75–92, Cambridge University Press, Cambridge, UK.

Nash, J. (1996) *Essays on Game Theory*. Edward Elgar, Cheltenham.

Nash, J. F. (1950) Equilibrium points in n-person games. *Proc. Natl. Acad. Sci.* **36**, 48–49.

Nash, J. F. (1951) Non-cooperative games. *Ann. Math.* **54**, 286–295.

Navarette, S. A. and Menge, B. A. (1997) The body size-population density relationship in tropical rocky intertidal communities. *J. Anim. Ecol.* **66**, 557–566.

Nelson, R. (1995) Recent evolutionary theorizing about economic change. *J. Econ. Lit.* **33**, 48–90.

Nelson, R. and Winter, S. (1982) *An Evolutionary Theory of Economic Change*. Harvard University Press, Cambridge, MA.

North, D. C. (1997) The contribution of the new institutional economics to the understanding of the transition problem. WIDER Annual Lectures 1, World Institute for Development Economics Research, Helsinki.

Nowak, M. A. (1990) An evolutionary stable strategy may be inaccessible. *J. Theor. Biol.* **142**, 237–241.

Nowak, M. A., Komarova, N. L., and Niyogi, P. (2001) Evolution of universal grammar. *Science* **291**, 114–118.

Nowak, M. A., Plotkin, J. B., and Jansen, V. A. A. (2000) The evolution of syntactic communication. *Nature* **404**, 495–498.

Nowak, M. A. and Sigmund, K. (1998) Evolution of indirect reciprocity by image scoring. *Nature* **393**, 573–577.

Nowak, M. A. and Sigmund, K. (2004) Evolutionary dynamics of biological games. *Science* **303**, 793–799.

Noy-Meir, I. (1975) Stability of grazing systems: An application of predator-prey graphs. *J. Ecol.* **63**, 459–483.

Orians, G. H. (1997) Evolved consequences of rarity. In *The Biology of Rarity*, eds. W. E. Kunin and K. J. Gaston, pp. 190–208, Chapman & Hall, London.

Ott, E., Sauer, T. D., and Yorke, J. A. (1994) *Coping with Chaos: Analysis of Chaotic Data and the Exploitation of Chaotic Systems*. Wiley, New York.

Page, K. M. and Nowak, M. A. (2002) Unifying evolutionary dynamics. *J. Theor. Biol.* **219**, 93–98.

Palander, T. (1935) *Beiträge zur Standortstheorie*. Almqvist & Wiksells, Uppsala (in German).

Palm, G. (1984) Evolutionary stable strategies and game dynamics for n-person games. *J. Math. Biol.* **19**, 329–334.

Parvinen, K. (1999) Evolution of migration in a metapopulation. *Bull. Math. Biol.* **61**, 531–550.

Parvinen, K. (2002) Evolutionary branching of dispersal strategies in structured metapopulations. *J. Math. Biol.* **45**, 106–124.

Parvinen, K. (2005) Evolutionary suicide. *Acta Biotheor.* **53**, 241–264.

Parvinen, K. (2006) Evolution of dispersal in a structured metapopulation model in discrete time. *Bull. Math. Biol.* **68**, 655–678.

Parvinen, K., Dieckmann, U., and Heino, M. (2006) Function-valued adaptive dynamics and the calculus of variations. *J. Math. Biol.* **52**, 1–26.

Peitgen, H.-O., Jürgens, H., and Saupe, D. (2004) *Chaos and Fractals: New Frontiers of Science*. Springer-Verlag, New York.

Pellmyr, O. and Leebens-Mack, J. (1999) Forty million years of mutualism: Evidence for Eocene origin of the yucca-yucca moth association. *Proc. Natl. Acad. Sci.* **96**, 9178–9183.

Pellmyr, O., Leebens-Mack, J., and Huth, C. J. (1996) Non-mutualistic yucca moths and their evolutionary consequences. *Nature* **380**, 155–156.

Peretto, P. (1998) Technological change and population growth. *J. Econ. Growth* **3**, 238–311.

Persson, L. (1985) Asymmetrical competition: Are larger animals competitively superior? *Am. Nat.* **126**, 261–266.

Persson, L., Leonardsson, K., de Roos, A. M., Gyllenberg, M., and Christensen, B. (1998) Ontogenetic scaling of foraging rates and the dynamics of a size-structured consumer-resource model. *Theor. Popul. Biol.* **54**, 270–293.

Petren, K., Grant, B. R., and Grant, P. R. (1999) A phylogeny of Darwin's finches based on microsatellite DNA length variation. *Proc. R. Soc. Lond.* B **266**, 321–329.

Pimentel, D. (1961) Animal population regulation by the genetic feedback mechanism. *Am. Nat.* **95**, 65–79.

Pimentel, D. (1968) Population regulation and genetic feedback. *Science* **159**, 1432–1437.

Pohley, H. J. (1983) Nonlinear ESS-models and frequency-dependent selection. *Biosystems* **16**, 87–100.

Polis, G. A. (1981) The evolution and dynamics of intraspecific predation. *Annu. Rev. Ecol. Syst.* **12**, 225–251.

Polis, G. A. (1988) Exploitation competition and the evolution of interference, cannibalism and intraguild predation in age/size-structured populations. In *Size-Structured Populations: Ecology and Evolution*, eds. B. Ebenman and L. Persson, pp. 185–202, Springer-Verlag, Heidelberg.

Porter, M. E. (1990) *The Competitive Advantage of Nations*. The Free Press, New York.

Poulin, R. and Grutter, A. S. (1996) Cleaning symbioses: Proximate and adaptive explanations. *Bioscience* **46**, 512–517.

Radinsky, L. B. (1987) *Evolution of Vertebrate Design*. The University of Chicago Press, Chicago.

Rainey, P. B. and Travisano, M. (1998) Adaptive radiation in a heterogeneous environment. *Nature* **394**, 69–72.

Ramasubramanian, K. and Sriram, M. S. (2000) A comparative study of computation of Lyapunov spectra with different algorithms. *Physica* D **139**, 72–86.

Ramsey, J., Bradshaw, H. D., Jr., and Schemske, D. W. (2003) Components of reproductive isolation in the monkeyflowers *Mimulus lewisii* and *M. cardinalis* (*Phrymaceae*). *Evolution* **57**, 1520–1534.

Rand, D. A., Wilson, H. B., and McGlade, J. M. (1994) Dynamics and evolution: Evolutionarily stable attractors, invasion exponents and phenotype dynamics. *Philos. T. Roy. Soc.* B **343**, 261–283.

Raup, D. M. (1981) Extinction: Bad genes or bad luck? *Acta Geologica Hispanica* **16**, 25–33.

Raup, D. M. (1991) *Extinction: Bad Genes or Bad Luck?* W. W. Norton, New York.

Raup, D. M. (1992) Large-body impact and extinction in the phanerozoic. *Paleobiology* **18**, 80–88.

Raup, D. M. and Sepkoski, J. J. (1982) Mass extinctions in the marine fossil record. *Science* **215**, 1501–1503.

Reed, J. and Stenseth, N. C. (1984) On evolutionarily stable strategies. *J. Theor. Biol.* **108**, 491–508.

Rice, W. R. and Hostert, E. E. (1993) Laboratory experiments on speciation: What have we learned in 40 years? *Evolution* **47**, 1637–1653.

Rinaldi, S. and Muratori, S. (1992a) Limit cycles in slow-fast forest-pest models. *Theor. Popul. Biol.* **41**, 26–43.

Rinaldi, S. and Muratori, S. (1992b) Slow-fast limit cycle in predator-prey models. *Ecol. Modelling* **61**, 287–308.

Rinaldi, S. and Scheffer, M. (2000) Geometric analysis of ecological models with slow and fast processes. *Ecosystems* **3**, 507–521.

Ritschl, H. (1927) Reine und historische Dynamik des Standortes der Erzeugungszweige. *Schmollers Jahrbuch* **51**, 813–870 (in German).

Ritzberger, K. (1995) The theory of normal form games from the differentiable viewpoint. *Int. J. Game Theory* **23**, 201–236.

Rizzoli-Larousse, ed. (2003) *Multimedial Encyclopedia*. R. C. S. Libri S. p. a., Milano (in Italian).

Roberts, G. and Sherratt, T. N. (1998) Development of cooperative relationships through increasing investment. *Nature* **394**, 175–179.

Roff, D. A. (1992) *The Evolution of Life Histories: Theory and Analysis*. Chapman & Hall, New York.

Roff, D. A. (1997) *Evolutionary Quantitative Genetics*. Chapman & Hall, New York.

Rohde, R. A. and Muller, R. A. (2005) Cycles in fossil diversity. *Nature* **434**, 208–210.

Romer, P. M. (1990) Endogenous technological change. *J. Polit. Econ.* **98**, 71–102.

Rosenzweig, M. L. (1971) Paradox of enrichment: Destabilization of exploitation ecosystems in ecological time. *Science* **171**, 385–387.

Rosenzweig, M. L. (1995) *Species Diversity in Space and Time.* Cambridge University Press, Cambridge, UK.

Rosenzweig, M. L., Brown, J. S., and Vincent, T. L. (1987) Red Queen and ESS: The coevolution of evolutionary rates. *Evol. Ecol.* **1**, 59–94.

Rosenzweig, M. L. and MacArthur, R. H. (1963) Graphical representation and stability conditions of predator-prey interactions. *Am. Nat.* **97**, 209–223.

Rosenzweig, M. L. and Schaffer, W. M. (1978a) Homage to Red Queen, I: Coevolutionary of predators and their victims. *Theor. Popul. Biol.* **9**, 135–157.

Rosenzweig, M. L. and Schaffer, W. M. (1978b) Homage to Red Queen, II: Coevolutionary response to enrichment of exploitation ecosystems. *Theor. Popul. Biol.* **14**, 158–163.

Rössler, O. E. (1976) An equation for continuous chaos. *Phys. Lett.* **57**, 397–398.

Roughgarden, J. (1979) *Theory of Population Genetics and Evolutionary Ecology.* MacMillan, London.

Roughgarden, J. (1983a) Coevolution between competitors. In *Coevolution*, eds. D. J. Futuyma and M. Slatkin, pp. 383–403, Sinauer Associates, Sunderland, MA.

Roughgarden, J. (1983b) The theory of coevolution. In *Coevolution*, eds. D. J. Futuyma and M. Slatkin, pp. 33–64, Sinauer Associates, Sunderland, MA.

Rozen, D. E., Schneider, D., and Lenski, R. E. (2005) Long-term experimental evolution in *Escherichia coli*, XIII: Phylogenetic history of a balanced polymorphism. *J. Mol. Evol.* **61**, 171–180.

Ruelle, D. and Takens, F. (1971) On the nature of turbolence. *Commun. Math. Phys.* **20**, 167–192.

Rundle, H. D., Nagel, L., Boughman, J. W., and Schluter, D. (2000) Natural selection and parallel speciation in sympatric sticklebacks. *Science* **287**, 306–308.

Rundle, H. D. and Schluter, D. (1998) Reinforcement of stickleback mate preferences: Sympatry breeds contempt. *Evolution* **52**, 200–208.

Sato, A., O'Huigin, C., Tichy, H., Grant, P. R., Grant, B. R., and Klein, J. (2001) On the origin of Darwin's finches. *Mol. Biol. Evol.* **18**, 299–311.

Saviotti, P. P. (1996) *Technological Evolution, Variety and the Economy.* Edward Elgar, Cheltenham.

Saviotti, P. P. (2001) Variety, growth and demand. *J. Evol. Econ.* **11**, 119–142.

Scheffer, M. (1998) *Ecology of Shallow Lakes.* Chapman & Hall, London.

Scheffer, M., Carpenter, S., Foley, J. A., Folke, C., and Walker, B. (2001) Catastrophic shifts in ecosystems. *Nature* **413**, 591–596.

Schemske, D. W. and Bradshaw, H. D., Jr. (1999) Pollinator preference and the evolution of floral traits in monkeyflowers (*Mimulus*). *Proc. Natl. Acad. Sci.* **96**, 11910–11915.

Schlaepfler, M. A., Runge, M. C., and Sherman, P. W. (2002) Ecological and evolutionary traps. *Trends Ecol. Evol.* **17**, 474–479.

Schlag, K. (1998) Why imitate, and if so, how? a bounded rational approach to multi-armed bandits. *J. Econ. Theory* **78**, 130–156.

Schliewen, U. K., Rassmann, K., Markmann, M., Markert, J., Kocher, T., and Tautz, D. (2001) Genetic and ecological divergence of a monophyletic cichlid species pair under fully sympatric conditions in Lake Ejagham, Cameroon. *Mol. Ecol.* **10**, 1471–1488.

Schliewen, U. K., Tautz, D., and Pääbo, S. (1994) Sympatric speciation suggested by monophyly of Crater Lake cichlids. *Nature* **368**, 629–632.

Schluter, D. (1988) Character displacement and the adaptive divergence of finches on islands and continents. *Am. Nat.* **131**, 799–824.

Schluter, D. (1994) Experimental evidence that competition promotes divergence in adaptive radiation. *Science* **266**, 798–801.

Schluter, D. (2000) *The Ecology of Adaptive Radiation*. Oxford University Press, Oxford, UK.

Schluter, D. and McPhail, J. D. (1992) Ecological character displacement and speciation in sticklebacks. *Am. Nat.* **140**, 85–108.

Schluter, D., Price, T. D., and Grant, P. R. (1985) Ecological character displacement in Darwin's finches. *Science* **227**, 1056–1059.

Schrage, M. (1995) Revolutionary evolutionist. *Wired* **3.07**, 120–123.

Schumpeter, J. A. (1912) *The Theory of Economic Development*. Dunker & Humbolt, Leibzig (in German; English translation by Harvard University Press, 1934).

Schumpeter, J. A. (1942) *Capitalism, Socialism, and Democracy*. Harper & Row, New York.

Schuster, P. and Sigmund, K. (1983) Replicator dynamics. *J. Theor. Biol.* **100**, 533–538.

Schwartz, I. B. and Smith, H. L. (1983) Infinite subharmonic bifurcations in an SEIR epidemic model. *J. Math. Biol.* **18**, 233–253.

Schwarz, D., Matta, B. M., Shakir-Botteri, N. L., and McPheron, B. A. (2005) Host shift to an invasive plant triggers rapid animal hybrid speciation. *Nature* **436**, 546–549.

Schwinning, S. and Fox, G. A. (1995) Population dynamic consequences of competitive symmetry in annual plants. *Oikos* **72**, 422–432.

Segerstrom, P. (1998) Endogenous growth without scale effects. *Am. Econ. Rev.* **88**, 1290–1310.

Sepkoski, J. J. (1984) A kinetic model of Phanerozoic taxonomic diversity, III: Post-Paleozoic families and mass extinctions. *Paleobiology* **10**, 246–267.

Shil'nikov, L. P. (1968) On the generation of periodic motion from trajectories doubly asymptotic to an equilibrium state of saddle type. *Math. USSR-Sb+* **6**, 427–437.

Sinervo, B. and Lively, C. M. (1996) The rock-paper-scissors game and the evolution of alternative male strategies. *Nature* **380**, 240–243.

Sinervo, B., Svensson, E., and Comendant, T. (2000) Density cycles and an offspring quantity and quality game driven by natural selection. *Nature* **406**, 985–988.

Smolin, L. (1997) *The Life of the Cosmos*. Oxford University Press, New York.

Soberon Mainero, J. and Martinez del Rio, C. (1985) Cheating and taking advantage in mutualistic associations. In *The Biology of Mutualism*, ed. D. H. Boucher, pp. 192–216, Croom Helm, London.

Sorhannus, U., Fenster, E. J., Burckle, L. H., , and Hoffmann, A. (1998) Cladogenetic and anagenetic changes in the morphology of *Rhizosolenia praeburgonii* Mukhina. *Hist. Biol.* **1**, 185–205.

Sorhannus, U., Fenster, E. J., Burckle, L. H., , and Hoffmann, A. (1999) Iterative evolution in the diatom genus *Rhizosolenia* Ehrenberg. *Lethaia* **24**, 39–44.

Soutar, A. and Isaacs, J. D. (1974) Abundance of pelagic fish during the 19th and 20th centuries as recorded in anaerobic sediment off California. *Fish. Bull.* **72**, 257–274.

Southwood, T. R. E. and Comins, H. N. (1976) A synoptic population model. *J. Anim. Ecol.* **45**, 949–965.

Stanley, S. M. (1974) Relative growth of the titanothere horn: A new approach to an old problem. *Evolution* **28**, 447–457.

Stearns, S. (1992) *The Evolution of Life Histories*. Oxford University Press, Oxford, UK.

Stebbins, G. L. (1959) The role of hybridization in evolution. *Proc. Am. Philos. Soc.* **103**, 231–251.

Stenseth, N. C., Falck, W., Bjornstad, O. N., and Krebs, C. J. (1997) Population regulation in snowshoe hare and canadian lynx: Asymmetric food web configurations between hare and lynx. *Proc. Natl. Acad. Sci.* **94**, 5147–5152.

Stenseth, N. C. and Maynard Smith, J. (1984) Coevolution in ecosystems: Red Queen evolution or stasis. *Evolution* **38**, 870–880.

Stephens, P. A. and Sutherland, W. J. (1999) Consequences of the Allee effect for behaviour, ecology and conservation. *Trends Ecol. Evol.* **14**, 401–405.

Stirling, A. (1998) On the economics and analysis of diversity. SPRU Electronic Working Paper Series n. 28.

Stone, H. M. I. (1998) On predator deterrence by pronounced shell ornament in epifaunal bivalves. *Palaeontology* **41**, 1051–1068.

Strogatz, S. H. (1994) *Nonlinear Dynamics and Chaos*. Addison-Wesley, Reading, MA.

Takada, T. and Kigami, J. (1991) The dynamical attainability of ESS in evolutionary games. *J. Math. Biol.* **29**, 513–529.

Taper, M. L. and Case, T. J. (1992) Models of character displacement and the theoretical robustness of taxon cycles. *Evolution* **46**, 317–333.

Taylor, E. B. and McPhail, J. D. (1999) Evolutionary history of an adaptive radiation in species pairs of threespine sticklebacks (*Gasterosteus*): Insights from mitchondrial DNA. *Biol. J. Linn. Soc.* **66**, 271–291.

Taylor, P. D. (1996) Inclusive fitness arguments in genetic models of behaviour. *J. Math. Biol.* **34**, 654–674.

Taylor, P. J. and Jonker, L. (1978) Evolutionarily stable strategies and game dynamics. *Math. Biosci.* **40**, 145–156.

Thom, R. (1972) *Structural Stability and Morphogenesis*. Benjamin, Reading, MA (in French; English translation by Benjamin, 1975).

Thompson, J. N. (1994) *The Coevolutionary Process*. Chicago University Press, Chicago.

Thompson, J. N. (1998) Rapid evolution as an ecological process. *Trends Ecol. Evol.* **13**, 329–332.

Tikhonov, A. M. (1952) Systems of differential equations containing small parameters at derivatives. *Mat. Sb.* **31**, 575–586 (in Russian).

Travis, J. (1990) The interplay of population dynamics and the evolutionary process. *Philos. T. Roy. Soc.* B **330**, 253–259.

Travis, J., Keen, W. H., and Juilianna, J. (1985) The effects of multiple factors on viability selection in *Hyla gratiosa* tadpoles. *Evolution* **39**, 1087–1099.

Travisano, M., Vasi, F., and Lenski, R. E. (1995) Long-term experimental evolution in *Escherichia coli*, III: Variation among replicate populations in correlated responses to novel environments. *Evolution* **49**, 189–200.

Tsuboi, M. (1984) Marine resources trends of purse seines—as main aims of sardine and mackerel fisheries—economical structure of purse seine fishery. Progress Report of Synthesis Investigations of the Japanese Fisheries, 1983 Project, Dai-Nihon Suisankai, March 1984.

Turchin, P. (2003a) *Complex Population Dynamics: A Theoretical/Empirical Synthesis*. Princeton University Press, Princeton, NJ.

Turchin, P. (2003b) *Historical Dynamics: Why States Rise and Fall*. Princeton University Press, Princeton, NJ.

Turchin, P. (2006) *War and Peace and War: The Life Cycles of Imperial Nations*. Pi Press, New York.

van Damme, D. and Pickford, M. (1995) The late Cenozoic ampullariidae (mollusca, gastropoda) of the Albertine Rift Valley (Uganda-Zaire). *Hydrobiologia* **316**, 1–32.

van den Bosch, F., de Roos, A. M., and Gabriel, W. (1988) Cannibalism as a life boat mechanism. *J. Math. Biol.* **26**, 619–633.

van der Laan, J. D. and Hogeweg, P. (1995) Predator-prey coevolution: Interactions across different timescales. *Proc. R. Soc. Lond.* B **259**, 35–42.

van Dooren, T. J. M. (1999) The evolutionary ecology of dominance-recessivity. *J. Theor. Biol.* **198**, 519–532.

van Dooren, T. J. M. (2000) The evolutionary dynamics of direct phenotypic overdominance: Emergence possible, loss probable. *Evolution* **54**, 1899–1914.

van Doorn, G. S. and Weissing, F. J. (2006) Sexual conflict and the evolution of female preferences for indicators of male quality. *Am. Nat.* **168**, 742–757.

Van Valen, L. (1973) A new evolutionary law. *Evol. Theory* **1**, 1–30.

Velicer, G., Kroos, L., and Lenski, R. E. (1998) Loss of social behaviors by *Myxococcus xanthus* during evolution in an unstructured habitat. *Proc. Natl. Acad. Sci.* **95**, 12376–12380.

Vincent, T. L. and Brown, J. S. (2005) *Evolutionary Game Theory, Natural Selection, and Darwinian Dynamics*. Cambridge University Press, Cambridge, UK.

Vincent, T. L., Cohen, Y., and Brown, J. S. (1993) Evolution via strategy dynamics. *Theor. Popul. Biol.* **44**, 149–176.

Volterra, V. (1926) Variazioni e fluttuazioni del numero d'individui in specie animali conviventi. *Mem. Accad. Naz. Lincei* **2**, 31–113 (in Italian).

von Neumann, J. and Morgenstern, O. (1953) *Theory of Games and Economic Behavior*. Princeton University Press, Princeton, NJ.

von Thünen, J. H. (1826) *Der isolierte Staat in Beziehung auf Landwirtschaft und Nationalökonomie*. Perthes, Hamburg (in German).

Vromen, J. J. (1995) *Economic Evolution: An Enquiry into the Foundations of New Institutional Economics*. Routledge, London.

Walker, J. A. (1997) Ecological morphology of lacustrine threespine stickleback *Gasterosteus aculeatus* L. (Gasterosteidae) body shape. *Biol. J. Linn. Soc.* **61**, 3–50.

Wallace, A. R. (1858) On the tendency of varieties to depart indefinitely from the original type. *Journal of the Proceedings of the Linnean Society, Zoology* **3**, 53–62.

Waxman, D. and Gavrilets, S. (2005) 20 questions on adaptive dynamics. *J. Evol. Biol.* **18**, 1139–1154.

Weber, A. (1909) *The Theory of Location of Industries*. The University of Chicago Press, Chicago.

Weinberg, W. (1908) On the demonstration of heredity in man. *Naturkunde in Württemberg* **64**, 368–382 (in German).

Weissing, F. J. (1996) Genetic versus phenotypic models of selection: Can genetics be neglected in a long-term perspective? *J. Math. Biol.* **34**, 533–555.

Wells, J. (2000) *Icons of Evolution: Science or Myth?* Regnery, Washington, DC.

Wilbur, H. M. (1984) Complex life cycles and community organization in amphibians. In *A New Ecology: Novel Approaches to Interactive Systems*, eds. P. W. Price, C. N. Slobodchikoff, and W. S. Gaud, pp. 196–224, Wiley, New York.

Williams, G. C. (1966) *Adaptation and Natural Selection: A Critique of Some Current Evolutionary Thought*. Princeton University Press, Princeton, NJ.

Williams, N. (1996) Streetcar carries evolution modelers around roadblocks. *Science* **271**, 1365–1366.

Wilson, D. S. (1975) The adequacy of body size as a niche difference. *Am. Nat.* **109**, 769–784.

Wilson, D. S. (1980) *The Natural Selection of Populations and Communities*. Benjamin Cummings, Menlo Park, CA.

Wilson, E. O., ed. (1988) *Biodiversity*. The National Academies Press, Washington, DC.

Wissel, C. and Stöcker, S. (1991) Extinction of populations by random influences. *Theor. Popul. Biol.* **39**, 315–328.

Witt, U. (2001) Learning to consume: A theory of wants and of growth of demand. *J. Evol. Econ.* **11**, 23–36.

Woese, C. R., Kandler, O., and Wheelis, M. L. (1990) Towards a natural system of organisms. *Proc. Natl. Acad. Sci.* **87**, 4576–4579.

Wolin, C. L. (1985) The population dynamics of mutualistic systems. In *The Biology of Mutualism*, ed. D. H. Boucher, pp. 248–269, Croom Helm, London.

Wray, G. A. (2001) Dating branches on the tree of life using DNA. *Gordon & Breach Publishing Group* **3**, 1–7.

Wright, S. (1931) Evolution in Mendelian populations. *Genetics* **16**, 97–159.

Wright, S. (1969) *Evolution and the Genetics of Populations*. The University of Chicago Press, Chicago.

Wynne-Edwards, V. C. (1962) *Animal Dispersion in Relation to Social Behaviour*. Oliver & Boyd, Edinburgh.

Yoshida, T., Jones, L. E., Ellner, S. P., Fussmann, G. F., and Hairston, N. G., Jr. (2003) Rapid evolution drives ecological dynamics in a predator-prey system. *Nature* **424**, 303–306.

Young, A. (1998) Growth without scale effects. *J. Polit. Econ.* **106**, 41–63.

Ziman, J. (2000) Evolutionary models for technological change. In *Technological Innovation as an Evolutionary Process*, ed. J. Ziman, pp. 41–52, Cambridge University Press, Cambridge, UK.

Zimmer, C. (2003) Rapid evolution can foil even the best-laid plans. *Science* **300**, 895.

Index

absorbing state, 277
accidental extinction, 13, 53, 67, 75
  probability, 92, 277
adaptive
  fitness landscape, 65
  phenotype, 14
  trait, 14
adaptive dynamics (AD), 46, 67, 74
  AD canonical equation (Chapters 2, 4–7,
    10), 68, 121, 143, 158, 174, 233
  AD canonical equation (Chapter 3), 74
  AD canonical equation (for general demo-
    graphic attractors; Chapter 9), 209
  AD canonical equation (slow-fast approxi-
    mation; Chapter 8), 195
  AD in economics, 119
age
  distribution, 44
  reproductive, 10, 50
  structure, 53, 232
age-structured demographic model, 174
agent
  artificial, 4
  economic, see economic agent
  risk-taking [-averse], 133
Allee effect, 202
allele, 6
  dominant, 8
  maternal [paternal], 47
  recessive, 8
amino acid, 6
ancestral condition, see evolutionary state (an-
    cestral)
archaeon, 5
asexual
  demographic model, 45, 74
  population, 69
  reproduction, 7, 44
  species, 7
attack rate, see consumer [cannibalistic] attack
    rate
attractor, 11, 244
  basin of attraction, 98, 244
  demographic, see demographic attractor
  evolutionary, see evolutionary attractor
autonomous

dynamical process, 11, 16
  ordinary differential equation (ODE), 88,
    125, 243
  recurrence, 50

bacterium, 5
bifurcation, 72, 110, 243
  analysis, 72, 110, 128, 267
  border-collision, 265
  branching, see branching bifurcation
  catastrophic, 261
  codimension, 250
  curve, 112, 249
  diagram, 113, 249
  extinction, see extinction bifurcation
  function, 270
  global, 251
  local, 250
  noncatastrophic, 261
  normal form, 253
  point, 112, 249
  sliding, see sliding bifurcation
  standard, 113
  undetermined, 261
bifurcation (codimension-1)
  boundary equilibrium, 163, 266
  collision of tangent points, 164, 268
  exchange of stability, 253
  flip, 238
  fold, 253
  grazing, 163, 266
  heteroclinic, 184, 251
  homoclinic, 146, 162, 223, 259
  homoclinic to saddle-node, 146, 259
  homoclinic to standard saddle, 259
  Hopf, 113, 146, 162, 210, 224, 239, 252
  period doubling, 238
  pitchfork, 252
  saddle-node, 112, 146, 155, 162, 178, 181,
    190, 250
  saddle-tangent-point connection, 164, 268
  supercritical [subcritical] Hopf, 254
  supercritical [subcritical] pitchfork, 254
  tangent of limit cycles, 146, 162, 252
  transcritical, 113, 140, 181, 210, 233, 252
bifurcation (codimension-2), 146, 160
  Bogdanov-Takens, 146, 163